# Genetically Modified Organisms in Agriculture

## Economics and Politics

# Genetically Modified Organisms in Agriculture

## Economics and Politics

**Edited by**
**Gerald C. Nelson**

University of Illinois, Urbana, USA
2001

**ACADEMIC PRESS**

A Harcourt Science and Technology Company

San Diego   San Francisco   New York   Boston   London   Sydney   Tokyo

Academic Press
*A Harcourt Science and Technology Company*
Harcourt Place, 32 Jamestown Road, London NW1 7BY, UK
http://www.academicpress.com

Academic Press
*A Harcourt Science and Technology Company*
525 B Street, Suite 1900, San Diego, California 92101-4495, USA
http://www.academicpress.com

ISBN 0-12-515422-4

Library of Congress Catalog Number: 00-110183

A catalogue for this book is available from the British Library

Typeset by M Rules, London
Printed and bound in Great Britain by MPG Books Ltd, Bodmin, Cornwall

01 02 03 04 05 06 MP 9 8 7 6 5 4 3 2 1

# Contents

## Part 1
## Analysis of Current GMO Benefits and Costs

## Part 2
## Perspectives on the Controversies

### Farm Perspectives

### Government Perspectives

### Developing Country Perspectives

### Environmental Perspectives

### Consumer Issues – Food Safety/Labeling

**Part 3
Special Topics**

# Contributors

Dennis Avery
   Center for Global Food Issues, PO Box 202, Churchville VA 24421, USA
Julie Babinard
   International Food Policy Research Institute, 2033 K Street NW,
   Washington DC 20006, USA
Nicole Ballenger
   Economic Research Service, 1800 M St. NW, Washington DC 20036-5831,
   USA
Mary Bohman
   Economic Research Service, 1800 M St. NW, Washington DC 20036-5831,
   USA
David Bullock
   Department of Agricultural and Consumer Economics, University of
   Illinois, 326 Mumford Hall, 1301 W. Gregory, Urbana IL 61801, USA
Richard Caplan
   US Public Interest Research Group (US PIRG), 218 D Street, SE, Washington
   DC 20003, USA
Marc J. Cohen
   International Food Policy Research Institute, 2033 K Street, NW
   Washington DC 20006, USA
Carrie Cunningham
   Economics and Management of Agrobiotechnology Center, University of
   Missouri, 129A Mumford Hall, MO 65211, Columbia, USA
Alessandro De Pinto
   408 W. Vermont Street, Urbana IL 61801, USA
José Benjamin Falck-Zepeda
   Department of Agricultural Economics, Auburn University, 202 Comer
   Hall, Auburn AL 36849-5406, USA
Gary Goldberg
   American Corn Growers Association, 3309 E. 66th St., Tulsa OK 74136, USA
Gene Grabowski
   Grocery Manufacturers of America, 1010 Wisconsin Ave NW, Ninth Floor,
   Washington DC 20007, USA
Tassos Haniotis
   European Commission, Cabinet Fischler, rue de la Loi 130, 1049 Brussels,
   Belgium
Michael Hansen
   Consumers Union, 101 Truman Avenue, Yonkers NY 10703, USA
Richard L. Hellmich
   United States Department of Agriculture, Agricultural Research Service,
   Corn Insects and Crop Genetics Research Unit, and Department of

Entomology, Iowa State University, 109 Genetics Laboratory, c/o Insectary, Ames IA 50011, USA

Lowell Hill
Department of Agricultural and Consumer Economics, University of Illinois, 326 Mumford Hall, 1301 W. Gregory, Urbana IL 61801, USA

Mark W. Jenner
American Farm Bureau Federation, 225 Touhy Avenue, Park Ridge IL 60068, USA

Timothy Josling
Institute for International Studies, Stanford University, Encina Hall East Wing, 616 Sera Street, Stanford CA 94305, USA

Malcolm Kane
Cambridge Food Control Ltd, 30 Brewery Road, Pampisford, Cambridge CB2 4EN, UK

Dennis Kucinich
US Representative, 10th District, Ohio

Roger Krueger
Monsanto, 800 North Lindbergh Boulevard, E2SE, St Louis MO 63167, USA

Gerald C. Nelson
Department of Agricultural and Consumer Economics, University of Illinois, 326 Mumford Hall, 1301 W. Gregory, Urbana IL 61801, USA

Robert G. Nelson
Department of Agricultural Economics, Auburn University, 202 Comer Hall, Auburn AL 36849-5406, USA

Elisavet I. Nitsi
Department of Agricultural and Consumer Economics, University of Illinois, 326 Mumford Hall, 1301 W. Gregory, Urbana IL 61801, USA

Per Pinstrup-Andersen
International Food Policy Research Institute, 2033 K Street NW, Washington DC 20006, USA

Donna Roberts
Economic Research Service, USDA, US Mission to the WTO, Geneva Switzerland

Mark W. Rosegrant
International Food Policy Research Institute, 2033 K Street NW, Washington DC 20006, USA

Vandana Shiva
Research Foundation for Science, Technology and Ecology, A-60 Hauz Khas, New Dehli 110 016, India

Blair D. Siegfried
Department of Entomology, 202 Plant Industry Building, University of Nebraska, Lincoln NE 68583 0812, USA

Greg Traxler
Department of Agricultural Economics, Auburn University, Comer Hall, Auburn AL 36849-5406, USA

Laurian Unnevehr
Department of Agricultural and Consumer Economics, University of Illinois, 326 Mumford Hall, 1301 W. Gregory, Urbana IL 61801, USA

Jack Widholm
Department of Crop Sciences, University of Illinois, 1201 W. Gregory Drive, Urbana IL 61801, USA

# Foreword

**Robert W. Herdt**

*Vice President for Program Administration, Rockefeller Foundation*

The safety and economic feasibility of genetically engineered crops and livestock, widely known as genetically modified organisms or GMOs, is an important issue for consumers, agricultural producers and the rest of the agricultural sector. The energy of the debate in the United States and Europe made this issue a natural focus for the inaugural activity of the European Union Center at the University of Illinois, recently founded with funding by the European Commission. At the same time the American Farm Bureau provided a grant to the Department of Agricultural and Consumer Economics at the University to conduct a thorough review of the issues, progress, and potential for genetic engineering technology in agriculture. The material presented in this book combines and extends the results of these efforts.

This debate, while most vigorous in Europe and the United States, has far wider implications. Food production in the developing world will have to more than double over the next 25 years. In the high production, intensively cultivated areas of China, India, and Indonesia, yields are near their genetic potential and crops with greater genetic yield potential will be needed to make that doubling possible. Food with improved nutritional potential, such as the so-called 'golden rice,' which has been engineered to contain beta-carotene, promises to prevent blindness in hundreds of thousands and debilitating infections in millions of children who depend on a rice diet. The developing nations have a vital interest in safe, effective ways to modify the genetics of their food crops, and the insights generated by the debate in the North will be useful for them as they seek to take advantage of the technology.

The University of Illinois EU Center director, Larry Neal, called on Lowell Hill to organize a conference titled, 'GMO Regulations: Food Safety or Trade Barrier?' in Chicago in October 1999. Both European and American perspectives were presented in a sequence of panels that began by comparing the scientific efforts in the US and Europe for creating GMOs, moved on to compare consumer attitudes toward GMO products in the two economies, confronted the trade issues from both sides of the Atlantic, and ended by discussing the institutional arrangements available for resolving at least the trade issue between the two great trading partners. Throughout the conference, representatives from each level of the 'food chain' from both Europe and the US were seated at small discussion tables, encouraged to articulate their own concerns but also to hear and respond to each other's concerns. The complexities of the issue became obvious to all, with a consensus gradually emerging on several key issues.

There was general agreement that educational efforts must receive at least as much attention as the technical strategies if there is to be widespread acceptance of the new development as part of everyone's daily diet. The same scientific rigor that enables GMOs to be created can and should be applied to testing their long-term effects on the environment and making the results quickly and widely known. Participants also agreed that there were legitimate and valid arguments on both sides of the issues and additional communication was essential to find common ground that would allow technology to be developed and adopted where such actions were in the best interest of society. Participants in the conference contributed four of the chapters in this book.

The research to provide some of the information needed to fill the communication void highlighted by the conference participants was generated by the team assembled by Gerald Nelson. Twelve chapters of the book contain the results of the intensive investigation conducted under the American Farm Bureau (AFB) grant. AFB had recognized the potential for these controversies to derail the (then) upcoming world trade talks in late 1999. The report to the AFB provided a comprehensive review of the issues and received favorable attention around the world. AFB and the research team recognized that all participants in the debates would benefit from improved access to accurate and objective information in order to make rational decision that will be culturally as well as economically acceptable to all. The authors of the AFB report also reported important first empirical results on the costs and benefits of already commercialized GMOs and highlighted the need for specific research needed to assess both market and nonmarket benefits and the costs of GMOs.

The need for better communication and greater understanding of GMO issues as well as of the technology that created them is evident. This book is a contribution toward meeting that need. It brings together not only the scientific aspects of GMOs, but also the economic implications, reports on consumer resistance or acceptance, regulatory responses, and the possible environmental hazards initially unforeseen. Each topic is treated in greater depth than was possible in a single conference, or a semester long seminar. A wider range of perspectives than simply the American and the European is also brought to bear on the ultimate issue and the ultimate challenges of providing safe and adequate nourishment for the world's population. This is an impressive outgrowth of initial efforts to explore this issue. The result is a solid base for continued research and dialogue among all parties concerned.

# Acknowledgements

Three organizations provided financial support for the earlier activities that culminated in this book. The American Farm Bureau Foundation for Agriculture (AFBFA) provided a grant to the University of Illinois at Urbana-Champaign (UIUC) for preparation of the initial version of what is now Part 1. The Office of Research, College of Agricultural, Consumer and Environmental Sciences (ACES), UIUC, provided support for its reproduction. The European Commission, through the EU Center at UIUC, provided support for the Fall 1999 Chicago conference on GMOs that is the source of several papers in Part 2.

Many individuals have made contributions to this work and it is impossible to include them all. However, I would like to thank the following individuals for comments and suggestions that have made this a better product.

University of Illinois at Urbana-Champaign: Bruce Chassy, Theodore Hymowitz, Andrew Isserman, David Onstad, Steven Pueppke, Kevin Steffey, and seminar participants, Department of Agricultural and Consumer Economics, UIUC.

American Farm Bureau Foundation for Agriculture: Terry Francl, John Hosemann, Mark Jenner, Marsha Purcell, and John Skorburg.

Economic Research Service, United States Department of Agriculture: Jorge Fernandez-Cornejo, Cassandra Klotz-Ingram, and Sharon Jans.

The **cover photographs** were taken by **John Robinson**, Robinson Farms, 828 East, 1350 North Road, Monticello, Illinois (jrob@pdnt.com). Both corn and soybean photos were grown on his fields in summer 2000.

# Disclaimer

The contents do not necessarily reflect the views of AFBFA, any units of UIUC, or any of the organizations that employ the authors. Responsibility for any errors or omissions remains with the authors. The use of a product name does not imply endorsement of the product by the authors.

# Part 1

## Analysis of Current GMO Benefits and Costs

# Introduction

**Gerald C. Nelson**

*University of Illinois, Urbana, Illinois, USA*

Although genetically modified organisms (GMOs)[1] in agriculture have been available only for a few years, their commercial use is expanding rapidly. GM crops in widespread use include corn (maize), soybeans, cotton, potatoes, and canola (rape). The first GM crops were commercialized by China in the early 1990s, with the introduction of virus-resistant tobacco and later a virus-resistant tomato. Among industrialized nations, the first commercial use of a genetically modified food product was the Flavr Savr tomato, a delayed-ripening tomato introduced in the US by Calgene in 1994 (James and Krattiger, 1996). Between 1996 and 1998, transgenic crop area increased fifteenfold, to almost 28 million hectares (James, 1998). By 2000 GMO crop area had expanded to 44.2 million hectares (James, 2000). Most GMO crops are grown in North America, but large areas of cultivation are found in Argentina, Mexico, and South Africa.

Earlier technological changes in agriculture, such as hybrid corn in the US and Green Revolution rice and wheat, had opponents. But hostility to crops produced using recombinant DNA technologies has arisen much more quickly and been much more in the public eye. This opposition, driven by concerns about consumer food safety, the environment, corporate control of agriculture, and ethics, is strongest in Europe, but resistance to its use in developing countries has also arisen. The European Union (EU) accepts imports of some versions of the two most widely grown GMOs – *Bt* corn and glyphosate-resistant (GR) soybeans. However, in 1999 the EU halted registration of new GM crop varieties for domestic cultivation and use. European policy is evolving. Alternatives range from ultimate acceptance to a ban on the use of GMO seeds and on imports of GMO products. Dramatic changes in EU food safety regulation are already underway. The regulatory environment in other countries is undergoing changes as well.

---

[1] We explain the term GMO in more detail in Chapter 2. Throughout this book we will use the terms GMO, GM crop, and transgenic crops as synonyms for crops produced using recombinant DNA (rDNA) technologies.

---

Genetically Modified Organisms in Agriculture
ISBN 0-12-515422-4

EU policy debates over the use of transgenic crops in the food chain and the countryside highlight the challenges facing national food safety, environmental and agricultural regulatory agencies and the international trading system. This challenge is likely to become apparent on three different fronts – how to deal with vocal consumer resistance, the speed of technological change, and restructuring of the agri-food industry. These changes pose national and international problems of efficiency, equity, and responsibility that must be addressed.

Thus far GMO issues have not led to open trade conflict, but recent policy decisions in the EU – most importantly the decision to place a moratorium on the regulatory approval process for GMOs, combined with widespread public sentiment against GMOs – make such a conflict almost inevitable without constructive dialogue and collective action among the principals. The successful conclusion of the Biosafety Protocol negotiations in Montreal in February 1999 may mark the beginning of such a dialogue, but many challenges remain.

## Goals of the Book

This book has several goals. First, it provides an overview of the gamut of GMO issues – biology, regulation, private and social economics, and politics. We argue that for most of the questions raised about GMOs, whether positive (contributions to farmer income and world food supply sustainability) or negative (food safety and environmental concerns), the answers need to be assessed at the level of the product of the technology rather than the technology itself. Because most of the commercial production today is of three GMO crops – glyphosate-resistant (GR) soybeans, *Bt* corn, and *Bt* cotton – we examine the biology of these crops in depth in Chapter 2. Chapter 3 presents an economist's approach to evaluating new technologies. Chapters 4 through 7 then apply that approach, assessing the market and nonmarket effects of these three crops.

The regulatory mechanisms for assessing the food and environmental safety of GM products are in a state of flux. We review how GMOs are regulated in the major importing countries and the roles currently played by international institutions with regulatory or standards setting authority such as the World Trade Organization (WTO), the Biodiversity Agreement, and the Codex Alimentarius.

This material is then used to inform a 'political economy' discussion of the subject of GMOs and the world trading system, which entails consideration not only of the issues themselves but also of the stakeholders and their positions. The stakeholders operate in the context of a regulatory framework at the national and international level and, in turn, try to influence this framework. The political marketplace brings together this set of interests and stimulates attempts at communication and persuasion. This marketplace also includes the international arena, where governmental and non-governmental bodies seek to develop multilateral institutions. These institutions in turn become the focus for attempts by countries to influence regulations in their own perceived interest.

We end this part of the book with our perspective on the possible futures for agricultural GMOs.

The political economy discussion of Chapter 8 sets the stage for Part 2 of this book. We invited leading individuals with different perspectives on the GMO controversies to contribute short chapters that present their views. Their voices provide a vivid illustration of the range of passions engendered by this set of new technologies.

The final part of this book presents a more in-depth look at selected issues. We provide a very brief history of agricultural biotechnology (Chapter 26), a more in-depth look at the biotechnology techniques, both those that involve transfer of novel genes and others, available to today's plant breeder (Chapter 27). We include chapters that provide the latest research on the monarch butterfly controversy (Chapter 28) and on the history of the beef hormone dispute between the EU and the US. Finally, we include a glossary of biotech terms and a list of EU biotech field trial approvals.

## Main Issues in the GMO Debate

Three sets of broad issues define the scope of the GMO debate – assessing costs and benefits of the technology and its products, formulating regulatory strategies to enhance human and environmental safety, and structuring legal institutions to encourage development of intellectual property. The first set of issues deals with the economic, social, and ethical benefits and costs associated with specific GMO products. The potential benefits include environmental improvements from reduced use of chemical inputs, plants with enhanced health characteristics, and more abundant food supplies. The potential costs include environmental and food safety hazards, as well as adverse distributional impacts if the technology were to favor large farmers or multinational corporations. Ethical concerns about the use of biotechnology in agriculture arise from the notion that genetic engineering methods extend the intrusion of humans into natural processes far beyond that of normal plant breeding. But there are also ethical considerations involved in deciding to repress a technology that offers humanitarian benefits. We explore the available evidence on the market and nonmarket benefits and costs of specific GMOs in Chapters 3 through 7. Ethical considerations are beyond the professional purview of economists (which the authors of the first part are) but various contributors to the second part of this book address ethical concerns.

The second set of questions is about regulatory responsibility. Have governments adequately assessed the possible health and environmental effects of GMOs, or has the process of adoption been rushed as a result of commercial pressures on companies responsible for the technologies? Should one wait until long-term studies of the effects of GMOs in the environment and in the diet can be concluded? Or is it enough to deduce from scientific studies what the long-term impacts might be? The international aspects of regulatory responsibility are also in a state of flux, as countries try to develop a bio-safety regime to go along with the trade regime established in the WTO. We review the regulatory process in the US and the EU in some detail in Chapter 9 and several contributors to Part 2 provide perspectives on the need for regulatory reforms.

The third set of issues surround the legal and effective ownership of genetic material. Developers of new plant varieties in industrialized countries have had some form of legal protection for the intellectual property embodied in these transformed products for many years. However, until the 1980s, most private-sector research was devoted to male-sterile hybrid technology in open-pollinating crops, principally corn. The nature of this technology is that cultivators must purchase seed each year to get the benefits from the technology. In essence, biology enforces intellectual property rights. The cost of developing GMO crops, recent changes in patent laws, the use of genetic markers, and the potential for genetic enforcement of legal rights (the 'terminator' technology) shift control of technology in the direction of the private sector. There is concern in some quarters that the nature of global agriculture and the relationship between farmers and other parts of the food system are undergoing radical change.

These three sets of issues lie at the heart of the GMO debate. Opinions differ sharply on the answers to the questions they pose, adding to the complexity of the debate. And the answers are likely to differ dramatically for different GMOs.

We continue, in Chapter 2, with a brief review of the development of the rDNA technology itself and the main traits currently available only in GMO crops. Two chapters in Part 3 provide more in-depth information on the history of and the science techniques available to agricultural biotechnology.

# 2

# Traits and Techniques of GMOs

**Gerald C. Nelson**

*University of Illinois, Urbana, Illinois, USA*

A simplistic, but useful interpretation of human development of agriculture is the search for genes that produce characteristics of value to humans. Domestication of crops has meant that plants with valuable characteristics to humans are encouraged, while competing characteristics, including those that might enhance survivability in the wild, are discouraged. Until the development of rDNA technologies, the extent of desirable gene combinations was limited by sexual reproduction. The power of genetic engineering is the ability to move genes between organisms that are not sexually compatible, creating novel organisms with hitherto unavailable bundles of desirable characteristics. Table 2.1 summarizes five categories of characteristics for which transgenic crops have either been developed or for which research is ongoing.

**Table 2.1**: Improved characteristics of transgenic crops

| Characteristic | Rationale | Examples |
|---|---|---|
| Herbicide tolerance | More efficient herbicide use and/or safer herbicide use | Glyphosate-tolerant soybeans, canola, corn |
| Disease/insect tolerance | Reduction in pesticide use and/or more efficient pest control | *Bt* cotton, corn, potatoes; virus-resistant papaya, tobacco, melon |
| Quality improvements | Development of new foods or sources of new products | Ripening-delayed tomato; soybean oil quality; carnation quality |
| Tolerance to biological stresses | Improved resistance to droughts, easier production in marginal areas, easier nitrogen fixation | Research on drought-tolerant corn |
| Productivity enhancements | Higher output per unit of land | High-yielding rice and corn |

*Source*: Overseas Development Institute (1999); James (1998).

Genetically Modified Organisms in Agriculture
ISBN 0-12-515422-4

Chapter 27 provides an extended discussion of the technologies available to today's plant breeder, including transfer of novel genetic material. In this chapter we summarize the process of creating a GMO. To make a transgenic crop, one or more genes of interest from another species are inserted into a plant cell along with promoter and marker genetic material.[1] The promoter material influences at what locations in the plant the desired trait is produced and at what levels. The genetic marker aids identification of successful transformations. In gene transfer experiments, only a small percentage of the recipient plant cells actually take up the novel genes, and many desirable traits are not easy to detect before the plant has fully developed. The genetic marker material causes the plant to produce a substance that can be detected soon after transformation. Examples include a substance that inactivates an antibiotic or a herbicide or that causes a color change in the presence of a cultivation medium.[2]

Successful transformations, called events, vary depending on the components of the genetic package and where the novel DNA is inserted. The insertion site may affect production of the desired trait and could affect other plant functions as well. After the novel genetic material has been inserted, the transformed cell is induced to grow an entire plant that expresses the property encoded by the new genetic material. This new plant is then incorporated into traditional breeding programs, to combine the new trait with other desirable traits in existing varieties.

Initial efforts at genetic engineering involved the insertion of a single gene. More recent efforts involve insertion of multiple genes to combine the traits added by each gene. This technique is called stacking.

It is also possible to use genetic modification techniques to protect intellectual property (for example, the so-called terminator technology). The basic idea is that novel genetic material would be activated only if an external agent were applied. The potential implementation of this technology that has received the most (negative) press is one in which seeds produced by a GM plant would be sterile. Other uses of the technology might be to activate the production of an insecticide only when there is an infestation, or to change the composition of the final crop depending on market prices.

The novel traits of GMOs can differ dramatically. Hence, it is useful to review in more detail the biology of the three most widely used GMOs – *Bt* corn and cotton, and glyphosate-resistant (GR) soybeans. While all are engineered to deal with pests of a commercial crop, they differ in the nature of the pest and in their functioning. *Bt* corn and cotton are designed to kill lepidopteran (the butterfly and moth order) corn and cotton pests (European and southwestern corn borers, cotton budworms and bollworms) while leaving most beneficial insects unaffected. GR soybeans allow the application of glyphosate (trade name Roundup), a broad-spectrum herbicide, directly to the soybean field, killing other plants while leaving the soybeans unharmed.

---

[1] Inserting new genetic material into plant cells is a difficult process because the cell wall is designed to exclude foreign material (Bains, 1998). Today, a number of techniques are used to perform this operation. The most common basic gene transfer methods are summarized in Chapter 27.

[2] The potential for negative side effects of this part of the technology is discussed in Chapter 7.

# The Biology of *Bt* Corn[3]

*Bt* corn hybrids were developed to deal primarily with crop damage from the European corn borer (ECB), *Ostrinia nubilalis*. This insect is the second most important pest species of corn in North America, after soil-inhabiting insects (rootworms, cutworms, wireworms) (Gianessi and Carpenter, 1999).[4] Losses resulting from European corn borer damage and control costs have been estimated to exceed $1 billion in some years. For example, losses during a 1995 outbreak in Minnesota alone exceeded $285 million. A recent four-year study in Iowa indicated average losses near 13 bushels per acre in both first and second generations of European corn borer, for total losses of about 25 bushels per acre (Ostlie *et al.*, 1997). In other years, however, losses from the ECB are small. The magnitude of loss depends on the number of larvae, which in turn depends on overwinter survival and seasonal reproduction.

*Bt* corn hybrids contain genetic material from one of several strains of *Bacillus thuringiensis* (*Bt*). *Bt* is a naturally occurring soil-borne bacterium found worldwide. *Bt* forms asexual reproductive cells, called spores, which enable it to survive in adverse conditions. During the process of spore formation, *Bt* also produces unique crystal-like or 'Cry' proteins. When eaten by a susceptible insect during its feeding stage of development (as larvae), the crystal acts as poison. The insect's digestive enzymes activate the toxin. The Cry proteins bind to specific receptors on the intestinal lining and rupture the cells. Insects stop feeding within two hours of a first bite and, if enough toxin is eaten, die within two or three days.

A unique feature of *Bt* as a pesticide is that a specific Cry protein is toxic only to specific groups of insects and has no effect on mammals. These characteristics make various *Bt* insecticides very desirable generally and crucial to the organic food industry. Scientists have identified more than 170 Cry proteins.[5] Of commercial importance are Cry proteins toxic to the Colorado potato beetle, corn earworm, tobacco budworm, and European corn borer. Most of the *Bt* corn hybrids produce only the Cry1Ab protein; a few produce the Cry1Ac protein or the Cry9C protein.[6]

For more than 30 years, various liquid and granular formulations of *Bt* have been available for treatment of European corn borer and other insect pests on a variety of crops. Table 2.2 contains a list of the *Bt* species approved by the US

---

[3] This section is based in part on Ostlie *et al.* (1997).

[4] The corn borer typically has two generations per crop year although some three- and four-generation varieties have appeared in the South. Full-grown larvae overwinter in a variety of vegetable material, and moths begin to appear when the temperature reaches 50°F. Their offspring feed on leaves before boring into the corn stalk. Moths emerge in midsummer, mate, and lay eggs on recently tasseled plants.

[5] See the following website for the latest scientific information on the Cry proteins: *http://www.biols.susx.ac.uk/Home/Neil_Crickmore/Bt/*. For information on the insecticidal activities of the different Cry proteins, see *http://www.glfc.forestry.ca/english/res/ Bt_HomePage/netintro.htm*.

[6] Other Cry toxins are known to be effective against mosquitoes, blackflies and some midges, the gypsy moth, wax moth larvae in honeycombs, the cabbage looper, and the boll weevil.

Environmental Protection Agency's (EPA) Office of Pesticide Programs as pesticide active ingredients.

**Table 2.2**: EPA-registered *Bt* strains

| EPA code | Name | Major targets | Number of products |
|---|---|---|---|
| 6400 | *Bt* Berliner | Lepidoptera and Diptera; tobacco hornworm, cabbage worm, mosquito | 1 |
| 6403 | *Bt* subsp. *aizawai* | Lepidoptera | 5 |
| 6426 | *Bt* subsp. *aizawai* GC-91 | Lepidoptera | 2 |
| 6401 | *Bt* subsp. *israelensis* | Mosquito control | 25 |
| 6476 | *Bt* subsp. *israelensis* EG2215 | Mosquito control | 1 |
| 6402 | *Bt* subsp. *kurstaki* | | 100 |
| 6407 | *Bt* subsp. *kurstaki* BMP123 | | 6 |
| 6420 | *Bt* subsp. *kurstaki* EG 2424 | | |
| 6424 | *Bt* subsp. *kurstaki* EG2348 | | 6 |
| 6423 | *Bt* subsp. *kurstaki* EG2371 | | 2 |
| 6447 | *Bt* subsp. *kurstaki* EG7673 Coleoptera toxin | | 3 |
| 6448 | *Bt* subsp. *kurstaki* EG7673 Lepidoptera toxin | | 2 |
| 6459 | *Bt* subsp. *kurstaki* EG7826 | Lepidoptera: loopers, armyworms, corn borers, budworms, webworms, moths, and caterpillars | 3 |
| 6453 | *Bt* subsp. *kurstaki* EG7841 | Lepidoptera | 2 |
| 6452 | *Bt* subsp. *kurstaki* M200 | Lepidoptera | 2 |
| 6405 | *Bt* subsp. *tenebrionis* | Coleoptera; Colorado potato beetle | 6 |

*Source*: Based on EPA websites *http://www.epa.gov/oppbppd1/biopesticides/ai/bacteria.htm* and *http://www.epa.gov/oppbppd1/biopesticides/biop_fr_ai.htm*.

Although conventional *Bt* insecticides may be as toxic as synthetic insecticides, their performance in the field is not always consistent. The toxic effect breaks down quickly after exposure to ultraviolet radiation, heat, or dry conditions. Conventional application techniques may give incomplete coverage of feeding sites. As a result, few farmers have used *Bt* insecticides to treat ECB. In fact, because of the challenge of monitoring infestation levels, the use of any insecticides for ECB is relatively rare except in areas where corn is grown continuously.[7]

Modifying a corn plant to produce its own *Bt* protein overcomes these problems. The protein is protected from rapid environmental degradation. Plants produce the protein in tissues where larvae feed, so coverage is not an issue. Finally, the protein is present whenever newly hatched larvae try to feed, so application timing is not a problem.

*Bt* corn improves ECB control dramatically. If timed well, synthetic insecticides typically provide control for 60 to 95% of first-generation larvae and 40 to 80% of second-generation larvae. In field tests, *Bt* corn hybrids (regardless of event) provide more than 99% control of first-generation ECB larvae. The level

---

[7] Giannessi and Carpenter (1999, p. 28) state 'approximately 5% of the nation's corn acres were sprayed for ECB control in 1995.'

of protection against late-season ECB infestations differs among *Bt* varieties. Because some of the late-season larvae initially colonize ears to feed on silks and developing kernels, they can avoid the toxin in *Bt* varieties that express the toxin only in green tissues and pollen. Table 2.3 lists all *Bt* crops registered by EPA as of July 1999.

**Table 2.3**: *Bt* crops registered by EPA as of July 2000

| Events/Products | Year registered | Expiration date | Toxin | Crop | Company |
|---|---|---|---|---|---|
| Newleaf | May 1995 | None | Cry3A | Potato | Monsanto / NatureMark |
| Newleaf plus | Dec. 1998 | None | Cry3A + Potato leaf roll virus resistance gene | Potato | Monsanto / NatureMark |
| Bollgard | Oct. 1995 | Jan. 2001 | Cry1Ac | Cotton | Monsanto |
| Event 176 (KnockOut) | Aug. 1995/ March 1998 | April 2001 | Cry1Ab | Field corn/ popcorn amendment | Novartis |
| Event 176 (NatureGard NGBt1) | Aug. 1995 | April 2001 | Cry1Ab | Field corn | Mycogen |
| BT11 (YieldGard) | Oct. 1996 | April 2001 | Cry1Ab | Field corn | Novartis |
| BT11 (Attribute) | March 1998 | April 2001 | Cry1Ab | Sweet corn | Novartis |
| MON810 (YieldGard) | Dec. 1996 | April 2001 | Cry1Ab | Field corn | Monsanto |
| DBT418 (Bt-Xtra) | March 1997 | April 2001 | Cry1Ac | Field corn | DeKalb (now part of Monsanto) |
| CBH-351 (StarLink) | May 1998 | May 2000 | Cry9C | Field corn | AgrEvo/PGS (now Aventis) |

*Source: http://www.epa.gov/oppbppd1/biopesticides/otherdocs/bt_position_paper_618.htm.*

Note: Bt-Xtra was not grown after 1999. DeKalb/Monsanto has voluntarily removed *Bt* corn hybrids containing the DBT418 event from the market. (Personal communication from Kevin Steffrey, Professor and Extension Specialist in Entomology, Department of Crop Sciences, UIUC.) CBH-351 (StarLink) was approved only for feed use. It was removed by Aventis after the 2000 controversy when traces were found in a number of food products.

From the perspective of the corn grower, *Bt* corn varieties have many desirable properties – ease of use, no exposure to noxious chemicals, and no safety issues for mammals. Many of these same characteristics also make *Bt* toxins desirable for the organic foods industry. The loss of their effectiveness would have potentially severe consequences for organic growers. We discuss the development of resistance and the organic industry in Chapter 7.

## The Biology of *Bt* Cotton

The cotton plant is susceptible to many insect pests, several of which are in the lepidopteran order. The most serious pest is the cotton bollworm (*Helicoverpa zea*), which has infested 8 to 10 million acres in the US cotton belt in recent years.[8] This insect has several names, including corn earworm. It reportedly prefers corn to cotton, moving to cotton after nearby corn plants have dried. Other important pests are tobacco budworm (*Heliothis virescens*), pink bollworm (*Pectinophora gossypiella*), found only in the western cotton growing regions, boll weevil, and plant bugs (lygus).

As with *Bt* corn, *Bt* cotton contains genetic material from *Bacillus thuringiensis*. The most receptive cotton variety to effect this transfer was an obsolete variety known as Coker 312. Monsanto scientists increased the expression of the *Bt* protein in Coker 312 a thousandfold to achieve the level necessary for commercially viable insect control. Monsanto then chose Delta and Pine Land Company (D&PL) as its seed partner to provide elite parent lines for the four generations of back-crossings necessary to replace the Coker traits with improved high-yielding characteristics (Falck *et al.*, 2000a).

The *Bt* gene transferred to cotton (given the trade name of Bollgard) expresses the Cry protein Cry1Ac, which has also been incorporated into a corn variety. The target insects are the cotton and pink bollworms and the tobacco budworm. These three pests accounted for an estimated $391 million in cotton losses and treatment expenses in the US in 1995, and $699 million in 1996 (Williams, 1996, 1997 as cited in Falck, *et al.*, 2000a). Early treatments for these pests included organochlorines such as DDT, benzene hexachloride, toxaphene, chlordane, and methoxychlor. These were later replaced by pyrethroids, e.g., bifenthrin, cyfluthrin, cypermethrin, esfenvalerate, lambda-cyhalothrin and tralomethrin (see Gianessi and Carpenter, 1999, for an extended discussion of the history of pesticide use in cotton). However, by the early 1990s resistance to pyrethroids had become a serious problem.

Two cotton varieties, NuCOTN 33[B] and NuCOTN 35[B], containing the Bollgard (*Bt*) gene were introduced commercially in 1996 through a licensing agreement between the gene discoverer, Monsanto, and the leading cotton germplasm firm in the US, Delta and Pine Land Company (D&PL). By the late 1990s the number of cotton varieties containing this gene had grown to around 50.

As Table 2.4 indicates, *Bt* control of tobacco budworm and pink bollworm is currently quite effective, although later generations of cotton bollworm (that affect cotton during the blooming period) are not controlled as well as early generations.

**Table 2.4**: Level of control of cotton pests by *Bt* Cotton in research plots

| Species | % Control |
| --- | --- |
| Tobacco budworm | 95 |
| Cotton bollworm pre-bloom | 90 |
| Cotton bollworm blooming | 70 |
| Pink bollworm | 99 |
| Cabbage looper | 95 |
| Beet armyworm | 25 |
| Fall armyworm | 20 or less |
| Saltmarsh caterpillar | 85 or more |
| Cotton leaf perforator | 85 or more |
| European corn borer | 85 or more |

*Source*: Moore *et al.* (1999) as referenced in Gianessi and Carpenter (1999).

---

[8] A close second is the boll weevil (a coleopteran), affecting about 5 million acres.

# The Biology of Glyphosate-resistant Soybeans[9]

Glyphosate, the active ingredient in Roundup, is a nonselective herbicide regis-tered by EPA for use on many food and nonfood field crops, as well as noncrop areas for which total vegetation control is desired. Glyphosate is effective on a wide spectrum of weeds. It is considered relatively safe because it breaks down rapidly and is largely inactivated on contact with soil. Research has found little likelihood of carcinogenic potential, and the product has mild or no effects on mammals.[10] The largest uses include hay and pasture, soybeans, and field corn.

Glyphosate controls weeds by inhibiting an enzyme (EPSP synthase) neces-sary for the plant's growth. GR crops contain the gene CP4 EPSP synthase isolated from *Agrobacterium* sp. strain CP4. This gene produces an enzyme suf-ficiently similar to the naturally occurring EPSP synthase to be able to carry out its function, but it differs enough not to be inhibited by glyphosate. Monsanto holds patents on both glyphosate (expired in 2000 in the US and earlier else-where) and the glyphosate-resistance gene. In 1996, Monsanto introduced the first glyphosate-resistant crop, Roundup Ready soybeans. In 1999, glyphosate-resistant varieties were available for five commercial crops – soybeans, cotton, corn, sugar beets, and canola – with several other glyphosate-resistant crops under development.

---

[9] This material is based largely on EPA Registration Eligibility Decision Fact Sheet: Glyphosate. Available at *http://www.epa.gov/oppsrrd1/REDs/factsheets/0178fact.pdf*.

[10] An article in a recent issue of *Cancer*, the journal of the American Cancer Society, sug-gests that exposure to glyphosate 'yielded increased risks for NHL [non-Hodgkin's lymphoma].' The authors state that with the rapidly increasing use of glyphosate since the time the study was carried out, 'glyphosate deserves further epidemiologic studies' (Hardel and Eriksson, 1999). Critiques of the article, written by Mark Cullen, Professor, Yale University School of Medicine and Dr Dimitrios Trichopoulos, Harvard School of Public Health and Dr Hans-Olov Adami, Karolinska Institute, argue that the conclusions cannot be supported by the evidence provided in the paper itself. Several potential problems are mentioned, including recall bias and collinearity. They point to the well-known fallibility of recall information and the fact that only four individuals with NHL could recall using pesticides. They point out the potential for collinearity problems; that is, exposure to one agricultural chemical is likely to imply exposure to multiple agricultural chemicals. According to representatives of the Monsanto Corporation who provided these critiques, the Swedish Chemicals Institute stated that this study would not change its regulatory appraisal of glyphosate.

# 3

# The Economics of Technology Adoption

**Gerald C. Nelson**
*University of Illinois, Urbana, Illinois, USA*

**David Bullock**
*University of Illinois, Urbana, Illinois, USA*

This chapter begins our in-depth analysis of the most widely used GMO crops – glyphosate-resistant (GR) soybeans, *Bt* corn, and *Bt* cotton. In this chapter we first review the economic theory of what determines whether a new technology is adopted by the private sector. The next two chapters review the effects of GR soybeans and *Bt* corn on profitability for an individual farmer and what effect the technology might have on world markets under different assumptions about the level of adoption. Chapter 6 examines the private profitability of *Bt* cotton.

The process of growing a crop involves combining resources that are natural (e.g., soil, water, temperature) and manufactured (e.g., fertilizer, tractor power, knowledge, improved seeds) to produce an economically useful product. 'Technical change' is a change in the production process that results in fewer or less expensive resources needed to produce the same amount of product or in enhanced characteristics of the product. Technical change might come in the form of more efficient tractors, seeds with modified characteristics, and improved knowledge about soil fertility management.

If the research that results in the technical change is done by the public sector (such as publicly funded universities and federal research laboratories), the results typically are made freely available. They become a public good. Innovators in the private sector are motivated by profit. They strive to create technical change that generates the largest increase in net revenue by lowering costs, increasing the amount of product per given amount of inputs, or by increasing the value of the product by increasing its quality. The creativity and skill embodied in the new technology is called intellectual property and in most countries is owned in some form by the inventor (see the discussion of US legal protection of intellectual property and recent changes in international protection for intellectual property in Chapter 9).

Genetically Modified Organisms in Agriculture
ISBN 0-12-515422-4

An innovation is useful only if it is used; commercial operators must have a reason to change their existing practices to adopt a new technology. Typically, the new technology must provide increased financial returns to the adopter, such as lower input costs, increased yields, more valuable product, and less managerial effort. For purely public-sector innovation, the adopter can use the results for free;[1] for private-sector innovation, the adopter must compensate the owner of the innovation. The private-sector innovator has several options for recouping development costs: a higher price for special seed, a specific technology fee, and mandatory sale of the output to the innovator. However, the private-sector innovator must set these fees so that some of the benefit of the new technology is transferred to the adopter.

The widespread adoption of a new technology can have diverse market and nonmarket effects. For example, prices of competitive products might fall, and use and prices of some inputs might increase, while use and prices of other inputs fall. Nonmarket effects are discussed in Chapter 7.

## Categories of Technical Change

It is useful to think of five categories of technical change.

### Technical change that increases the biological maximum yield

The Green Revolution rice technology developed by the International Rice Research Institute is an example of a technical change that increased the maximum amount of rice that can be produced in ideal agronomic conditions, the biological maximum. With this type of technical change, the economically optimum (farm level) yield is usually higher as well.

### Technical change that increases the economic optimum yield

Even if a technical change does not increase the crop's biological maximum, it can increase farm yield. For example, as will be discussed in more detail in Chapter 4, before the arrival of *Bt* corn technology, many farmers made little or no effort to control the European corn borer (ECB). While methods for controlling the ECB have long existed (pesticides can be sprayed or, at an extreme, the pests could be eliminated by hand), for most farmers these alternatives cost more than the value of the corn they saved. As a result, most corn producers in the past elected simply to do nothing to control the ECB. *Bt* corn technology provides farmers with an economical way to control ECB. So, while the new technology does not increase the biological maximum yield, farm yield grows, because the new technology increases farmer profit.

---

[1] For purely public innovation, funds to pay for the research come from tax revenues. Increasingly, public–private research partnerships involve funds from both public and private sectors, with various mechanisms to deal with ownership of intellectual property.

## Input-switching technical change

GR soybeans are an example of this type of technical change. Several herbicides exist that are good substitutes for glyphosate. Any weed that can be controlled with glyphosate can also be controlled with other herbicides. GR technology itself cannot increase maximum obtainable yields per acre. The GR technology leads neither to an increase in maximum biological yield nor to an increase in the economic optimum yield, given the levels of other inputs. The benefit from GR technology is that it often allows the farmer to control the weeds more cheaply, because glyphosate is relatively inexpensive to manufacture. Glyphosate is also very convenient to use, and so can reduce managerial costs. For many producers, it is economically profitable to plant GR soybeans and then substitute glyphosate for other herbicides to control weeds.[2]

While input-switching technical change need not increase yield, it can lead to an increase in total area planted to the crop in question, because the costs of growing the crop have fallen. This increase in total area planted can lead to an increase in the total quantity produced, even though no producer has higher yields. But this total production increase can come only at the expense of declining production of other crops.

## Quality-enhancing technical change

Some types of technical change alter the characteristics of the final product made from the crop. Examples include high-oil corn, rice with high levels of beta carotene (the precursor to vitamin A, essential to the development of normal eyesight and often deficient in diets in the developing world), and vegetables with enhanced nutraceutical content. Consumers benefit from the enhanced quality of the product and are willing to pay more for it. Producers will grow more of the enhanced crop if a price premium is paid for it. The benefit to adopters and innovators is higher output price and/or larger volume.

## Risk-reducing technical change

New varieties that are drought or flood tolerant are examples of risk-reducing technical change. This change increases the profitability in marginal areas for which the production environment is less friendly to agriculture. It is likely to increase aggregate production and lower prices. The effects on other crops depend on what, if anything, was grown on the marginal lands.

---

[2] No data are available on whether GR soybeans affect economically optimal yields. There are reports of a small reduction in yields of GR soybeans relative to otherwise identical non-glyphosate-resistant hybrids, but the industry contends this difference is based on statistics that are not directly comparable. In any case most observers expect any gap to close over time.

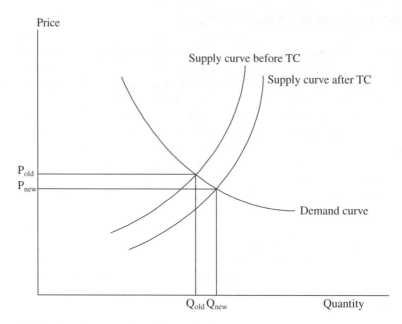

**Figure 3.1**: Supply-enhancing technical change (TC)

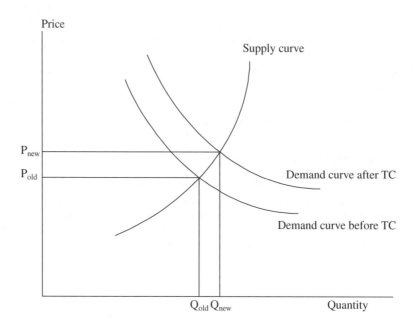

**Figure 3.2**: Quality-enhancing technical change (TC)

## Market Effects of Technology Adoption

The model used in Chapter 5 to simulate world market effects of GR soybeans and *Bt* corn simulates the effects of technical change as shifts in supply and/or demand curves for market goods. In this section, we explain how technical change is incorporated into the model and give a qualitative explanation of some of the basic results.

The first three types of technical change described above, and the last (risk-reducing), can be illustrated in a supply and demand diagram by a downward and rightward shift in the supply curve, as shown in Figure 3.1. Whether the change increases economically optimal yield or decreases costs, given any price of output, producers will choose to supply a greater quantity.

After the shift in the supply curve, producers supply more of the good and receive a lower price per unit. Producer incomes might rise or fall after such a technical change; for while costs per unit of production are lower, so is the price received for each unit. Consumers are typically made better off because of the price decline.[3]

The effect of a quality-enhancing technical change is illustrated in Figure 3.2 by a shift in the demand curve. If a new technology makes a product taste better or have better nutritional properties, demand shifts out because the product is worth more to consumers. As a result, the market price of the good increases. Unlike the other technical changes discussed above, this change improves the well-being of consumers. This improved well-being comes about not because the price of the good is lower (indeed, it might be higher), but because the good is now actually better. In addition, this type of technical change raises the income of producers; they sell more of the good and at a higher price.

With this overview of how new technologies affect market outcomes, we turn to a careful assessment of the market effects of *Bt* corn and GR soybeans.

---

[3] The magnitude of the consumer benefit depends on how much of the price of the final product is determined by the price of the commodity with technical change. Often the value of an agricultural commodity is a fraction of the value of the final product. For example, the value of the wheat in a loaf of bread in the US is less than 10% of the final price; the remaining value added comes from transportation, milling, baking, and retail. Even a large change in the price of wheat would result in only a small change in the price of the bread.

# 4

# GMO Adoption and Private Cost Savings: GR Soybeans and *Bt* Corn

**David Bullock**
*University of Illinois, Urbana, Illinois, USA*

**Elisavet I. Nitsi**
*University of Illinois, Urbana, Illinois, USA*

In this chapter, we apply the principles of technology adoption discussed in the previous chapter to the two most widely grown GM crops – glyphosate-resistant (GR) soybeans and *Bt* corn. For each crop, we start with a discussion of the theoretical issues that confront an evaluation of its profitability. We then use currently available data to assess the economic gains and their distribution.

It is useful to summarize briefly the process of commercializing a new GMO. A technology company identifies a gene that, when inserted in a crop, will add a commercially valuable characteristic. After the gene is inserted, the individual plant is added to the range of varietal material available to the regular plant breeding process. Sexual reproduction involving the modified plant becomes the mechanism to incorporate the gene into any number of crop varieties with varying characteristics. By virtue of its research, the technology company can gain a patent on the combined novel gene and its incorporation into a particular crop. It may choose to develop commercial seed varieties using the gene, or more commonly, license the technology to companies that develop seed varieties for commercial use. These new commercial varieties that include the novel gene are what we commonly refer to as GMOs.

Genetically Modified Organisms in Agriculture
ISBN 0-12-515422-4

## GR Soybeans

Figure 4.1 presents a model of the economic effects of the introduction of GR (Roundup Ready) soybean technology.[1] The model is based on the same principles underlying Figures 3.1 and 3.2 of the previous chapter. In this model several interrelated markets appear, since prices in many markets potentially are affected by the introduction of GR soybeans. In Figure 4.1 we focus on the price effects in seven principal markets: in Figure 4.1 the bottom panel is market A, the market for the right to use the patented technology to produce GR soybean seeds (called the GR license market below); the middle row's panels are the input markets (B, other herbicides; C, glyphosate; D, conventional soybean seeds; and E, GR soybean seeds); and in the top row's panels are output markets (F, soybeans; and G, corn). All markets except the GR license market are assumed to be competitive; that is an individual buyer or seller has no effect over market prices.[2]

We model the market for the GR soybean technology as oligopolistic, rather than as monopolistic as have other studies (e.g., Falck-Zepeda *et al.*, 2000b, for *Bt* cotton; Moschini *et al.*, 2000). Monsanto developed the GR technology and holds a patent on it. In the early 1990s, before GR soybeans were sold commercially, Monsanto sold permanent rights to produce GR soybean seeds to the seed companies Pioneer Hi-Bred International, Inc. (now owned by Dupont) and Novartis (Carson, 2000).[3] Monsanto also leases the rights to produce GR seeds to other seed producers. But because Pioneer and Novartis can legally produce as much GR seed as they wish, Monsanto is constrained in how much it can charge other seed companies. The current GR license market is therefore not a monopoly, but rather an oligopoly, with three firms owning rights. Essentially, Monsanto captured some of its monopoly profits ahead of time by selling away its monopoly rights.

Oligopoly theory, especially in a dynamic setting, is more difficult to present in simple diagrams than is monopoly theory. However, the results predicted by oligopoly theory are qualitatively similar to those predicted by monopoly theory. The equilibrium price under oligopoly is lower than under a monopoly, but higher than under pure competition. Therefore we simply assume that the oligopolistic price for the technology at $p_{\text{license}}^{\text{oli}}$ is above the competitive price and below the monopoly price.

---

[1] To keep the presentation simple, Figure 4.1 does not consider the effects of simultaneous introduction of glyphosate-resistant corn technology, *Bt* corn technology, or any other GMO technology.

[2] It is important to note that the patent on glyphosate itself (not GR soybeans) expired in Fall 2000. Competitors are expected to enter the glyphosate market, causing the price of glyphosate to decrease. Therefore, in Figure 4.1 it is assumed that glyphosate is sold in a competitive market.

[3] Monsanto, Novartis, and DuPont (which now owns Pioneer) are currently engaged in legal disputes regarding the original contracts they signed about the purchase of GR technology. Details about these legal disputes are not publicly available (Carson, 2000).

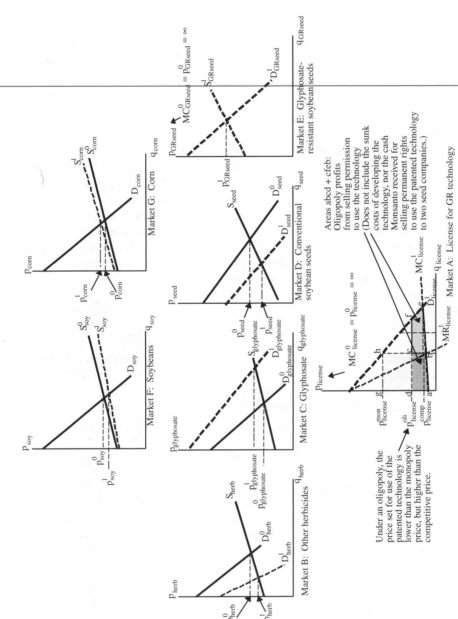

**Figure 4.1**: The economic effects of GR soybean technology

The economic equilibrium established before the introduction of the GR technology is denoted by the superscript 0: $p^0_{soy}$ is the initial price of soybeans, $p^0_{corn}$ is the initial price of corn, etc. Before the introduction of a new technology, the marginal cost of purchasing the technology license is infinite, as is the marginal cost of granting permission to use it: $MC^0_{license} = p^0_{license} = \infty$. (In other words, because the technology does not exist, it simply cannot be purchased for any amount of money.) Similarly, the cost of producing GR seed is infinite before the technology has been developed, as is the price for buying GR seed: $MC^0_{GRseed} = p^0_{GRseed} = \infty$. Equilibrium prices in the input and output markets are established where supply and demand curves intersect.

After the introduction of GR technology, the biotechnology companies' marginal costs of supplying the technology license are quite low, as shown by the curve $MC^1_{license}$ in Figure 4.1A.[4] Seed producers' demand for a license is shown by the curve $D^1_{license}$ and the corresponding marginal revenue curve is $MR^1_{license}$. If only one firm supplied the patented technology (a monopoly), it would charge seed companies $p^{mon}_{license}$ for each unit of GR seeds sold. At the quantity demanded at this price, marginal cost and marginal revenue are equal. Areas abcd + cdgh in Figure 4.1A represent a monopoly patent holder's revenues minus non-R&D costs. If this area were greater than R&D costs (not shown), a monopoly patent holder would make a profit. Because the technology is in the hands of an oligopoly, we assume that the equilibrium price, $p^{oli}_{license}$ is lower than the monopoly price $p^{mon}_{license}$ but higher than the competitive price $p^{comp}_{license}$.

After the GR technology is introduced, the marginal cost of producing GR seeds is no longer infinite, and $S^1_{GRseed}$ is the supply curve of GR seeds. Farmer demand for the GR seed is $D^1_{GRseed}$, and the price is $p^1_{GRseed}$ in this competitive market. The introduction of GR seed lowers farmers' demand curve for conventional seed from $D^0_{seed}$ to $D^1_{seed}$, causing the price of conventional seed to drop from $p^0_{seed}$ to $p^1_{seed}$.

The introduction of GR technology raises demand for glyphosate from $D^0_{gly}$ to $D^1_{gly}$ in Figure 4.1C because many farmers use it in combination with GR soybean seeds. The effect of the demand shift is to raise the price of glyphosate from $p^0_{gly}$ to $p^1_{gly}$.[5]

Because glyphosate and other herbicides are substitutes, demand for other herbicides shifts left after the introduction of GR technology. The effect is to lower the price of other herbicides from $p^0_{herb}$ to $p^1_{herb}$ in Figure 4.1B.

In the soybean market, the GR technology lowers many farmers' costs of producing soybeans (unless this were true, there would be no incentive for farmers to purchase GR soybeans). This causes farmers to move acreage from other uses, such as corn, into soybeans. The soybean supply curve shifts out, and

---

[4] These marginal costs do not include the costs of the research and development of the technology — costs that are undoubtedly quite high, but are 'sunk' at this point in time.

[5] The increase in competition in the supply of glyphosate since the patent ran out has resulted in lower prices for glyphosate. To keep the presentation simple, we assume that the patent had already run out before the introduction of glyphosate-resistant technology.

the price of soybeans drops from $p^0_{soy}$ to $p^1_{soy}$.[6] Because farmers have moved land out of corn and into soybeans, the supply curve for corn shifts back, less corn is produced, and the price of corn rises from $p^0_{corn}$ to $p^1_{corn}$.

The magnitude of the price changes depends on the actual shapes of the supply and demand curves. For instance, if the glyphosate industry can produce glyphosate at roughly constant returns to scale (which would be the case if new glyphosate-producing factories can be built and new workers can be hired at the same costs as in existing factories), the supply curve $S_{gly}$ would be flat, and the increase in the demand for glyphosate would have no effect on its price. It also should be remembered that we have ignored the introduction of technology for GR corn. We expect the marginal costs of production in both industries to fall. However, additional acreage can be added to corn and soybeans only by taking it out of other crops, and the opportunities are fairly limited. Therefore, we might expect only small increases in the quantities of corn and soybeans produced, and correspondingly small drops in their prices.[7]

## Summary of the theoretical results

1. *While the patent on the technology to produce GR soybean seeds remains in effect, it is not likely that all farmers will ever choose to use GR technology.*

Until the patent on GR technology expires, only three companies have the right to produce GR seed. Monsanto holds the patent, and so may lease the right to seed companies. Novartis and Pioneer have already purchased the permanent right to sell GR seed. If the objective of this oligopoly were to maximize the number of farmers who adopt GR technology, Monsanto could simply charge a very low price to seed companies to lease this technology, or Novartis and Pioneer could charge a very low premium for their GR seed. Because many seed companies compete to sell seeds to farmers, a lower price set by the patent holder for the technology use would lead to a smaller 'price premium' for GR seeds, and many or most farmers would be expected to plant GR seeds. However, if we assume that each firm in the oligopoly desires to recoup R&D costs and maximize profits, not maximize the number of customers, such a strategy would not be optimal. Rather, Monsanto, Novartis, and Pioneer should be expected to act as oligopolists, charging a fairly high price, such as $p^{oli}_{license}$, for the technology, and thus creating a price premium for GR seeds. The higher the premium, the more farmers would find it optimal to use the conventional

---

[6] In this analysis, we ignore the existence of government commodities policies. In fact, the prices received by US corn and soybean farmers cannot drop below the floor set by the nonrecourse loan program established by government policy. Thus, in certain situations the US farm price of soybeans would not fall even if soybean acreage were increased. Of course, if we think of the supply and demand diagrams of this model as representing world supply and demand, then an increase in the quantity of acres planted to soybeans could indeed cause the world price to fall, since the US nonrecourse loan rate applies only to domestic production.

[7] We also expect increased returns to owners of land where soybeans and corn can be grown.

technology – the costs savings they achieve with the GR seeds would be out-weighed by the high premium charged for these seeds.

2. *Introduction of the GR soybean technology should cause demand for other herbi-cides to shift in and prices of these herbicides to fall.*

3. *The effect of the introduction of GR technology on the income of farmers is ambigu-ous from the point of view of economic theory.*

The introduction of GR soybean technology lowers soybean production costs. But this cost reduction causes the supply curve for soybeans to shift down and out, leading to more soybeans produced, and a lower price of soybeans. The combination of lower costs *and* lower prices may ultimately increase or decrease farm income.[8]

4. *Economic theory predicts that the introduction of GR technology causes prices for some consumer goods to drop and others to rise.*

Lowering costs of soybean production will tend to cause US farmers to plant more land to soybeans and less to corn, causing the price of soybeans to fall and the price of corn to rise (though government nonrecourse loan policies prevent prices from falling below certain levels – see footnote 6). If GR technology for both corn and soybeans is introduced, farmers take land out of other uses and put it into corn and soybeans. This effect tends to cause corn and soybean prices to drop, but the prices of some other crops to rise.[9]

## Empirical analysis

The average savings from the use of GR soybeans come from four sources. First, glyphosate is a relatively inexpensive herbicide effective on a broad range of weeds. If GR soybeans are planted, then glyphosate replaces more expensive selective herbicide 'cocktails.' Second, as illustrated in Figure 4.1B, the intro-duction of GR soybeans decreases the demand and price of nonglyphosate herbicides, lowering production costs even for those farmers who do not use GR technology. In fact, a marked reduction in nonglyphosate herbicide prices occurred after 1996, the first year that GR soybeans were marketed. For example, the price of Synchrony, a widely used herbicide, was $16.65 per quart in 1995, but only $8.39 per quart in 1999.

---

[8] We do not discuss here the important possibility that many consumers might consider genetically modified and non-genetically modified grains to be different products. For consumers who worry that consumption of genetically modified grains might entail health or other risks, GM and non-GM crops are two different goods. The demand for product differentiation has led to a price 'premium' for non-genetically modified grains. These issues are discussed in more depth in Chapter 11. See also Bullock *et al.* (2000).

[9] Using more sophisticated economic theory, it can be shown that the effect of the tech-nology change on consumer well-being in general is positive because the 'direct effects' of lower costs passed along to the consumer tend to outweigh the 'indirect effects' of taking land and other resources away from the production of other crops.

Third, because glyphosate is easy to use, GR soybeans save management resources. Most nonglyphosate herbicides are relatively selective in the weeds they can control, so use of these herbicides requires farmers to scout fields and identify weed types, and often to mix and spray a number of different kinds of herbicides. Because glyphosate controls a broad spectrum of weeds, it can be relatively easy to use, and scouting for and identifying weeds are not as important. Fourth, glyphosate can kill larger weeds than can nonglyphosate herbicides, which provides farmers who spray postemergence herbicides a wider time window in which to do so.[10]

## Methodology

To estimate the effect of the introduction of GR technology on the average cost of soybean production in the United States, we used data generated by a survey mailed to 1398 farms across eight midwestern states (Pike *et al.*, 1997) about weed and pest problems. The information was analyzed in a software package called the Ohio Herbicide Selector Program (Ohio State University, 1999).[11] The Pike survey results provided information about the presence and severity of various types of weeds on each farm surveyed. Farms were surveyed in Illinois, Indiana, Iowa, Michigan, Minnesota, Missouri, Ohio, and Wisconsin – the eight midwestern states in which corn and soybeans are principal crops. The one-page survey asked farmers to categorize various weeds as either 0 = 'not present,' 1 = 'occurs occasionally, may not warrant control measures,' 2 = 'present, but not difficult to control,' or 3 = 'present in high populations or is difficult to control.' We considered weeds rated 2 or 3 to be weeds that the farmer would commonly need to spray with a herbicide, and weeds rated 0 or 1 to be weeds the farmer would decide not to spray. For each farm, we developed a list of weeds that the farmer would typically desire to control with a herbicide. For some farms, this list was quite short with no or only a few weeds; for other farms, this list included almost all of the weeds in the survey.[12]

The Ohio Herbicide Selector Program allows farmers to enter up to six kinds of weeds they wish to control. Control options include different mixes of

---

[10] For example, if a field receives a lot of rain after planting, it may be too muddy to enter until weeds have grown too large to spray with conventional herbicides. Because glyphosate kills larger weeds, the window of opportunity to control weeds is wider.

[11] Ideally, a study of the economic effects of a new technology would be conducted using field experiments in many locations to obtain data on the effect of the technology on yield response to other inputs. These data could then be used to estimate the economic consequences of the introduction of the technology. Unfortunately, very few data exist that describe the effects of glyphosate-resistant soybean technology on yields and input use over time, given how growing conditions and field characteristics vary geographically. Therefore, we conducted our empirical study using survey data on the geographic distribution of weeds, and using the Ohio Herbicide Selector Program as a guide to how yield responds to the presence of various types of weeds.

[12] The Ohio Herbicide Selector program would allow us to enter data on only six weeds per farm. For farms reporting more than six weeds, we included the six weeds that were most commonly ranked as 2 or 3 across all farms in the survey.

herbicides as well as GR soybeans. The software calculates the costs and degree of effectiveness of various herbicide plans that use one or more commercial herbicides, possibly at different points in the growing season. The software considers whether conventional-till or no-till practices are used. For each herbicide plan, the software determines the cost per acre (including labor and application costs) and whether control of each of the listed weeds would be 'Excellent,' 'Good,' 'Fair,' or 'Poor.' If control of all weeds was rated either 'Excellent' or 'Good', we assumed that the herbicide plan prevented the weeds from affecting yield.[13] For 1340 out of the 1398 farms, the least expensive glyphosate and conventional (i.e., nonglyphosate in postemergence) herbicide plans controlled weeds sufficiently such that the weeds had no effect on yields. For farms on which bindweed or shattercane were a problem, no herbicide plans achieved better than a 'Fair' performance, but both the least expensive GR and the least expensive non-GR plans achieved the same 'Fair' performance; therefore, yield losses from weeds were identical, and we did not need to adjust the cost differential of the plans to account for yield losses. Only five of the 1398 farms faced a choice of higher yields with a more expensive plan. Because this occurrence was so infrequent, and because differences in the costs of the more expensive and least expensive plans were small, we did not attempt to adjust cost differences for yield loss differentials when considering the costs of different herbicide plans.[14] We also did not adjust for savings associated with reduced managerial effort when growing GR soybeans.[15]

For every farm in the data set, we used the software to calculate per-acre costs for both GR and conventional soybeans for 1999, finding the least expensive GR and conventional management plans. This enabled us to estimate for each farm the average 1999 cost savings (or losses) that accrued from the use of GR or conventional technology. Then we calculated the per-acre costs only for non-GR soybeans for the year 1995, because there was no GR soybean production in 1995. This enabled us to estimate the impact of the lower 1999 herbicide prices on costs of production.

## Results

Our results show that the cost of using herbicides has decreased for most farmers since the introduction of GR technology, even those not growing GR soybeans. For conventional-tillage cropping systems using non-GM soybeans,

---

[13] Personal communication from Mark Loux, Department of Horticulture and Crop Sciences, The Ohio State University.

[14] See Fernandez-Cornejo *et al.* (1999) for statistical analysis of a large-scale sample of GR soybean use in 1997 that finds a small, but statistically significant increase in yield between GR and non-GR soybeans.

[15] In addition to reductions in input and management costs, many farmers get satisfaction from having 'clean' fields, i.e., fields with few visible weeds. Therefore, the amount of herbicide that would maximize profits is not always the amount that makes the farmer happiest. We do not include these nonmonetary benefits in our empirical analysis.

the herbicide costs fell $5.00 per acre between 1995 and 1999.[16] For no-till acres, the decline was $6.76 per acre.

Soybean producers using GR soybean seed in 1999 paid a premium of approximately $6.25 per acre. On a farm in our sample with average weed conditions, the conventional-tillage costs using GR seeds were $4.07 per acre lower than for conventional seeds. For these farmers, use of the GR technology cost $2.18 more per acre ($6.25 – $4.07). Total variable costs for growing soybeans in the United States in 1997 were approximately $78.86 per acre; therefore on the 'average farm' in our sample, use of GR technology increased variable costs by 2.18/78.86 = 2.8%. In a similar calculation for no-till acreage, variable production costs using GR technology are $1.94 per acre (or by about 2.5%) higher than conventional technology. In all, for 71% of the farms in our sample, the price premium on GR soybean seed was greater than the non-management cost savings from using GR technology.

Several factors explain the apparent contradiction between the observed rapid growth in US GR soybean adoption (which stopped in 2000) and our estimates of negative gains to the average farmer in our sample. First, our calculations omit some important management cost savings from applying GR technology (the 'convenience' factor). If we were to take these into account, our estimate of the number of farms willing to pay the price premium for GR technology would increase. In addition, the companies that own rights to the technology are not trying to maximize the number of farms that use the technology. Rather, they set the license fee to maximize their returns. There is no reason to suggest that this fee would make GR soybean use profitable on all farms. Our results show that, even ignoring management cost savings, 29% of the farmers in our sample would find the adoption of GR technology to be profitable, in some cases by more than $20.00 per acre. Whether a farmer can make money by adopting GR technology will depend very much on the particular weed situation of the farm.

To estimate the effects of the introduction of GR technology on per-acre costs for the farms in our sample, we subtracted the 1995 per-acre herbicide costs from the minimum of (1) the $6.25 per-acre price premium plus 1999 per-acre GR herbicide costs, and (2) 1999 non-GR per-acre herbicide costs. Results were that some farms in our sample faced lower costs in 1999 than in 1995 because nonglyphosate herbicide prices dropped, some farms faced lower costs because the glyphosate-resistant system worked inexpensively and effectively on their weed situation. Some farms in our sample actually faced higher herbicide costs in 1999 than in 1995. But this was not the typical case. Our final estimate is that the introduction of GR technology saved the average farm in our sample $5.00 per acre on conventional-till acres. (This $5.00 is a minimum cost saving, since we have not included the glyphosate-resistant management cost savings in our estimate.)

---

[16] This number includes any differences in application and labor costs. Also, as discussed above, the glyphosate-resistant and nonglyphosate-resistant plans had similar effects on yields, so we did not need to account for a yield differential when estimating this cost differential.

Total variable costs for growing soybeans in the United States in 1997 were approximately $78.86 per acre; therefore, the percentage decrease in costs per acre (and per bushel, assuming no yield reduction from the GR technology) due to the introduction of GR technology is estimated as $5.00 per $78.86 = 6.34% on conventional-till acres. Similar methods were used to estimate a percentage decrease in variable costs of 8.57% per no-till acre. These percentages are used in Chapter 5 to estimate the effects of GR technology on world prices and quantities demanded and supplied.

## *Bt* Corn

### Theoretical analysis

A theoretical model of the economic effects of the introduction of *Bt* corn technology is illustrated in Figures 4.2 to 4.4. Prices in many markets, including livestock and meat markets, are potentially affected by the introduction of the

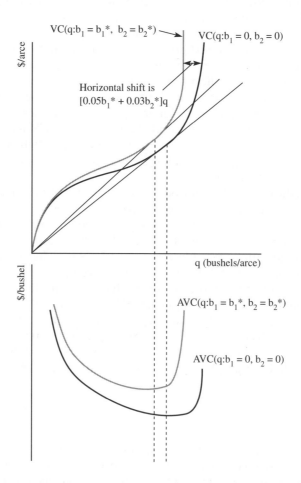

**Figure 4.2**: Economic effects of ECB infestation

technology. In these figures, we focus on the cost effects that the new technology has on the market for corn itself.

The top panel of Figure 4.2 shows how the variable costs of producing corn increase with quantity.[17] In the United States, different subspecies of European corn borer (ECB) produce one, two, three, or four generations per growing season, but two-generation corn borers dominate the central Corn Belt (Mason *et al.*, 1996). The yield loss caused by ECB depends on the level of infestation in each generation, which in turn depends heavily on weather. While research estimates vary, generally it has been demonstrated that during the whorl stage of corn growth (the time when first-generation borers do damage), corn yield is reduced by 5–6% per borer per plant. During ear development (when second-generation borers do damage), each borer causes a yield loss of 2–4% (Bessin, 1996; Bode and Calvin, 1990; Calvin, 1995; Mason *et al.*, 1996; Monsanto, 1998; Ostlie *et al.*, 1997). In Figure 4.2, variables $b_1$ and $b_2$ denote the number of first- and second-generation borers per corn plant, respectively. The dark curve labeled $VC(q; b_1 = 0, b_2 = 0)$ shows variable costs with no corn borers. The light curve labeled $VC(q; b_1 = b_1^*, b_2 = b_2^*)$ shows variable costs per bushel with $b_1^*$ first-generation and $b_2^*$ second-generation borers per plant. Per-acre yield is reduced by the quantity $[0.05b_1^* + 0.03b_2^*]q$, causing the variable cost curve to shift horizontally by that quantity. This in turn shifts up the average variable cost curve in the bottom panel of Figure 4.2 from the dark curve labeled AVC ($q$; $b_1 = 0, b_2 = 0$) to the light curve labeled $AVC(q; b_1 = b_1^*, b_2 = b_2^*)$.

When the farmer is making management decisions, the number of corn borers during the upcoming growing season is unknown. Therefore the values of $b_1$ and $b_2$ are not known with certainty. Let $E\{b_1\}$ be the expected number of first-generation borers per plant and $E\{b_2\}$ the expected number of second-generation borers per plant.[18] In Figure 4.3 the expected horizontal shift in the variable cost curve is then $[0.05E\{b_1\} + 0.03E\{b_2\}]q$, and the expected variable cost curve for non-*Bt* corn is shown by the light curve EVC($q$).

With *Bt* corn, ECB larvae have very little effect on yield.[19] The 1999 price premium for YieldGard (MON810 event) hybrid corn seed, one of the most popular *Bt* corn varieties, was approximately $8.00 per acre. So, the variable cost curve with *Bt* corn, labeled $VC_{Bt}(q)$, is shifted up by $8.00 from $VC(q; b_1 = 0, b_2 = 0)$. When *Bt* corn is planted, $AVC_{Bt}(q)$ is both the average variable cost curve and the expected average variable cost curve, since *Bt* is assumed to remove all yield uncertainty caused by ECB. When conventional corn is planted, EAVC($q$) is the expected average variable cost curve.

The price of corn determines the most profitable technology. If the expected price of corn is lower than the price at which $AVC_{Bt}(q)$ intersects EAVC($q$) (lower panel, Figure 4.3), the farmer maximizes profit by planting conventional corn. If

---

[17] In Figure 4.2, we assume that rainfall, temperature, and other weather variables take on average values and that the prices of inputs such as seed, fertilizer, and herbicides do not change.

[18] Roughly speaking, the expected number of borers is the average number over many years.

[19] For example, the MON810 event known by the trade name YieldGard prevents approximately 96% of yield loss from corn borers (Monsanto, 1998).

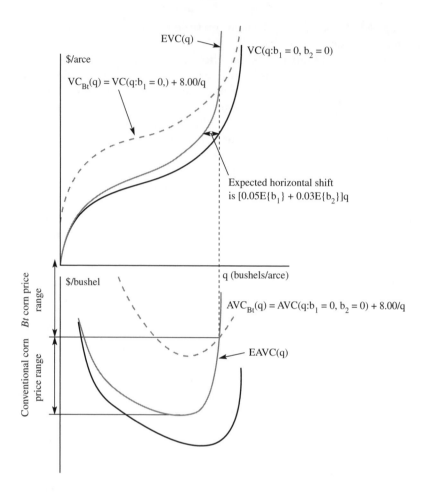

$EVC(q)$

$\$/arce$

$VC(q:b_1 = 0, b_2 = 0)$

$VC_{Bt}(q) = VC(q:b_1 = 0,) + 8.00/q$

Expected horizontal shift
is $[0.05E\{b_1\} + 0.03E\{b_2\}]q$

$q$ (bushels/arce)

$\$/bushel$

$AVC_{Bt}(q) = AVC(q:b_1 = 0, b_2 = 0) + 8.00/q$

$EAVC(q)$

*Bt* corn price range

Conventional corn price range

**Figure 4.3**: Economics of *Bt* corn adoption

the price is higher, *Bt* corn gives the highest profit. The height of $AVC_{Bt}(q)$ is determined in part by the price premium for *Bt* corn seed. In setting this premium, the patent holder or holders can influence how many farmers adopt *Bt* technology. Figure 4.3 demonstrates that the higher the premium for *Bt* seed, the higher expected corn prices must be for farmers to adopt the new technology. Conversely, given some price premium for seed, the higher the expected price of corn, the more likely it is farmers will adopt *Bt* technology.

Because ECB infestations tend to be heavier in some geographic areas than in others (typically, infestation levels are higher in the western Corn Belt than in the eastern Corn Belt), and because corn yields tend to be higher in some geographic areas than others, cost savings from *Bt* corn also vary geographically. *Bt* corn offers more value in areas with high ECB infestation levels and high yields. We can use this information to derive the aggregate demand curve for *Bt* corn technology.

Figure 4.4 shows that the number of farmers adopting *Bt* corn technology depends on the price premium charged for the use of that technology. The horizontal axis shows purchases of *Bt* corn seed. Farmers with ECB infestations

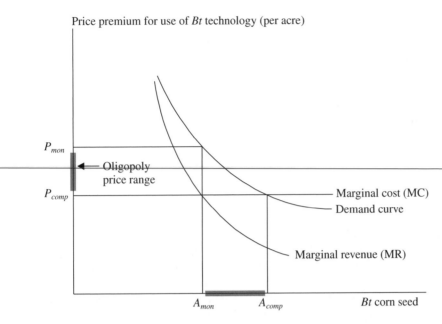

Price premium for use of *Bt* technology (per acre)

$P_{mon}$

Oligopoly price range

$P_{comp}$

Marginal cost (MC)

Demand curve

Marginal revenue (MR)

$A_{mon}$    $A_{comp}$    *Bt* corn seed

**Figure 4.4**: The *Bt* corn premium market

generate demand for the seed. One can also view the x-axis as an ordering of area planted with highest infestations of ECB and/or the highest yields (where *Bt* corn adoption is most profitable) to lowest. If the price premium were zero, then all farmers would adopt *Bt* technology. Conversely, a high price premium would cause only those farmers who traditionally face high ECB infestation levels to adopt the technology.

The demand curve in Figure 4.4 shows the maximum amount farmers are willing to pay for *Bt* technology. The amount they actually pay depends on the competitiveness of the market. In a competitive market, the price charged is equal to the marginal cost to supply the technology. Once the large fixed costs of developing the technology have been incurred, the additional variable cost of producing *Bt* corn instead of conventional corn is quite low, as shown by level MC in Figure 4.4. In a competitive market the price premium for *Bt* corn would equal the marginal cost of producing *Bt* corn instead of conventional corn, and there would be a very high adoption rate of *Bt* technology.

However, like GR soybeans, the *Bt* license market is oligopolist. Only a few companies license *Bt* technology to seed companies. Therefore, we might expect the companies to charge a price premium higher than zero but lower than the monopoly price premium (where marginal revenue equals marginal cost in Figure 4.4). The region between $A_{mon}$ and $A_{comp}$ in Figure 4.4 is the amount of *Bt* corn seed not purchased (and therefore area planted) if the technology were supplied by a monopolist. Because the market is oligopolistic, the actual *Bt* seed sales lie somewhere in that range.

It is of interest to compare the differences in benefits to non-adopters of GR soybeans and *Bt* corn. For all soybean growers, the introduction of the GR technology meant lower herbicide costs. *Bt* corn technology does not offer the same potential savings to nonadopters. Few economically feasible ECB control methods currently exist for most corn farmers.[20] For pesticide treatment to be effective, timing must be precise. Spraying just a few days early or late can result in little effect on the ECB population. The high costs of scouting, especially for second-generation borers, make it economically undesirable for the great majority of corn farmers to spray for ECB (Hyde *et al.*, 1999). As a result, for most farmers there are no cost-effective substitutes for *Bt* corn.

In summary, as with GR technology, the market for the supply of *Bt* technology is not very competitive. We expect the price premium to be set high enough to keep some farmers from adopting it. Unlike GR technology, we expect that growers who do not adopt *Bt* corn technology will not benefit from its introduction.

## Empirical results

The theoretical model of the effects of *Bt* corn technology developed above presents several challenges for an empirical investigation. First- and second-generation ECB, respectively, lower yields by about 5% and 3% per borer per plant. But the level of infestation varies greatly from year to year, depending on many factors, including weather. The challenge is to estimate the number of larvae of each ECB generation per plant when *Bt* technology is not used and the yield and profit damage compared to the results when *Bt* technology is used.

To estimate the average number of first- and second-generation borers per corn plant in the US Corn Belt, we used data displayed in Figure 4.5. These data come from extension service surveys conducted by the Universities of Illinois, Minnesota, and Wisconsin during the fall seasons 1943 through 1997. Researchers took samples of the numbers of fifth-instar, second-generation borers per corn stalk in fields around their respective states.[21]

It is clear from Figure 4.5 that infestation levels vary greatly from year to year and across geographic sites in any given year. Unfortunately, because the severity of winter also varies from year to year, the number of borers attempting to overwinter does not directly correspond to the number that survives into the next spring. So, high infestation levels in one year do not necessarily lead to high levels the next year. Because few on-farm surveys of first-generation ECB infestation levels have been conducted, less historical data exist on the levels of first-generation infestations than on the levels of second-generation infestations.

Hill *et al.* (1973) presented comparative data on first-generation and second-generation infestation levels across the north-central United States for the years 1960 through 1969. Calvin (1995) used the Hill data to run a regression relating the number of first-generation larvae ($x$) to the average number of second-generation larvae per plant ($y$) and found $y = 0.46 + 3.45x$. When first-generation

---

[20] We should be clear here that this discussion is about feed corn, not sweet corn. ECB infestation alters the appearance of sweet corn, making it much less desirable for most consumers. Therefore sweet corn producers tend to spray often to control ECB. Sweet corn acres make up a very small percentage of total corn acres.

[21] European corn borer larvae pass through five growth stages, or instars. Only second-generation ECB in the fifth instar attempt to overwinter by boring into corn stalk stubble or other vegetative material. Borers in other instars or generations do not survive the winter.

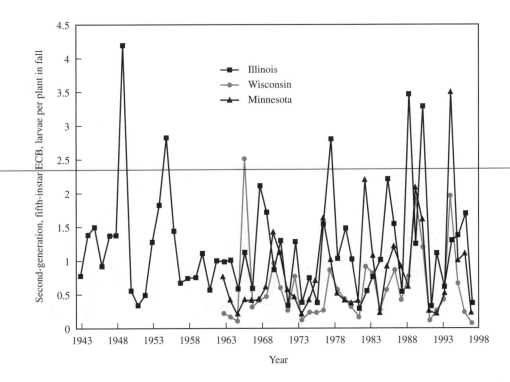

**Figure 4.5**: Fifth-instar, second-generation ECB (larvae per plant), average per state

*Sources*: Iowa State University (1999); Rice (1999); Steffey and Gray (1997); Steffey (1998); Briggs and Guse (1986); National Center for Food and Agricultural Policy (1999).

borer levels are near zero, second-generation borers still average about 0.46 per plant. For every first-generation borer, there are an additional 3.45 second-generation borers per plant. We used this estimated relationship and data on second-generation larvae per plant (depicted in Figure 4.5) to estimate the average number of first-generation corn borers in Illinois for the years 1943 to 1998, and in Minnesota and Wisconsin for 1960 to 1998. Then we estimated percentage yield losses by assuming that yield losses were 5% per first-generation borer per plant, and 3% per second-generation borer per plant.

As shown in Figure 4.6, the estimated mean percentage yield loss from ECB infestation in three midwestern states during the years 1943 through 1998 was 3.53%. The estimated percentage yield loss varied greatly from year to year and state to state. In many years and states it is near zero. In other years and states it is as high as 18%.

These results can be used, albeit imperfectly, to assess the nationwide effect of *Bt* corn use. US corn yields for the years 1994 through 1998 were 138.6, 113.5, 127.1, 126.7, and 134.4 bushels per acre, respectively, for a mean of 128.06 bushels per acre (National Agricultural Statistics Service, various years). Assuming that ECB infestation reduces per-acre corn yields on average by 3.53%, we estimate that US yields would have averaged approximately 128.06/(1 − 0.0353) = 132.76 bushels per acre had all corn planted been *Bt* corn, an increase of 4.70 bushels per acre.

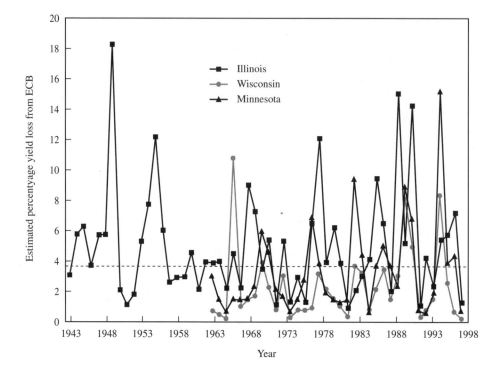

**Figure 4.6**: Estimated yield losses from ECB in three Midwestern states, 1943 to 1998 (mean yield loss = 3.53%)

*Source*: Estimated from data in Figure 4.5, using the formula second-generation ECB larvae = 0.46 + 3.4 (first-generation larvae) – and assuming yield loss from first-generation ECB is 5% per larva per stalk from first-generation ECB and a 3% per larva per stalk from second-generation ECB.

**Table 4.1**: Experiments comparing *Bt* and conventional corn yields

| Researcher | Year | Location | Corn yield gain, Bt versus non-Bt (bushels per acre) |
| --- | --- | --- | --- |
| Northrup-King/Novartis | 1995 | Locations across Corn Belt | 10.4 |
| Northrup-King/Novartis | 1996 | Locations across Corn Belt | 8.0 |
| Northrup-King/Novartis | 1997 | Locations across Corn Belt | 9.4 |
| Monsanto | 1997 | 310 locations across Corn Belt | 10.8 |
| Pioneer | 1997 | Locations across Corn Belt | 17.0 |
| Asgrow | 1997 | 6 locations across Corn Belt | 11.5 |
| Iowa State University | 1997 | 14 locations across Iowa | 9.6 |
| Iowa State University | 1998 | 14 locations across Iowa | 2.9 |
| Monsanto | 1998 | 310 locations across Corn Belt | 2.4 |
| Northrup-King/Novartis | 1998 | Locations across Corn Belt | 4.6 |

*Source*: Gianessi and Carpenter (1999).

Examination of Table 4.1 and Figure 4.6 shows that our methods of estimating expected yield gains from *Bt* corn give results reasonably consistent with the small amount of experimental data from the past few years.[22] For example, our estimation methods predict that yield gains in 1995 had all corn been *Bt* corn would have been 6.10, 10.37, and 19.98 bushels per acre in Illinois, Minnesota, and Wisconsin. This result can be compared to the first line in Table 4.1 reporting an agronomic experiment conducted in 1995 across the US Corn Belt that found an average yield gain from *Bt* corn of 10.4 bushels per acre.

Our estimate is that the European corn borer causes a yield loss of 4.70 bushels per acre to the typical Corn Belt farmer in a typical year. While *Bt* corn does not control ECB perfectly, the StarLink (now withdrawn) and YieldGard corn hybrids provide 98% control of first- and second-generation ECB. Other *Bt* hybrids provide 98% control of first-generation ECB but only 50 to 75% of second-generation larvae (Rice and Pilcher, 1998).

The 1999 price premium for YieldGard was about $8.00 per acre. With P as the price of corn and y as the yield loss from ECB, it is profitable to use YieldGard seeds only if the value of the crop saved (0.98*P*y) is greater than $8.00. With a corn price of $2.50, only farmers with expected yield losses from ECB greater than $8.00/($2.50*0.98) = 3.27 bushels have an economic incentive to purchase YieldGard. If the price of corn is $2.00, the breakeven yield gain from *Bt* corn is 4.08 bushels. Because our estimate of expected yield loss from ECB is 4.70 bushels, we conclude that use of YieldGard is profitable for the 'typical farmer.' Of course, as with GR technology, there is no 'typical farmer' using *Bt* technology. Farmers who have high average ECB infestations will find it profitable to pay the *Bt* premium. Those with relatively light average ECB infestations will not.

*Bt* technology has the potential to raise average yields to all corn farmers who experience yield loss from ECB. With the current price premium for *Bt* seed, it is not profitable for every farmer to adopt *Bt* technology. It is not possible, given current data limitations on how corn borer infestations vary geographically, to estimate the beneficial effect of *Bt* technology on average yields for different parts of the United States.

## Summary

We have conducted a theoretical and empirical analysis of the private cost savings to farmers from adopting glyphosate-resistant soybean and *Bt* corn technologies. Our theoretical analysis of glyphosate-resistant soybean technology makes clear that even farmers who do not adopt the technology are likely to benefit from its introduction, since the technology lowers the demand and thereby the prices of nonglyphosate herbicides. Our analysis predicted that

---

[22] Had experimental data on yield gains from *Bt* corn been available from many years of experiments, there would have been no need to use our indirect methods. But as can be seen in Table 4.1, experimental data are only available for a few years. ECB infestations in those years are not necessarily representative of either historical or future ECB infestations.

because the market for glyphosate-resistant technology is oligopolistic, the price premium for GR seeds would be high enough to keep many farmers from adopting the technology. Our empirical analysis found cost savings in US herbicide markets. We estimated that on average weed control costs were $5.00 per acre lower in 1999 than in 1995 on conventional-till acreage for the farms in our sample. The production cost savings from glyphosate-resistant soybean technology depended on the type of weeds plaguing the farm. For some farms, the technology lowered herbicide and labor costs by over $20.00 per acre, clearly paying for the $6.50 per acre seed premium placed on glyphosate-resistant soybean seeds. For most farms (71%), however, the reduction in herbicide and labor costs was under $6.50 per acre, indicating that unless the technology reduced management costs sufficiently to cover the seed price premium, glyphosate-resistant soybean technology could not pay for itself on those farms.

We also developed a theoretical and empirical analysis of cost savings from *Bt* corn. Costs saving from *Bt* corn are a stochastic variable, depending on the level of ECB infestation, which cannot be determined ahead of time. We estimated that *Bt* corn technology increases expected corn yields by about 4.70 bushels per acre, which is greater than our calculated break-even point for the average farm in our sample of 4.08 bushels per acre. Farmers who generally face high ECB infestations clearly have more to gain from adopting *Bt* technology than farmers who face low infestation levels.

# 5

# Simulation of World Market Effects: The 2010 World Market With and Without *Bt* Corn and GR Soybeans

**Mark W. Rosegrant**
*International Food Policy Research Institute (IFPRI), Washington DC, USA*

## Introduction

This chapter assesses the effect of the current generation of *Bt* toxin corn and glyphosate-resistant soybeans on corn and soybean production in the United States and the world, and world prices and US exports of these commodities. The analysis is undertaken by incorporating estimates of farm-level impacts and future adoption rates of *Bt* toxin corn (hereafter *Bt* corn) and glyphosate-resistant soybeans (GR soybeans) into a global food supply, demand, and trade model, and simulating the outcomes to the year 2010. These results are compared to a simulation that limits adoption of the current generation of *Bt* corn and GR soybeans to the levels attained in 1998.

The model utilized here is IFPRI's International Model for Policy Analysis of Agricultural Commodities and Trade (IMPACT), which offers an integrated methodology for analyzing alternative scenarios for global food demand, supply, and trade. IMPACT covers 36 countries and regions (which account for virtually all of world food production and consumption), and 16 commodities, including all cereals, soybeans, roots and tubers, meats, milk, eggs, oils, oilcakes, and meals. IMPACT models a competitive agricultural market for crops and livestock. It is specified as a set of country or regional sub-models, within each of which supply, demand and prices for agricultural commodities are determined. The country and regional agricultural sub-models are linked through trade, a specification that captures the interdependence of countries and commodities in the global agricultural markets. The model uses a system of supply

Genetically Modified Organisms in Agriculture
ISBN 0-12-515422-4

and demand elasticities, incorporated into a series of linear and nonlinear equations, to approximate the underlying production and demand functions. Sectoral growth multipliers are used to determine the intersectoral effects of changes in income in agricultural and nonagricultural sectors. World agricultural commodity prices are determined annually at levels that clear international markets. Demand is a function of prices, income and population growth. Growth in crop production in each country is determined by the area and yield response functions. Harvested area is specified as a response to the crop's own price, the prices of other competing crops, and the growth trend in harvested area. Yield is a function of the commodity price, the prices of inputs (such as fertilizer and labor), and a trend factor reflecting productivity improvements. Annual production is then estimated as the product of its area and yield. Future productivity growth is estimated by its component sources, including crop management research, conventional plant breeding, wide-crossing and hybridization breeding, and biotechnology and transgenic breeding. Other sources of growth considered include private sector agricultural research and development, agricultural extension and education, markets, infrastructure, and irrigation.

A wide range of factors with potentially significant impacts on future developments in the world food situation can be modeled within IMPACT. They include population and income growth, the rate of growth in crop and livestock yield and production, feed ratios for livestock, agricultural research, irrigation and other investment, price policies for commodities, and elasticities of supply and demand. For any specification of these underlying factors, IMPACT generates projections for crop area, yield, production, demand by food, feed and other uses, prices, and trade; and for livestock numbers, yield, production, demand, prices, and trade.

In the rest of this chapter, the baseline projections for production and trade of corn and soybean are briefly presented, then the alternative 'full-adoption' scenarios for *Bt* corn and GR soybean are described, and the impact of these scenarios on production and US exports of corn and soybean are assessed.

The baseline, no-adoption scenario assumes that world adoption of *Bt* corn and GR soybeans is frozen at 1997 levels. In the full-adoption scenario, only biological and economic factors limit adoption. Chapter 4 describes the details of full adoption. It reports the estimates of farm-level impacts of GR soybeans and *Bt* corn, and assumptions regarding speed and final extent of adoption of these varieties. The assumptions underlying the full-adoption scenario are summarized in Table 5.1, which provides details on three corn simulations and one soybean simulation. Each row represents one simulation. These simulations assume that adoption is limited only by economic and technological factors, not by political factors. The remaining columns indicate assumptions about farm-level impacts and how rapidly and completely the GMO is adopted in different parts of the world.

An assumption common to all simulations is that percentage cost savings from GMO adoption are the same everywhere and are based on cost savings in the United States. These scenarios assume no political constraints to adoption. GR soybeans are estimated to reduce production costs by $8 per acre, or a reduction of 5% of total cost. Under the GR soybean scenario, the United States and Canada are projected to fully adopt by 2000, with the EU and

Argentina and Brazil also adopting rapidly. China and the rest of the world are projected to adopt more slowly, but to achieve high adoption levels by 2010 (Table 5.1).

For *Bt* corn full adoption, three simulations are included: high farm-level impacts (8.1% increase in crop yield), medium impacts (3.2% yield increase), and low impacts (1.8% yield increase). The United States and North America are projected to achieve the most rapid adoption of *Bt* corn, followed by Latin America and China, with Africa and the rest of the world lagging behind (Table 5.1).

**Table 5.1**: World market scenario descriptions

| Cost savings/ yield increase from adoption | Speed and location of adoption (linear adoption rate) | | | | | |
| --- | --- | --- | --- | --- | --- | --- |
| | US/North America | EU | Argentina/ Brazil | Africa | China | Other countries |
| *Glyphosate-resistant soybeans* | | | | | | |
| $8 per adopted acre; 5% of total cost | 0 in 1995 to 100% in 2000 | 0 in 2000 to 100% in 2002 | 0 in 1996 to 100% in 2002 | 0 in 2000 to 80% in 2010 | 0 in 2000 to 100% in 2010 | 0 in 2000 to 80% in 2010 |
| Bt *corn* | | | | | | |
| Yield increase assumptions for adopters: | 0 in 1996 to 60% in 2002 | 0 in 1999 to 60% in 2005 | 0 in 1998 to 30% in 2008 | 0 in 2000 to 30% in 2015 | 0 in 2000 to 40% in 2010 | 0 in 2000 to 40% in 2010 |
| High: 8.1% Medium: 3.2% Low: 1.8% | | | | | | |

*Notes on assumptions*: Assumptions on when GR soybean adoption starts: United States had 12% of total soybean area in GMOs in 1997, so we assume 0 starts in 1995. Argentina had 22% of area in 1997.

Assumptions on when *Bt* corn adoption starts: US area in 1997 was 9% in *Bt* corn. Areas in other countries were negligible in that year. Adoption in Africa and Asia is assumed to be hindered by the lack of a seed production and delivery system. All adoption values assume a 20% refuge requirement.

## No-adoption Scenario Projections

Table 5.2 shows the baseline production projections of corn and soybeans for selected regions and the world, and for production in 2010 with no adoption. World production of soybeans is projected to grow by one-third between 1995 and 2010, by a total of 43.4 million metric tons (mt). Soybean production in the United States is projected to grow by more than 15 million mt between 1995 and 2010, or nearly 25%. The share of the United States in total world production will decline slightly, from 48% to 45% because of rapid growth in production elsewhere in the world. Soybean production in Latin America is projected to grow particularly rapidly, by 18 million mt, or 45%. In China, soybean production is projected to grow by 6.3 million mt, or 44%.

**Table 5.2**: No-adoption scenario, total production for soybeans and corn (million metric tons)

| Region | Soybeans | | Corn | |
|---|---|---|---|---|
| | 1995 actual | 2010 | 1995 actual | 2010 |
| China | 14.3 | 20.6 | 113.3 | 164.9 |
| Latin America | 40.3 | 58.3 | 71.0 | 100.5 |
| Sub-Saharan Africa | 0.5 | 0.7 | 26.2 | 40.3 |
| European Union | 1.1 | 1.3 | 32.0 | 35.3 |
| United States | 63.2 | 78.7 | 226.7 | 267.0 |
| World | 131.3 | 174.7 | 558.2 | 723.7 |

*Source*: IFPRI IMPACT.

Global corn production is projected to increase nearly as rapidly as soybean production, by 165.5 million mt, or 30%, between 1995 and 2010 (Table 5.2). Compared to the developing regions shown in Table 5.2, US corn production will grow relatively slowly in percentage terms, growing only 18% by 2010. Nevertheless, this represents a substantial increase of 40.3 million mt. Production growth in China and Latin America will be dramatic, at 46% and 42%, respectively, between 1995 and 2010. In absolute terms, corn production in China is projected to increase by 51.6 million mt, greater even than in the United States. Rapid growth in corn production is driven mainly by the demand for feed for the rapidly expanding livestock sector.

In spite of China's rapid growth in soybean and corn production, it will become an increasingly large importer of these commodities because of more rapid growth in demand. As can be seen in Table 5.3, China is projected to increase soybean imports from 2.6 million mt in 1995 to 5.4 million mt in 2010, and to boost corn imports from 5 million mt to 18.1 million mt. Much of these imports will come from the United States, which is projected to increase soybean exports from 21.9 million mt to 27.4 million mt, and corn exports from 49 million mt to 60.1 million mt. Latin America will also increase soybean exports substantially, while reducing corn imports. The European Union will show increases in soybean imports (Table 5.3).

**Table 5.3**: No-adoption scenario, net trade for soybeans and corn (million metric tons)

| Region | Soybeans | | Corn | |
|---|---|---|---|---|
| | 1995 actual | 2010 | 1995 actual | 2010 |
| China | −2.6 | −5.4 | −5.0 | −18.1 |
| Latin America | 3.9 | 7.2 | −4.9 | −3.8 |
| Sub-Saharan Africa | 0.0 | 0.0 | −1.0 | 0.0 |
| European Union | −14.2 | −17.0 | −1.6 | −1.0 |
| United States | 21.9 | 27.4 | 49.0 | 60.1 |
| World | 0.0 | 0.0 | 0.0 | 0.0 |

*Source*: IFPRI IMPACT.

A negative number means net imports.

With production growth for soybeans and corn balanced against rapid demand growth, particularly for animal feed in developing countries, world price trends for these two commodities are projected to be considerably stronger than in recent years. Compared to large declines in real world prices during the past 15 years, the real world price of soybeans is projected to decline only from $249 per mt in 1995 to $248 per mt in 2010. The real world price of corn is projected to increase slightly, from $116 to $120 per mt.

## Full-adoption Scenario Projections

In Tables 5.4 and 5.5, we report the results of five full-adoption simulations – complete adoption of *Bt* corn alone at three levels of yield increase, adoption of GR soybeans alone, and adoption of both GMOs with intermediate *Bt* corn-yield increase. The simulations assume no consumer resistance or restrictions on imports anywhere. Adoption is limited only by profitability, the lack of a seed distribution industry in parts of the developing world, and refuge requirements for *Bt* corn.

As can be seen in Table 5.4, under the high farm-level impact scenario, adoption of *Bt* corn has a significant impact on US and world production, US exports, and world prices of corn. US production is projected to increase by 3.3% in 2010 compared to the no-adoption scenario, and exports to increase by 4.9%. For the world as a whole, corn production is projected to increase by 1.9%, resulting in a 4.2% fall in the world price of corn. An interesting secondary effect from the adoption of *Bt* corn can also be seen for soybeans. The increase in corn yield and production reduces the world price of corn, thus inducing some substitution of soybean area for corn area, and increasing production and US exports of soybeans.

**Table 5.4**: Full-adoption simulations; production, US exports, and world price

| | Corn (% change from no adoption) | | | | Soybean (% change from no adoption) | | | |
|---|---|---|---|---|---|---|---|---|
| | US production | World production | US exports | World price | US production | World production | US exports | World price |
| Scenario 1: Corn (a) | 3.3 | 1.9 | 4.9 | −4.2 | 0.3 | 0.2 | 0.5 | −0.4 |
| Scenario 2: Corn (b) | 1.1 | 0.6 | 1.8 | −1.7 | 0.1 | 0.1 | 0.1 | 0.0 |
| Scenario 3: Corn (c) | 0.5 | 0.3 | 0.9 | −0.8 | 0.0 | 0.0 | 0.0 | 0.0 |
| Scenario 4: Soybeans | 0.0 | 0.0 | 0.0 | 0.0 | 0.3 | 0.4 | 0.2 | −0.4 |
| Scenario 5: Corn (b) and soybeans | 1.2 | 0.6 | 1.8 | −1.7 | 0.4 | 0.5 | 0.4 | −0.6 |

*Source*: IMPACT model runs; see Table 5.1 for detailed adoption assumptions.

*Notes*: Scenarios are described in detail in Table 5.1. Corn (a)–(c) represent high-, medium-, and low-impact estimates.

**Table 5.5**: Full-adoption scenarios for *Bt* corn and GR soybeans: production in other regions

| | Corn production (% change from no adoption) | | | | Soybean production (% change from no adoption) | | | |
|---|---|---|---|---|---|---|---|---|
| | China | Latin America | Sub-Saharan Africa | European Union | China | Latin America | Sub-Saharan Africa | European Union |
| Scenario 1: Corn (a) | 0.8 | 1.4 | 0.0 | 3.1 | 0.0 | 0.2 | 0.0 | 0.0 |
| Scenario 2: Corn (b) | 0.2 | 0.5 | 0.0 | 1.1 | 0.0 | 0.1 | 0.0 | 0.0 |
| Scenario 3: Corn (c) | 0.0 | 0.3 | 0.0 | 0.5 | 0.0 | 0.0 | 0.0 | 0.0 |
| Scenario 4: Soybeans | 0.0 | 0.0 | 0.0 | 0.0 | 0.5 | 0.3 | 0.6 | 0.5 |
| Scenario 5: Corn (b) and soybeans | 0.2 | 0.5 | 0.0 | 1.1 | 0.5 | 0.5 | 0.6 | 0.5 |

*Source*: IMPACT model runs; see Table 5.1 for detailed adoption assumptions.

*Notes*: Scenarios are described in detail in Table 5.1 Corn (a)–(c) represent high-, medium-, and low-impact estimates.

For other regions of the world, the impact of *Bt* corn is smaller, as would be expected, given slower rates of adoption than in the United States. Under the high-impact *Bt* corn scenario, the European Union is projected to increase corn production by 3.1% in 2010 compared to the no-adoption scenario, Latin America by 1.4%, and China by 0.8%. There is projected to be no increase in corn production in sub-Saharan Africa, where the production impact of the decline in world prices offsets the small increases due to slow adoption of *Bt* corn (Table 5.5).

The medium- and low-impact *Bt* corn scenarios have a comparable pattern of effects, but of lower magnitude. Under the medium-impact scenario, US corn production is projected to increase by 1.1%, and corn exports by 1.8%. The low-impact scenario cuts these effects by about one-half again. Small positive secondary effects on US soybean production are still seen in the medium-impact scenario, but virtually disappear under the low-impact scenario (Table 5.4). For other regions, the impacts on corn production for medium- and low-impact scenarios decline by comparable proportions relative to the high-impact scenarios (Table 5.5).

As was described above, GR soybeans are estimated to have smaller farm-level impacts than *Bt* corn, and thus the long-run effect on global food markets of adoption is also smaller. For the United States, the GR soybean scenario (Scenario 4, Table 5.4) is projected to result in a 0.3% increase in US soybean production, and 0.2% in US soybean exports. World soybean production is projected to increase by 0.4%, and the world price to decline by 0.4% (Table 5.4). Soybean production is also estimated to increase by 0.5% each in China and the European Union, by 0.3% in Latin America, and by 0.6% in sub-Saharan Africa (Table 5.5). The final scenario (Scenario 5) combines the assumptions of the medium-impact *Bt* corn scenario with the assumptions of the soybean scenario. The estimated impacts in 2010 are essentially additive from the individual scenarios.

The fact that these two technologies have relatively small effects should perhaps not be surprising. These technological improvements involve increasing productivity by making lower cost inputs more effective. Neither has any effect on the biological yield maxima of the crops and for both, (more expensive) technologies already existed to control the problems addressed by the GMO.

## Summary

The farm-level analysis of the market effects of the two widely used food GMOs, *Bt* corn and GR soybeans, suggests that for some farmers, but not all, adoption is profitable (see Chapter 4). For corn growers, adoption is profitable if expected infestation levels and the output price are together high enough so that the increase in yields offsets the additional price of the GMO seed. All soybean growers have benefited from increased competition that has driven down the price of competing herbicides. Technology-producing firms, especially Monsanto, have benefited from licensing and technology fees. In addition, Monsanto has seen large short-run profits from increased sale of glyphosate.

Moreover, the results shown indicate that the US will continue to be the leading exporter of corn and soybean to the year 2010, capturing a significant share of increased imports to rapidly growing developing countries such as China. The continued adoption of *Bt* corn and GR soybeans would further enhance US production and exports. Under the estimated farm-level impacts utilized in this study, the impacts of *Bt* corn and GR soybeans on global food markets are nevertheless relatively moderate. However, the high-impact *Bt* corn scenario shows that continued improvement in varieties to further boost on-farm crop yields and reduce production costs could significantly increase US production and exports as well as world production of corn and soybeans.

Nevertheless, relative to the size of the global market, the impact of these technologies to date has been minor. Even with full adoption, unfettered by political constraints, the effects on world prices will be relatively small. A world with full adoption instead of no adoption would see small price declines. For consumers of goods made from these two GMOs, there is very little benefit from their adoption, mainly because the GMOs confer no enhanced consumer characteristics. Furthermore, they are inputs into final consumer goods and contribute only a fraction of the cost of the final product. Even with full adoption, consumers would see little change in prices. The current generation of GR soybean and *Bt* corn technology will continue primarily to benefit farmers.

# 6

# Cotton GMO Adoption and Private Profitability[1]

**José Benjamin Falck-Zepeda**
*Department of Agricultural Economics, Comer Hall, Auburn, Alabama, USA*

**Greg Traxler**
*Department of Agricultural Economics, Comer Hall, Auburn, Alabama, USA*

**Robert G. Nelson**
*Department of Agricultural Economics, Comer Hall, Auburn, Alabama, USA*

## Introduction

As discussed in Chapter 2, when larvae of various lepidopteran insects feed on the cotton plant, they reduce the amount of cotton produced and possibly also its quality. To control these insects farmers have a choice of using conventional cotton varieties and treating them with pesticides or using *Bt* cotton and treating only for insects not susceptible to the Cry protein. All farmers using *Bt* cotton pay a technology fee to Monsanto and a premium on seed cost to Delta and Pine Land (D&PL). In 1996 and 1997, the technology fee was $32.00 per acre and the seed premium over conventional varieties was $2 per acre. Beginning in the 1998 crop season, Monsanto and D&PL included the technology fee and seed premium in the price of the bag. They still charged $32 per acre, but adjusted the

[1] Financial support for the research was provided by the Natural Resource Conservation and Management Branch and the Market and Trade Economics Division of USDA Economic Research Service and by USDA National Research Initiative Award #99-35400-7869. This draws heavily on Falck-Zepeda (1999 and 2000a).

Genetically Modified Organisms in Agriculture
ISBN 0-12-515422-4

price of the bag based on the seed drop rate (somewhat similar to seeding rates). The decision to adopt the *Bt* technology depends on a farmer's assessment of the cost reduction from reduced pesticide treatments compared to the additional costs of the technology fee and seed premium.

In this chapter we review the economic benefits and costs of *Bt* cotton, with detailed analysis of rent creation and distribution in 1996 and 1997 using survey data. We provide preliminary surplus estimates for *Bt* cotton in 1998. We then compare the preliminary results from 1998 both to estimates from 1996, and to the estimate from the 1997 planting season presented in Falck *et al.* (2000b). Finally, we discuss some of the implications of the estimated distribution of rents on farmer and society welfare and impacts of biotechnology varieties in the US and abroad.

## Background

*Bt* cotton was planted on 4.2 million acres in the US in 1999, up from 1.8, 2.3 and 2.7 million acres in 1996, 1997 and 1998 respectively (Table 6.1). The *Bt* technology has also been adopted by farmers abroad. *Bt* cotton was planted on 514 000 acres outside the US in 1998, up from 202 000 acres in 1997 (Table 6.2). In 1997 and 1998, Australia had the highest acreage planted to *Bt* cotton with 165 000 and 200 000 acres respectively. In 1998, the second largest acreage was planted in China, representing 156 000 acres. *Bt* cotton is also grown in Mexico and South Africa, and has been tested successfully in Argentina (Videla *et al.*, 1999).

**Table 6.1**: Adoption of *Bt* cotton in the United States, selected states (%)

|  | 1996 | 1997 | 1998 | 1999 |
|---|---|---|---|---|
| Alabama | 74 | 70 | 69 | 70 |
| Arizona | 23 | 61 | 71 | 70 |
| Arkansas NE | 1 | 12 | 1 | 4 |
| Arkansas SE | 38 | 13 | 30 | 40 |
| Florida | 43 | 60 | 56 | 60 |
| Georgia | 30 | 38 | 47 | 59 |
| Louisiana | 15 | 38 | 60 | 62 |
| Mississippi | 39 | 43 | 60 | 68 |
| New Mexico | * | 1 | 35 | 29 |
| North Carolina | * | 6 | 12 | 20 |
| Tennessee | 1 | 3 | 19 | 65 |
| US | 14 | 17 | 25 | 31 |
| Total US *Bt* (000 acres) | 1 851 | 2 294 | 2 732 | 4 234 |
| Total US (000 acres) | 13 052 | 13 462 | 10 713 | 13 601 |

*Source*: Williams (1997, 1998, 1999, 2000).

*Note*: * = less than 0.5% adoption.

In the US adoption of *Bt* cotton varieties and of other genetically modified (GM) varieties has varied across states, from 1% in Virginia to 80% in Florida

(Table 6.3) in 1998.[2] Some regions with low adoption of *Bt* varieties such as North Carolina, Texas, Tennessee, and Virginia have higher adoption rates of other GM varieties such as BXN and Roundup Ready resistant cotton varieties. This may indicate that in these regions the tobacco budworm–cotton bollworm (BBW) complex is not economically important or that the available varieties are not appropriate for the region.

In 1999 adoption of *Bt* and stacked *Bt*/Roundup Ready cotton varieties increased to 32% of the total acreage planted to cotton in the United States. Adoption of BXN and Roundup Ready cotton varieties has increased modestly in 1999. Adoption of the *Bt* gene within states has changed over time, perhaps due to changing pest population pressures, differences in insect resistance to insecticides, and the increase in the number of varieties available to farmers. Adoption of the *Bt* gene including stacked varieties in all of the selected states in Table 6.3 varied from 8% in Texas to 68% and 75% in Florida and Alabama respectively.

**Table 6.2**: International Plantings of *Bt* cotton, 1996–1998 (000 acres)

|  | *1996* | *1997* | *1998* | *1999* |
|---|---|---|---|---|
| Australia |  | 165 | 200 | 309 |
| China |  | 0 | 156 | 740 |
| South Africa |  | 0 | 29 | 25 |
| Mexico |  | 37 | 111 | 50 |
| Argentina |  | 0 | 20 | 20 |
| Total |  | 202 | 514 | 1 144 |
| US | 1 851 | 2 272 | 2 411 |  |
| Total world acres planted to cotton (a) | 83 500 | 82 800 | 80 600 |  |

*Source*: James (1998), except (a) USDA/ERS (1999).

**Table 6.3**: Adoption of cotton GM varieties in the United States, selected states, 1998 and 1999 (%)

| *State* | *Bt cotton* | | *stacked Bt/RR* | | *BXN cotton* | | *Roundup Ready cotton* | | *Total* | |
|---|---|---|---|---|---|---|---|---|---|---|
|  | *1998* | *1999* | *1998* | *1999* | *1998* | *1999* | *1998* | *1999* | *1998* | *1999* |
| Alabama | 59 | 38 | 2 | 37 | * | 0 | 11 | 3 | 72 | 78 |
| Arkansas | 14 | 17 | * | 5 | 28 | 41 | 2 | 2 | 44 | 65 |
| Florida | 80 | 32 | * | 36 | * | 1 | 6 | 18 | 87 | 87 |
| Georgia | 30 | 18 | 18 | 33 | 5 | 2 | 17 | 22 | 70 | 75 |
| Louisiana | 69 | 57 | 2 | 10 | 8 | 9 | 1 | 1 | 80 | 77 |
| Mississippi | 52 | 40 | 7 | 25 | 7 | 15 | 2 | 1 | 69 | 81 |
| New Mexico | * | 26 | * | 2 | 11 | 0 | 2 | 0 | 13 | 28 |
| North Carolina | 2 | 3 | 2 | 33 | 4 | 13 | 30 | 25 | 38 | 74 |
| Tennessee | 5 | 13 | 2 | 46 | 40 | 10 | 16 | 11 | 63 | 80 |
| Texas | 7 | 5 | 1 | 3 | 0 | 1 | 27 | 35 | 34 | 44 |
| Virginia | 1 | 4 | * | 12 | 3 | 11 | 19 | 27 | 24 | 54 |
| US | 18 | 16 | 3 | 16 | 6 | 8 | 17 | 20 | 45 | 60 |

*Source*: USDA/AMS (1999, 2000). Note the *Bt* cotton estimates differ from those in Table 6.1 because of different sources.
*Note*: * = less than 0.5% adoption.

---

[2] Adoption estimates from the USDA/AMS publication 'Cotton Varieties Planted' differ in some states from the estimates made by entomologists in Williams (1997, 1998, 1999, 2000) and by James (1998).

## Previous Studies of the Economics of *Bt* Cotton

GM cotton is the most widely studied of the transgenic crops. Table 4 in Falck *et al.* (1999) presents a four-page list of studies made in the US and abroad evaluating *Bt* cotton under different biotic conditions and locations. We present here a brief discussion of a sample of these studies.

In 1995, farmers in Alabama had one of the worst cotton insect losses in history due to the pyrethroid-resistant BBW complex. One year later, with widespread use of *Bt* cotton, Alabama had the lowest amount of insecticide applications on record (Smith, 1997). Although the reductions in insecticide application cannot be attributed entirely to the *Bt* technology, it played a role in this significant reduction in insecticide application. Smith (1998) indicated that in 1997 *Bt* varieties were treated between two and four times, whereas conventional varieties were treated six to eight times. A majority of the applications on *Bt* cotton in 1997 were used to control pests of cotton not susceptible to the *Bt* toxin, such as stink and other plant bugs, and fall armyworms. In addition to the difference in the number of applications, there was a difference in the cost per application and a difference in lint yield. Average application cost for conventional varieties ranged between $10 and $15 an acre, whereas application costs on *Bt* varieties averaged $3–7 per acre.

Using farm survey data from Mississippi, Gibson *et al.* (1997) estimated an average net return for *Bt* cotton in 1996 of $246.30 per acre. For non-*Bt* cotton average net returns were $230.08 per acre, a difference of $16.23 per acre in favor of *Bt* cotton. To decrease the possibility of insect resistance to *Bt*, producers had to set aside areas, called refugia (or refuges), to support non-resistant pest populations. Producers had two options for the refugia. In the first option, producers could plant 4% of the area to non-*Bt* cotton. This acreage could not be treated with any pesticide targeted to control the BBW complex. This was called the 96/4 option. In the second option, producers could plant 20% of the area with non-*Bt* cotton and treat the area with any pesticide except foliar *Bt* spray. This was called the 80/20 option. A comparison of the average net returns without considering the refugia set aside would not capture total economic costs. The authors of this study made an adjustment for refugia, estimating that average per acre net returns for the surveyed refugia were $230.08 for the 80/20 option, whereas for the 96/4 option average per acre net returns were $120.60. Incorporating the net returns for the refugia to estimate total economic costs of *Bt* cotton yields an adjusted average difference in per acre net returns of $11.17 in favor of *Bt* cotton varieties.

Carlson *et al.* (1997)[3] surveyed approximately 300 producers in Alabama, Georgia, South Carolina, and North Carolina for the 1996 planting season to evaluate the impact of *Bt* variety adoption in the southeast. The researchers reported that average yields were 11.4% higher in fields planted to *Bt* varieties than conventional varieties. Application costs were reduced by an average of $33 and insecticide applications were reduced almost 72%. Additional profits from the adoption of *Bt* cotton varied between $59 and $111 per acre.

---

[3] An electronic version of this paper is Marra *et al.* (1996).

In 1998, Mullins and Mills (1999) surveyed 109 sites where growers, consultants, and university/extension researchers kept records of costs and yields of similarly managed and closely located *Bt* and conventional fields in southern and southeastern states. Average yields for *Bt* fields were 37 lb per acre higher than for fields planted with conventional varieties. The overall number of insecticide applications on *Bt* fields was lower than in conventional fields, yet the number of insecticide application for pests not controlled by *Bt* varieties was higher. Average insecticide control costs were $15.43 lower on *Bt* fields, after paying the technology fee. The net economic advantage of *Bt* varieties was approximately $40 per acre.

In Mexico, Magaña *et al.* (1999) found in one hectare plots that yields of transgenic *Bt* cotton were between 13% and 29% higher for transgenics, with no significant differences in operating expenses between varieties, and with a difference in gross returns that varied between 4% and 30%.

## The Empirical Model

We modeled the introduction of *Bt* cotton in 1996 as occurring in a large open economy with no technology spillovers, and assumed linear supply and demand curves and a parallel shift in supply from the new technology (Alston *et al.*, 1995, p. 213). Figure 6.1 illustrates this approach.

In Figure 6.1, adoption of the *Bt* gene technology in the US induces a shift in the supply curve from $S_{US,0}$ to $S_{US,1}$. Because the US is a major net exporter of cotton in world markets and thus affects directly world prices, the shift in the US supply curve will induce a shift in the excess supply curve from $ES_{US,0}$ to $ES_{US,1}$. A shift in the $S_{US}$ and the $ES_{US}$ curves induces a decrease in price from $P_0$ to $P_1$.

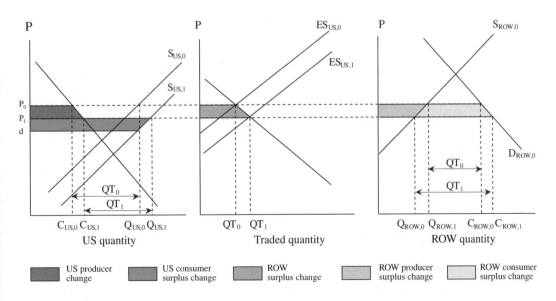

**Figure 6.1**: Effects of *Bt* cotton adoption

In this graph the distance between $P_0$ and d is a measurement of the vertical shift induced by the technology.

For 1997 and 1998, *Bt* cotton was available for planting in other countries besides the US. We modified the 1996 model to include the technology spillovers in 1997 and 1998. The main difference between models is that a separate *k* (the vertical shift of the supply function estimated as a proportion of the initial equilibrium price) is estimated for the US and the rest of the world (ROW), and is inserted in the formulas. This implies that the Z (a variable that measures the reduction in price relative to the initial equilibrium price that is due to the shift in the supply curve) estimated in the no-technology spillovers model differs from one that allows for spillovers. The graph and formulas differ for the 1996 and 1997/1998 models and are presented in Falck-Zepeda *et al.* (2000a, 2000b).

US and rest of the world (ROW) supply and demand are modeled as:

US supply: $Q_{US} = \alpha_{US} + \beta_{US}(P + k) = (\alpha_{US} + \beta_{US} k) + \beta_{US}P$
US demand: $C_{US} = \gamma_{US} - \delta_{US}P$
ROW supply: $Q_{ROW} = \alpha_{ROW} + \beta_{ROW}P$
ROW demand: $C_{ROW} = \gamma_{ROW} - \delta_{ROW}P$

where *k* is the vertical (price) shift in the supply function due to the introduction of the technology, and P is the world price of cotton. $Q_{US}$ and $C_{US}$ are quantities produced and consumed in the US, $Q_{ROW}$ and $C_{ROW}$ are quantities produced and consumed in the rest of the world. Trade equilibrium, $QT_0 = C_{ROW,0} - Q_{ROW,0} = Q_{US,0} - C_{US,0}$, is assumed in the model. The technology-induced cotton supply shift was estimated for each of the 30 cotton-producing regions of the US using data on yield, cost savings net of increased seed costs, and adoption rates.[4]

---

[4] A counterfactual price reduction must be calculated to isolate the effect of the technology-induced supply shift from other exogenous changes in supply and demand. This price change differs from the observed change in world price in that it represents what the world price would have been in 1996 if all supply and demand conditions had been identical except for the introduction of the new technology. Following Alston *et al.* (1995) and Pinstrup-Andersen *et al.* (1976), the counterfactual world price ($P_0$), and the relative price change (Z) can be calculated in elasticity form as:

$(P_0 = P_1 / \{1 - [\varepsilon_{US} K / [\varepsilon_{US} + S_{US} \eta_{US} + (1 - S_{US}) \eta^{EB}]\}$, and
$Z = -(P_1 - P_0)/P_0 = \varepsilon_{US} K / [\varepsilon_{US} + S_{US} \eta_{US} + (1 - S_{US}) \eta^{EB}]$,

where $K = k / P_0$ converts the absolute price shift to a percentage reduction in price, $\varepsilon_{US}$ is the US elasticity of supply for cotton lint, $\eta_{US}$ is the absolute value of the US demand elasticity, $\eta^{EB}$ is the absolute value of the elasticity of export demand (the ROW excess demand elasticity), and $S_{US}$ is the share of US production consumed domestically. Because all US regions face the same world price, the relative price change is the same and, by invoking the Law of One Price, regional prices differ only by transportation costs.

The formulas for changes in domestic and ROW producer and consumer surpluses were:

$$\Delta CS_{US} = P_0 C_{US,0} Z (1 + 0.5 Z \eta_{US}),$$
$$\Delta PS_{US} = P_0 Q_{US,0} (K - Z) (1 + 0.5 Z \varepsilon_{US}),$$
$$\Delta CS_{ROW} = P_0 C_{ROW,0} Z (1 + 0.5 Z \eta_{ROW}),$$
$$\Delta PS_{ROW} = -P_0 Q_{ROW,0} Z (1 + 0.5 Z \varepsilon_{ROW}),$$
$$\Delta ROWS = \Delta CS_{ROW} + \Delta PS_{ROW},$$

where $\Delta CS_{US}$ is the change in consumer surplus in the US, $\Delta PS_{US}$ is the change in producer surplus in the US, $\Delta CS_{ROW}$ is the change in consumer surplus in the rest of the world, $\Delta PS_{ROW}$ is the change in producer surplus for the foreign sector, $\Delta ROWS$ is the change in rest of the world surplus, $P_0$ is the pre-innovation price, $\eta_{ROW}$ is the absolute value of ROW demand elasticity, and $\varepsilon_{ROW}$ is the ROW supply elasticity.

Because there is a single supplier of the genetic technology, we measure monopoly profits produced by the downstream seed input supplier and the biotechnology firm. A complete description of the model and the methodology used can be seen in Falck-Zepeda *et al.* (2000a, 2000b).

Data for the estimation of economic rents in 1996 and 1997 come from a survey of consultants made by Plexus Marketing Research Inc. and Timber Mill Research Inc. in 1996 and 1997. Consultants were compensated for the collection of data on costs and yields of paired on-farm *Bt* and non-*Bt* fields. In both cases, fields were selected so that they had similar agronomic, production, and crop rotation histories. For the preliminary estimates for 1998 we used results from the survey presented by Mullins and Mills (1999). Since this survey is not comprehensive, we used the average of yield and cost changes of the states with available information to provide an estimate for the states with no information.

Because there is limited information on cotton farming in the rest of the world, we assumed that non-US cotton farmers obtained the same efficiency as that obtained by US farmers. Results from a study by Magaña *et al.* (1999) in Mexico indicate that *Bt* varieties were relatively well adapted to the environmental and biocide conditions in a state in Mexico.

To estimate monopoly profit accruing to Monsanto and D&PL we assumed that the seed multiplication process was identical for transgenic and conventional varieties. We assumed that the market for conventional seed cotton is competitive, so that the conventional market price, $P_c$, is equivalent to the marginal cost of producing seed. In other words, we considered that any increase in price above the price of conventional seed ($P_c$) contributes to monopoly profits. Thus total monopoly profits can be estimated by the formula: $Q_{Bt} (P_{Bt} - P_c)$. In this formula $Q_{Bt}$ and $P_{Bt}$ are the quantity and price of *Bt* seed, and $P_c$ is the marginal cost of producing seed or the price of conventional varieties. These figures represent gross *Bt* revenue – no administrative, marketing, or intellectual property rights (IPR) enforcement costs were deducted.

To estimate the allocation of monopoly profits between Monsanto and D&PL we used information contained in the D&PL and Monsanto annual reports for 1997 as used in Falck-Zepeda *et al.* (1999). Estimates from these reports indicated that Monsanto passed approximately $5.11 per acre back to D&PL for using their germplasm. Along with the $2 per acre seed premium, this provided D&PL

with gross revenue per acre of $7.11. Monsanto's share of gross profit was estimated by subtracting $5.11 from the $32 technology fee, resulting in gross revenue of $26.89 per acre for Monsanto.

The observed world cotton price ($P_1$) was (per pound) $0.63 in 1996, $0.84 in 1997, and $0.53 in 1998. The share of US production consumed domestically was 0.59 in 1996, and 0.6 in 1997 and 1998. The prices and share information come from the USDA/ERS publication *Cotton and Wool: Situation and Outlook* for various years. The elasticities used in this study came from previous studies. We used a value of 0.84 estimated by Taylor (1993) for the elasticity of supply, $\varepsilon_{USA}$. We used a value of 0.101 from Kinnucan and Miao (1999) for the domestic demand elasticity ($\eta_{USA}$). For the export elasticity, $\eta_{EB}$, we used an estimate of 1.62 from Duffy *et al.* (1990). For $\varepsilon_{ROW}$ and $\eta_{ROW}$, the ROW supply and demand elasticities, we used the estimates of 0.15 and 0.13, respectively, from Sullivan *et al.* (1989).

## Results

Results for the 1996 cotton crop year, using the Plexus survey data, are presented in Table 6.4. In that year *Bt* cotton created $134 million in additional surplus. Farmers received $58.2 million, representing 43% of total surplus. Monsanto gained $49.7 million and D&PL gained $13.1 million, representing 37% and 10% of total surplus, respectively. US consumers gained almost $7.6 million. Consumers in the rest of the world gained $12.8 million, but non-US producers lost $7.6 million due to lower prices.

**Table 6.4**: Estimates of economic surplus due to the introduction of *Bt* cotton in 1996

|  | Mean values ($ million) | Percent of total |
| --- | --- | --- |
| US consumer surplus | 7.6 | 6 |
| US farmer surplus | 58.2 | 43 |
| Monsanto | 49.8 | 37 |
| Delta and Pine Land | 13.2 | 10 |
| ROW producer surplus | (7.6) | – |
| ROW consumer surplus | 12.9 | – |
| Net ROW surplus | 5.3 | 4 |
| Total world surplus | 134.0 | 100 |

Data used to estimate rents for 1998 are presented in Table 6.5. The last column in this table presents the regional estimate of farmer surplus. In our estimates most states gained from the planting of *Bt* cotton in 1998, except for northeast Arkansas, California, Missouri and some regions in Texas. Results of the estimation of economic surplus for the 1998 planting using different efficiency assumptions are presented in Table 6.6. Columns 2 and 3 present results for observed yield and cost changes in the Mullins and Mills survey. Total surplus in 1998 amounted to $213 million. US consumers captured $14.0 million and US farmers gained $97.2 million. In contrast ROW producers lost $14.0 million, whereas ROW consumers gained $23.4 million. Results of our estimations are percentage-wise similar to the results obtained in 1996 using the Enhanced Marketing Data (EMD) and other sources of data.

In order to examine the sensitivity of our estimates to the cost and yield changes data, we estimate rent distribution with 50% drops in cost and/or yields for all states. Thus, columns 4 and 5 present the results of decreasing observed yield changes for each region by 50%. Columns 6 and 7 present results where cost changes are reduced for each region by 50%, and columns 8 and 9 present results for decreasing both yield and cost changes by 50%. Results of the sensitivity analysis presented in columns 4 through 7 indicate that as expected our results are more sensitive to the yield change assumption.

In the 50% decrease in the yield change and cost change scenarios, farmers' share of rents decrease to 38% and 32% respectively, whereas the innovators' share increases to 52% and 61% for the same scenarios. A drastic reduction in either the yield or cost makes *Bt* cotton less attractive to farmers in some regions.

**Table 6.5**: Percent adoption, total acres, yield difference, cost reduction and farmer surplus, 1998

| State/Region | Percent adoption Bt cotton[a] | Total acres planted to cotton (000 acres)[a] | Percent yield difference (Bt – non-Bt)[b] | Percent pesticide cost reduction (Bt-Non-Bt)[b] | 1998 Farmer surplus (000 $) |
|---|---|---|---|---|---|
| Alabama Central | 73 | 80 | 4 | 9 | 2 132 |
| Alabama North | 75 | 185 | 4 | 11 | 4 617 |
| Alabama South | 62 | 190 | 4 | 9 | 3 927 |
| Arizona | 71 | 250 | 4 | 6 | 10 004 |
| Arkansas Northeast | 1 | 513 | 11 | 6 | −831 |
| Arkansas Southeast | 30 | 342 | 3 | 8 | 3 293 |
| California | * | 826 | 4 | 5 | −3 012 |
| Florida | 56 | 80 | 4 | 11 | 1 753 |
| Georgia | 47 | 1300 | 4 | 8 | 27 174 |
| Louisiana | 60 | 540 | 3 | 10 | 10 976 |
| Mississippi Delta | 44 | 565 | 3 | 8 | 8 687 |
| Mississippi Hills | 82 | 365 | 3 | 9 | 10 798 |
| Missouri | 1 | 350 | 4 | 10 | −547 |
| New Mexico | 35 | 60 | 4 | 10 | 1 351 |
| North Carolina | 12 | 695 | 8 | 7 | 2 428 |
| Oklahoma | 10 | 120 | 4 | 11 | 217 |
| South Carolina | 40 | 280 | 8 | 5 | 5 634 |
| Texas Coastal Bend | 12 | 425 | 4 | 10 | 1 050 |
| Texas Far West | 22 | 159 | 4 | 11 | 914 |
| Texas High Plains | * | 1900 | 4 | 11 | −2 844 |
| Texas Lower Rio Grande | 5 | 215 | 4 | 10 | −7 |
| Texas North Central | 23 | 240 | 4 | 15 | 1 353 |
| Texas Rolling Plains | 17 | 180 | 4 | 15 | 633 |
| Texas South Central | 29 | 140 | 4 | 7 | 1 317 |
| Texas South Rolling Plains | 35 | 170 | 4 | 18 | 1 539 |
| Tennessee | 19 | 450 | 11 | 5 | 4 628 |
| Virginia | 3 | 91 | 8 | 6 | 2 |
| Total | | | | | $97 188 |

[a]*Source*: 1999 Beltwide Cotton Conference Proceedings.

[b]*Source*: Mullins and Mills (1999).

*Note*: *= less than 0.5 % adoption.

**Table 6.6:** Estimates of economic surplus due to the planting of Bt cotton in 1998

| | Results using survey yield and cost changes[a] | | 50% yield decrease | | 50% cost change decrease | | Decrease both cost and yield change by 50% | |
|---|---|---|---|---|---|---|---|---|
| | Value ($ million) | Percent of total | Value ($ million) | Percent of total | Value ($ million) | Percent of total | Value ($ million) | Percent of total |
| US consumer surplus | 14.0 | 7 | 9.9 | 6 | 6.9 | 4 | 3.6 | 3 |
| US farmer surplus | 97.2 | 46 | 68.3 | 38 | 49.3 | 32 | 26.4 | 21 |
| Monsanto | 73.5 | 34 | 73.5 | 41 | 73.5 | 48 | 73.5 | 59 |
| Delta and Pineland/Other | 19.4 | 9 | 19.4 | 11 | 19.4 | 13 | 19.4 | 15 |
| ROW producer surplus | (14.0) | | -9.9 | | -6.9 | | -3.6 | |
| ROW consumer surplus | 23.4 | | 16.5 | | 11.5 | | 6.1 | |
| Net ROW surplus | 9.3 | 4 | 6.6 | 4 | 4.6 | 3 | 2.4 | 2 |
| Total world surplus | 213.4 | 100 | 177.7 | 100 | 153.7 | 100 | 125.4 | 100 |

[a]Table assumes rest of the world (ROW) farmers have the same efficiency as US farmers. Results from assuming lower efficiencies yielded similar results.

With a 50% reduction in *both* cost and yield increases, the innovators' share increases to 74% from the previous scenarios, whereas farmers' share decreases to 21%, consumers' share decreases to 3% and ROW net share decreases to 2%.

It is important to note that these are additional benefits to farmers because of the adoption of the technology. Farmers are capturing an extra 21% of additional rent in the worst case scenario due to the adoption of the technology. We should also note that those former benefits are net of adoption costs, while industry benefits are 'gross'. Additional marketing and enforcement costs have not been subtracted from Monsanto/Delta and Pineland benefits.

## Discussion

Results of our estimations of distribution of rents in 1996, for 1997 (presented in Table 6.7) and for 1998, are similar. Farmers and the monopolist innovators share the additional rents created by the innovation almost in equal proportion. The only occasion where there is a departure from this result is when yield and cost decreases are reduced enough so that it becomes hard to justify using *Bt* in some regions, as shown in our simulation of decreasing by half yield and/or cost changes. As with any other *ex-post* analysis our estimations do not take into consideration risk preferences and premiums. From the standpoint of arriving at a true estimate of the value of *Bt*, there is a need to evaluate gains/losses of adoption in an *ex-ante* analysis where risk is taken into consideration. We concur with entomologists (Benedict, 1996) that *Bt* cotton is a risk management tool, and thus even if a region may have 'lost' in an *ex-post* analysis, in an *ex-ante* analysis this would have been a winning strategy.

**Table 6.7**: Economic surplus results 1997 cotton crop (ROW producers obtain same yield increases and cost decreases as US producers)

|  | Values ($ million) | Percent of total |
| --- | --- | --- |
| US consumer surplus | 14.0 | 7 |
| US farmer surplus | 80.0 | 42 |
| Monsanto | 67.1 | 35 |
| Delta and Pineland/Other | 17.7 | 9 |
| ROW producer surplus | (12.1) | |
| ROW consumer surplus | 23.4 | |
| Net ROW surplus | 11.2 | 6 |
| Total World Surplus | 190.1 | 100 |

*Source*: Falck-Zepeda *et al.* (2000a).

## Summary and Conclusions

We used a data set from Plexus Marketing Research Inc. and Timber Mill Research Inc. to estimate rent creation and distribution from the adoption of *Bt* cotton in 1996. Our results indicate that farmers gain between 43% and 59% of all rents created from the introduction and adoption of *Bt* cotton. The innovators D&PL and Monsanto gain between 26% and 47% of all rents in 1996.

Preliminary results from the estimation of rent creation and distribution for 1998 indicate that farmers and the innovators share almost equally rents created by adopting *Bt* cotton. Farmers gain 43% of total rents whereas the innovators gain 47% of total rents. Regionally there were winners and losers from the adoption of *Bt* cotton in 1998. Regions with low adoption such as California and Missouri lost because farmers suffered a reduction in cotton lint prices without having the benefits of the technology.

We performed a sensitivity analysis to evaluate results by reducing the yield and/or cost change assumptions by half. In the worst case scenario, where yield increases and cost reductions were reduced by 50%, farmers still were able to capture 21% of the total rents, whereas the innovators gained 74% of total rents.

Over the three years that have been analyzed by the authors, results have been fairly consistent. Farmers share almost equally with the innovators the rents created by the technology even when a monopolistic structure for the input market is assumed.

# 7

# GMO Adoption and Nonmarket Effects

**Gerald C. Nelson**
University of Illinois, Urbana, Illinois, USA

**Alessandro De Pinto**
University of Illinois, Urbana, Illinois, USA

The goal of both traditional plant breeding and genetic engineering is to identify desirable genetic traits and combine them in a crop variety that can be used in agriculture. (See Chapter 27 for an extended discussion of techniques available to the modern plant breeder, including genetic modification and others.) Desirable traits can be broadly divided into two classes – agronomic characteristics of the plant and quality characteristics of the product. Agronomic characteristics include yield, resistance to diseases, insects, and herbicides, and ability to thrive under adverse environmental conditions. Quality characteristics include processing, preservation, nutrition, and flavor of the final product. Traditional breeding efforts combine varieties of the same species and screen for offspring with desired characteristics. Such efforts can introduce traits found only in close relatives. Genetic modification techniques expand the range of achievable genetic variation.

As with any new product a new GMO has the potential for unintended consequences, negative and positive. In this chapter, we review three sets of issues associated with the adoption of GMOs – environmental safety, food safety, and the indirect impact of the spread of the technology on developing countries. Each issue raises questions about existing regulatory approaches (reviewed in Chapter 9). Together they represent a significant challenge to the domestic and international structure of regulatory control. Ethical issues further complicate the debate.

Genetically Modified Organisms in Agriculture
ISBN 0-12-515422-4

The chapter starts with an overview of potential environmental and food safety side effects. We then review the available evidence for GR soybeans, *Bt* corn, and *Bt* cotton. We conclude with a discussion of intellectual property rights concerns of developing countries that arise from granting patents on GMOs. We recognize the importance of ethical concerns but have no expertise to contribute to that discussion.[1]

## Potential Environmental and Food Safety Side Effects[2]

Generally, both positive and negative side effects of GMOs are hypothetical today because only a small number are in large-scale commercial production (these include GR soybeans and corn, *Bt* corn and cotton, tomatoes, potatoes, and canola (rape)). We address first potential environmental benefits and costs and then turn to food safety issues.

### Potential environmental benefits

The first commercial GMO crops were designed primarily to deal with pests, either weeds (GR soybeans) or lepidopteran insects (*Bt* corn and cotton). To the extent that they reduce overall pesticide application, they reduce the potential for collateral damage, such as to nontarget species, including humans. Even if the effect of the technology is primarily to substitute one pesticide for another (such as GR crops), the net effect might be to reduce negative environmental consequences. For example, this would occur if the new pesticide is less toxic to nontarget species, affects fewer species or degrades more quickly.

Herbicide-resistant crops also can contribute to soil-related environmental improvements. With less need for weed control, fewer passes of the tractor through the field are needed. Total soil movement, and hence soil erosion, is reduced. Reduced work in the field results in less soil compaction, which in turn corresponds to more oxygen in the topsoil and better water conditions in the root zone. Finally, fewer trips over the field reduce fuel consumption.

Finally, if a GMO increases the productivity of a crop it is possible that less land will be devoted to its cultivation. In areas where there is pressure to convert environmentally sensitive land to agriculture, increased productivity elsewhere may reduce this pressure. (See Chapter 20 for a vigorous statement of these benefits.) As with all the other potential environmental benefits and costs, this effect depends on many factors that have nothing to do with the GMO.

### Potential environmental costs

We have divided these costs into two categories – short and long run. Short-run costs, such as off-site mortality to nontarget organisms (e.g., the monarch butterfly, see Chapter 28 for details) are usually reversible by ending the use of the

---

[1] A useful reference in this area is Thompson (1997).

[2] Perspectives on the environmental benefits and costs are contained in Chapters 19 and 20.

GMO. Long-run costs, such as development of pest resistance or enhanced survivability (such as virus-resistant crops), have two characteristics – they usually develop relatively slowly and usually are not reversible.

## Short-run off-site environmental costs

### Risks for nontarget organisms
The primary short-run negative environmental side effect of GMOs designed to resist pests is that nontarget organisms might be affected as well. For example, the target species of *Bt* corn are a few lepidopterans, but other members of this order are killed if they consume *Bt* toxins. In addition, other species (e.g., birds, fish, mammals, arthropods) that rely on affected target or nontarget pests might suffer population declines if their food supply is reduced.[3] As long as populations are not completely eliminated by the GMO, they can recover if use of the GMO is eliminated. Hence, these costs are reversible.

## Long-run sustainability costs
Three long-run, sustainability concerns associated with GMOs have received the most attention – resistance development in target and nontarget populations, the possibility that with enhanced survival traits the plant might become a weed, and flow of the novel genetic material to other species changing local or global ecosystems. Another sustainability concern arises from the use of antibiotic resistance-inducing marker genes.

### Selection of resistant insects and weeds
Development of resistance in target species is probably the single most important sustainability concern associated with GMOs engineered to enhance pest management. In a changing environment, natural selection mechanisms encourage the development of populations to deal with the changes. If a plant produces a compound generally toxic to an insect species, any individual of that species not susceptible to the toxin has a reproductive advantage. Because insects have a short lifespan and high reproductive capability, resistance can move quickly through the population. The speed of resistance development is influenced by the nature of the toxin, the number of individuals with resistance-generating genetic material, and whether the genetic material is recessive or dominant (whether both parents must have it or just one). Weeds are slower to evolve resistance to pesticides than insects because they are less mobile and usually reproduce more slowly. However, many weeds have developed resistance to one or more herbicides.[4]

---

[3] For example, preliminary, unpublished research suggests that because of reduced ECB densities in *Bt* corn, the populations of *Macrocentrus grandii* (a wasp parasitoid) and *Nosema pyrausta* (a microsporidian disease pathogen) might be reduced by 95% or greater because of the relative scarcity of their host (personal communication from Kevin Steffey, Professor and Extension Specialist in Entomology, Department of Crop Sciences, UIUC).

[4] See the results of the 1998 International Survey of Herbicide-Resistant Weeds, available at *http://www.weedscience.com/*, accessed October 9, 1999.

*Genetic flow*

Genetic flow is the sexual transmission of genetic material from one species to another. Two concerns have been raised for current GMOs – movement of herbicide- and virus-resistance genes to wild relatives of the GM crop and antibiotic resistance marker genes to microorganisms in the stomachs of animals that eat a GM crop. When plants can exchange genes, novel genes might be passed from crops to weeds and disturb local ecosystems. This is particularly true for crops grown where wild relatives are found. If a herbicide- or virus-resistant plant passes the resistance to wild relatives, weed control becomes more difficult. If an insect-resistant plant passes the resistance to wild relatives, resistance management becomes much more difficult. Of the current commercial GM crops, canola (rape) has been the primary crop of concern. Canola is grown in regions where weeds with some degree of sexual compatibility exist such as field mustard (*Brassica rapa*), wild mustard (*Sinapsis arvensis* L.), hoary mustard (*Hirschfeldia incana* L.) and wild radish (*Raphanus raphanistrum* L.) (Chèvre *et al.*, 2000). Other GM crops where the potential for genetic flow has been researched include sugar beets (*Beta vulgaris* ssp. *vulgaris*) (Pohl-Orf *et al.*, 2000) and various cucurbits (cucumbers, squashes).

*Weediness of crops*

A commercial crop becomes a weed if it survives beyond its economic life (for example, over winter) or disburses seeds that can germinate and interfere with an alternate crop in the next growing period. For example, if a GR plant sprouts in a field where the intended crop is another GR plant, the effectiveness of glyphosate is reduced. Another possibility is that the crop itself would move into wild ecosystems and disrupt them.

Of the current commercial GM crops, weediness seems to be a problem only with canola because some viable seed is typically lost in the harvesting process. Corn[5] appears as a volunteer in some fields and roadsides, but it never has been able to establish itself outside of cultivation (Gould, 1968). Some relatives of corn are successful wild plants but have no pronounced weedy tendencies (Galinat, 1988). Soybean plants[6] are annuals and do not survive from one growing season to the next (Hymowitz and Singh, 1987). Some soybean seeds may be lost in harvesting and survive to sprout in the next crop year but there is no evidence to suggest that these plants are able to establish themselves.

*Loss of antibiotic effectiveness*

As indicated in Chapter 2, the most common genetic markers used to indicate successful uptake of the novel genetic material cause the plant to produce a substance that inactivates an antibiotic or a herbicide.

As can be seen in Table 7.1, the kanamycin resistance gene (known both as kan[r] and NptII) is widely used. Plant cells that express the kanamycin resistance gene survive and replicate on laboratory media in the presence of any of several

---

[5] Extracted from *http://www.aphis.usda.gov/biotech/corn.html*.
[6] Extracted from *http://www.aphis.usda.gov/biotech/soybean.html*.

**Table 7.1**: Selection markers in GM crops

| Crop | Selection markers |
|------|-------------------|
| Corn | Bla, cat, NptII, PAT enzyme, gox gene, CP4 EPSPS |
| Tomato | NptII |
| Canola | NptII |
| Cotton | NptII |
| Potato | NptII, CP4 EPSPS |
| Soybean | NptII |
| Squash | NptII |
| Sugar beet | GUS |
| Papaya | NptII |
| Radicchio | NptII, PAT |

*Source*: Derived from petitions to APHIS (Animal and Plant Health Inspection Service) for determination of nonregulated status published in the *Federal Register*. The petition includes a description of the product and selection markers (*http://www.gpo.ucop.edu/search/default.html*).

*Notes*: Antibiotic resistance: NptII (neomycin phosphotransferase II, also known as kan$^r$); PAT (phosphinothricin-*N*-acetyltransferase).

Glyphosate resistance: gox gene (glyphosate oxidoreductase), CP4 EPSPS (5-enolpyruvylshikimate-3-phosphate synthase).

Other selection markers: GUS (β-D-glucuronidase), bla (beta-lactamase), cat (chloramphenicol acetyltransferase gene).

antibiotics.[7] By linking the selectable marker gene to another gene that specifies a desired trait, scientists can identify plants that have taken up and expressed the desired genes. The kanamycin resistance gene has been used as a selection marker in more than 30 crops. Once the desired plant variety has been selected, the kanamycin resistance gene serves no further useful purpose, although it continues to produce the enzyme responsible for the resistance.

The continued existence of the gene in the crop and the enzyme it produces raise two concerns. First, antibiotics intended for therapeutic use might be inadvertently inactivated. Second, the genetic material that confers antibiotic resistance might flow to microorganisms in the gut of animals consuming the GMO (genetic flow).

In 1990, the Food and Drug Administration (FDA) was first asked to evaluate the use of the kan$^r$ gene by Calgene in its Flavr Savr tomato. In 1993, Calgene asked FDA to change its request to a food additive petition under section 409 of the Federal Food, Drug, and Cosmetic Act (see Chapter 9 for more details on FDA regulatory powers). FDA evaluated the use of the gene not only in tomatoes but also in cotton and canola.

In 1994, FDA decided that the product of this gene, APH(3′)II, was safe as a food additive. FDA looked explicitly at both food and feed use of products containing the kan$^r$ gene, both for the likelihood of inactivation of therapeutic antibiotics and for the potential for genetic flow to microorganisms. There are examples of resistance gene exchange in the stomach and intestines, some of them across genus and more distant lines. It is also the case that about 10% of the

---

[7] In addition to kanamycin, the enzyme also inactivates the following antibiotics: neomycin, paromomycin, ribostamycin, gentamicins A and B, as well as butirosins. *http://vm.cfsan.fda.gov/~dms/OPA-ARMG.HTML#1*, accessed September 23, 1999.

human population harbors bacteria carrying the APH resistance gene. The essence of the decision to approve commercial use was that the frequency of *in vivo* transfer is very low, and the natural incidence of resistance genes in the environment is much higher. FDA thus regarded the risk as real but small compared to other known mechanisms of dissemination of genes that occur frequently in nature.[8] It concluded that neither potential problem was likely to take place.[9]

## Unintended Food Safety Effects[10]

Virtually all breeding techniques, not just those involving novel DNA, have the potential to create unexpected food safety effects. These effects can be positive or negative. Mutations unrelated to the desired modification might be induced; undesirable traits might be introduced along with the desired traits; newly introduced DNA might inactivate a host gene or alter control of its expression; or the introduced gene product or a metabolic product affected by the genetic change may interact with other cellular products. Three categories of potential side effects affect all crop-breeding techniques – changes in known toxicants, changes in nutrient levels and composition and in allergenicity. A fourth category – introduction of new substances – is exclusive to genetic engineering.

### Changes in known toxicants[11]

Plants produce toxic substances as part of their natural defense mechanisms. Toxicants and antinutritional factors such as neurotoxins, alkaloids, protease inhibitors, and hemolytic agents are found in crops as common as cereals, legumes, and Cucurbitaceae. Commonly, these toxicants are either present at levels that do not pose a threat to human health or are broken down to harmless forms during processing. The process of breeding or genetic manipulation can

---

[8] Personal communication from Bruce Chassy, Professor, Department of Food Science and Human Nutrition, UIUC.

[9] For more details, see FDA's review document at *http://vm.cfsan.fda.gov/~dms/OPA-ARMG.HTML#1*, accessed September 23, 1999. The UK Advisory Committee on Novel Foods and Processes (ACNFP) expressed some concern about the potential that the antibiotic resistance gene would flow to microorganisms in connection with Ciba-Geigy's application for approval of *Bt* corn in the EU. The concern was with respect to the use of unprocessed corn fed to farm animals. The EU scientific committees on pesticides, animal feedingstuffs and food decided that this risk was minimal, and the permission for commercial use was granted. See *Food Safety Information Bulletin*, United Kingdom Ministry of Agriculture, Fisheries and Food, No. 81, January 1997. Available at *http://www.maff.gov.uk*, accessed September 23, 1999.

[10] This section draws heavily on two sources: FDA's 1992 Statement of Policy: Foods Derived from New Plant Varieties, *Federal Register*, May 29, 1992 (reproduced at *http://vm.cfsan.fda.gov/~lrd/fr92529b.html*) and Maryanski (1995).

[11] Sources for this section are Kessler *et al.* (1992) and FDA's 1992 Statement of Policy: Foods Derived from New Plant Varieties, *Federal Register*, May 29, 1992, *http://vm.cfsan.fda.gov/~lrd/fr92529b.html*, accessed October 1, 1999.

cause toxicants ordinarily present at safe levels to be produced at high levels. A common regulatory function is to make sure that new plant varieties do not have levels of toxicants significantly higher than in other edible varieties of the same species.[12]

## Changes in nutrient levels

Another unintended consequence of genetic modification of the plant may be a significant alteration in levels of important nutrients. In addition, there may be changes in availability of a nutrient due to changes in its form or the presence of increased levels of other constituents that affect absorption or metabolism of nutrients.

## Allergenicity

All food allergens are proteins. However, only a small fraction of the thousands of proteins in the diet have been found to be food allergens. Genetic manipulation may transfer proteins known to be allergenic from one food source to another. Testing for this potential risk is fairly straightforward, as the chemical structure of known allergens is available and various testing methods have been developed.

However, when a new protein that has never been a part of the diet, such as a *Bt* Cry protein, is added to foods, methods of testing for potential allergenicity are less well developed. It is possible to compare the molecular structure to known allergens but this does not ensure that some segment of the population will not develop an allergic reaction.

## Introduction of new substances

Because plant breeders using genetic engineering techniques are able to introduce essentially any trait or substance whose molecular genetic identity is known into virtually any plant, it is possible to unintentionally introduce a protein that differs significantly in structure or function, or to modify a carbohydrate, fat, or oil, so that it differs significantly in composition from such substances currently found in the crop. Such changes can be either positive or negative from a food safety perspective.

## Food safety effects from agronomic changes

Changes in agronomic characteristics might also result in unintended food safety effects. For example, toxicants such as aflatoxin and fumonisin are more likely to occur in plants that have suffered mechanical damage, say from insect feeding. Food produced from GM plants that reduce insect damage will also be less likely to contain these toxicants.

---

[12] FDA's Center for Food Safety and Applied Nutrition has an extensive database on poisonous plants at *http://vm.cfsan.fda.gov/~djw/readme.html*.

## Structural and Income Distribution Concerns

Proponents of GM technology see the promise of great benefits for developing countries (see Chapter 17). Examples include crops that perform well in adverse conditions or that have enhanced nutritional characteristics. At the same time, concern about potential for negative consequences of GMOs for developing countries has grown (see Chapter 18). In many ways, these concerns mirror those expressed about crops developed in earlier Green Revolutions such as hybrid corn and high-yielding rice and wheat. There were concerns about excessive reliance on modern technology, dependence on modern inputs sold by large corporations, and undesirable changes in traditional ways of doing things. In fact, early adopters, and those better integrated with other sectors, benefited initially, but benefits quickly became widespread. Changes in the structure of the farm sector and the input supply industries resulted. Though generally welcomed, the adoption of the new varieties caused tensions and required new institutions and policies to capture the benefits without imposing unacceptable costs.[13]

In one crucial way, GMOs differ from the most important Green Revolution crops. Technologies developed in both earlier Green Revolutions were often distributed by the public sector. Intellectual property rights were mainly an issue with privately developed hybrid corn and biology effectively enforced those rights. In contrast, most GMOs have been developed by the private sector and are currently grown in developed countries. Furthermore, the number of firms engaged in this development has shrunk dramatically in the last five years' increasing the potential for exercise of monopoly power.

The private sector has made sure that intellectual property issues are a high priority in national and international fora. In addition, it has proposed the development of biological innovations that would make it easier to enforce intellectual property. The terminator technology, described in Chapter 27, has received especially negative publicity in developing countries. Small farmers in developing countries often rely on saved seed, which would not be possible with purchased seeds containing the terminator technology. Also, in many developing countries, commercial seed industries are poorly developed and regular delivery of seed is not always reliable. Finally, unintentional cross-fertilization with plants in neighboring fields might make the neighbor's seeds unviable. These concerns do not apply to currently available GMOs because the technology is not available for commercial use, but in the absence of concrete evidence, it is easy to speculate about the downside effects. Monsanto announced October 4, 1999, that it is 'making a public commitment not to commercialize sterile seed technologies, such as the one dubbed "Terminator."'[14]

A related issue is that some of the genetic stock used to develop GMOs or other patented varieties is from plants found in developing countries. Once the genetic material is incorporated to improve plants, the developer can patent

---

[13] See Fitzgerald (1990) for a fascinating account of the interactions of science, economics, and politics in the research on hybrid corn at the University of Illinois, Urbana-Champaign, in the early part of the 20th century.

[14] See *http://www.monsanto.com/monsanto/gurt/default.htm*, accessed October 9, 1999.

and control the use of the new varieties. Developing countries object that they do not have control over these benefits from their natural resource base.

*Bt* corn and cotton and GR soybeans have been in use for several years. In the next section, we review the available evidence of nonmarket benefits and costs for these crops.

## Nonmarket Effects of *Bt* Corn

### Genetic flow and weediness

The evidence suggests that weediness of *Bt* corn is not an issue.[15] Commercial corn varieties, including *Bt* varieties, have shown no ability to establish wild populations.

Genetic flow is not a problem where *Bt* corn is currently grown because there are no wild relatives with sexual compatibility. In regions of the world where wild relatives exist, principally Central America, genes of corn might escape from fields in two ways – by pollen transfer or physical movement of grain.[16] If viable pollen of transgenic plants can be transferred by wind to any receptive plants within the 30-minute period of pollen viability, an escape of genetic material could take place. This potential transfer becomes more unlikely as distance increases from the transgenic plants. Physical transport is certainly possible, but germination and growth would have to take place near plants receptive to corn pollen.

If the *Bt* protein gene did become established in wild relatives, it seems unlikely that it would have any effect on the survivability of wild relatives, but we are aware of no research that directly addresses this issue.

### Risks for nontarget organisms

The best publicized nontarget organism at potential risk from *Bt* corn is the monarch butterfly (*Danaus plexippus*). The concern is that monarchs will ingest the toxins from corn pollen blown onto the leaves of milkweed, their primary food. Chapter 28 provides an in-depth review of the research on this topic. We summarize that review here.

There is little question that the Cry proteins in *Bt* corn are toxic to monarchs. However, there remain many unanswered questions about overall effect of *Bt* corn on monarch populations. Answers depend on the extent of corn varieties that express *Bt* toxins in pollen, the amount and type of toxin in the pollen, and the coincidence of pollen production and monarch feeding in the area.

The initial evidence suggests that the threat to monarchs as a species is not great. Monarchs are not equally sensitive to all currently used Cry proteins. Cry1Ab is the most toxic to monarchs and so the question is how widespread are the *Bt* corn varieties that produce this toxin: Event 176, *Bt*11, and MON810.

---

[15] Weediness of GR corn is also not likely to be a problem. As discussed above, cultivated corn has shown no tendency to survive as wild stands anywhere it is grown.

[16] Extracted from *http://www.aphis.usda.gov/biotech/corn.html.*

Event 176 is the most dangerous for monarchs because it expresses Cry1Ab at higher levels due to a pollen-specific gene promoter. Sales of varieties with this *Bt* event represented only 2% of total corn grown in the US in 1999.

Results from studies reported in Chapter 28 suggest that there is some overlap between pollen production and adjacent monarch feeding in some parts of the US but of short duration. Pollen deposition on milkweed depends heavily on weather conditions. One study reported that an average of 30% of the pollen potentially available for deposition actually stayed on the milkweed.

One issue that has received no attention in either the popular press or the scientific literature is the possibility that monarchs might develop resistance to one or more *Bt* toxins.

In addition to the monarch butterfly, other nontarget lepidopterans are potentially at risk. One of the few published research findings in this area reports a field study of the effects of *Bt* corn pollen on the black swallowtail *Papilio polyxenes* (Wraight *et al.*, 2000). The research failed to detect any effect of a pollen from a widely cultivated variety of *Bt* corn (Event 810) on either survivorship or larval mass of this insect. The study concludes that '*Bt* pollen of the variety tested is unlikely to affect wild populations of black swallowtails' (Wraight *et al.*, 2000, p. 7700).

Finally, it is of interest to note that as part of its refuge requirements for the 2000 season, EPA required companies selling *Bt* corn to encourage farmers to plant refuges in ways that would reduce the likelihood of *Bt* pollen affecting monarchs.

## Development of resistance[17]

Generation of *Bt*-toxin-resistant pests is the primary environmental threat posed by widespread commercialization of *Bt* plant varieties. Long-term exposure to an insecticide is a key factor in the selection pressure on both target pests and other susceptible insects. Target insect pests, such as the European corn borer, are killed when they consume the *Bt* toxin expressed in a *Bt* plant variety. It has been shown in the laboratory that the European corn borer (ECB) can develop resistance to the Cry1Ab protein used in first-generation *Bt* corn (Huang *et al.*, 1997). The preponderance of evidence currently suggests that the trait is recessive,

---

[17] See also the discussion about EPA regulations in Chapter 9. The primary reference documents for this section are:

- Scientific Advisory Panel on *Bacillus thuringiensis (Bt)* Plant-Pesticides, February 9–10, 1998. Transmittal of the final report of the Federal Insecticide, Fungicide, and Rodenticide Act (FIFRA) Scientific Advisory Panel on *Bacillus thuringiensis (Bt)* Plant-Pesticides and Resistance Management, Meeting held February 9–10, 1998. (Docket Number: OPPTS-00231), *http://www.epa.gov/pesticides/SAP/1998/february/finalfeb.pdf*.
- US Environmental Protection Agency, 1998. White Paper on *Bt* Plant-Pesticide Resistance Management. US EPA, Biopesticides and Pollution Prevention Division (EPA Publication 739-S-98-001).
- EPA and USDA Position Paper on Insect Resistance Management in *Bt* Crops, 1999, *http://www.epa.gov/pesticides/biopesticides/otherdocs/bt_position_paper_618.htm*.

which slows the rate of establishment. Current resistance management strategies are predicated on that fact.[18]

At present, no proven strategy exists for preventing the emergence of resistant insects. The strategy adopted by the industry, with support from EPA, has been to combine high-dose varieties with refuges.[19] The refuge (sometimes called refugia) concept is a simple one. If fields containing the non-*Bt* crop are planted near fields with the *Bt* crop, pests that are not resistant to *Bt* toxin will survive to reproduce, thereby diluting the selective effect of *Bt* toxin on the next generation.

Although the refuge concept is generally accepted in the scientific community, there is not unanimous agreement on its effectiveness and the methodology of implementation. The US government has allowed *Bt* corn seed to be sold, based on the promise that technology licenses will mandate farmer implementation of refuges either along the perimeters of their fields or as blocks within the field.

In 1995, when the first Cry1Ab (Event 176) field corn registration was issued, no scientific consensus existed on how to establish refuges to manage resistance in the two primary target pests, ECB and southwestern corn borer.[20] Therefore, EPA did not require any specific refuge size, except for *Bt* crops cultivated in cotton areas. Before 1997, the refuge concept was included only as a recommendation in a resistance management plan provided by the producing company to the farmers. For instance, growers were instructed by Monsanto/NatureMark to maintain at least 20% of farm potato acres as non-*Bt*-expressing potatoes that could be treated with conventional insecticides. Similarly, both Monsanto and DeKalb Genetics mandated certain activities under their technology agreements with corn growers even if EPA had not made any such requirement for the conditional registration. The technology agreements required growers to implement either a 5% untreated non-*Bt* refuge or a 20% treated non-*Bt* refuge. Novartis and Mycogen did not require implementation of any specific refuge option.

In 1996 and 1997, grower guides and technical bulletins issued by Mycogen and Novartis indicated their commitment to development of long-term resistance management strategies through the support of research efforts, but they did not mandate or even recommend a particular refuge option.

Two documents modified the EPA's position about requiring refuges. In December 1997, the USDA NC-205 regional research committee on ecology and management of the European corn borer and other stalk-boring Lepidoptera published its insect resistance management (IRM) recommendations in '*Bt* Corn and European Corn Borer: Long-Term Success Through Resistance

---

[18] A recent study (Huang *et al.*, 1999) found that resistance appears to be inherited as an incompletely dominant autosomal gene and might spread more rapidly than if it were completely recessive. Other researchers have questioned this result.

[19] The FIFRA Scientific Advisory Panel Subpanel on *Bt* Plant-Pesticides and Resistance Management defined high dose as '25 times the toxin needed to kill susceptible larvae.' They identified five 'imperfect' ways to assess this level. See *http://www.epa.gov/pesticides/SAP/1998/february/finalfeb.pdf*, accessed October 1, 1999.

[20] Corn earworm (CEW) is a secondary pest of field corn, but a primary pest of sweet corn. CEW is also a major pest of cotton and is known as the cotton bollworm. See Chapter 6 for a discussion of the cotton bollworm and *Bt* cotton.

Management,' (Ostlie *et al.*, 1997) and later in an October 1998 supplement to the report.

As of 2000, NC-205 recommends a 20 to 30% untreated refuge or 40% treated refuge planted within close proximity of *Bt* corn. This recommendation would apply to most of the Corn Belt east of the High Plains region. However, NC-205 indicated that further research regarding the efficacy of a 20% sprayed refuge was needed, especially in higher risk areas such as the High Plains region, where insecticide use has been historically high.

In light of these findings, EPA mandated specific structured refuge options for Novartis Event 176 (Cry1Ab) popcorn and AgrEvo CBH351 (Cry9C) field corn registration in March and May 1998, respectively.[21]

**Table 7.2**: *Bt* field corn and popcorn refuge requirements for 1999

| Company/Crop | Year registered | Refuge requirement |
|---|---|---|
| AgrEvo CBH351 (Cry9C) field corn | 1998 | 20 to 30% unsprayed refuge or, if treated with non-*Bt* insecticides, a 40% refuge planted within 1500 to 2000 feet of *Bt* corn fields |
| Monsanto Event 176 (Cry1Ab) popcorn | 1998 | 20 to 30% unsprayed refuge, or, if treated with non-*Bt* insecticides, a 40% refuge planted within 0.5 miles of *Bt* corn fields |
| Monsanto MON810 (Cry1Ab) field corn | 1996 | 10% unsprayed refuge, or 20% sprayed refuge within close proximity of *Bt* corn fields |
| DeKalb DBT418 (Cry1Ac) *Bt* field corn | 1997 | 10% unsprayed refuge, or 20% sprayed refuge within close proximity of *Bt* corn fields |
| Novartis Event 176 (Cry1Ab) field corn | 1995 | 20% non-*Bt* corn refuge that may be treated with non-*Bt* insecticides |
| Novartis BT11 (Cry1Ab) | 1996 | 20% non-*Bt* corn refuge that may be treated with non-*Bt* insecticides |
| Mycogen Event 176 (Cry1Ab) | 1995 | 20% untreated non-*Bt* corn refuge; or, if treated with non-*Bt* insecticides, a 40% non-*Bt* corn refuge |

*Source*: EPA and USDA Position Paper on Insect Resistance Management in *Bt* Crops, 1999, *http://www.epa.gov/pesticides/biopesticides/otherdocs/bt_position_paper_618.htm.*

*Note*: DeKalb's BtXtra will not be sold after 1999. The company is voluntarily removing *Bt* corn hybrids containing the DBT418 event from the market. (Personal communication from Kevin Steffey, Professor and Extension Specialist in Entomology, Department of Crop Sciences, UIUC.)

The Aquevo (now Aventis) CBH351 is known as StarLink. It was removed from the market after the 2000 season.

In April 1999, several firms in the biotech industry[22] and the National Corn Growers Association submitted to EPA a revised resistance management

---

[21] The agency registered BT11 (Cry1Ab) in sweet corn (*Bt* sweet corn) in March 1998. No specific refuge requirements were mandated for *Bt* sweet corn because harvesting occurs before insects mature, approximately 21 days after silking. Growers must destroy any Cry1Ab sweet corn stalks that remain in the fields within a month after harvest.

[22] Novartis, Pioneer Hi-Bred International, Mycogen Seeds/Dow Agrosciences, and Monsanto.

strategy for *Bt* corn.[23] It mandated a minimum 20% refuge area except in cotton-growing areas in the South, where the minimum would be 50%. Maximum refuge distance from *Bt* fields would generally be half a mile with at least a quarter of a mile preferred. Growers would have the option of applying conventional insecticide treatments to the non-*Bt* corn refuge; however, they are specifically instructed to do so only if the level of pest pressure meets or exceeds economic thresholds. See Table 7.2 for detailed *Bt* corn refuge requirements for 1999. For 2000, the requirements are similar except that 'Registrants must ensure that growers plant a minimum structured refuge of at least 20 percent non-*Bt* corn. For *Bt* corn grown in cotton areas, registrants must ensure that farmers plant at least 50 percent non-*Bt* corn in these areas. . . . For certain products where *Bt* is not expressed at a high dose, there will be sales and planting restrictions on specific products in certain limited growing areas.' (*http://www.epa.gov/pesticides/biopesticides/otherdocs/bt_corn_ltr.htm*, accessed August 27, 2000).

We have been unable to find any systematic evidence as to the degree of compliance with the refuge requirements or recommendations. However, both the implied complexity of adding refuge management to the existing challenges of managing hybrid corn cultivation and the anecdotal evidence suggest that compliance has been significantly less than complete. For example, the following sentence is part of the industry submission to EPA described above, 'The *Bt* corn industry is not aware of any growers that actually planted a 40 percent sprayed refuge.'

## Biopesticides, *Bt*, and the organic industry

Organic farming became one of the fastest growing segments of US agriculture during the 1990s and although still small relative to the total food market, demand for organic foods has grown tremendously in recent years. Total US retail organic food sales rose from $178 million in 1980 to $1 billion in 1990 and reached $5 billion by 1998 (Figure 7.1). The growing number and variety of retail outlets that are offering organic foods, the growth in demand for organic products, the substantial efforts made by the government to harmonize certification procedures, and the support received by the state administrations will most likely ensure continued growth of the organic farming sector.

Organic foods are distinguished from conventionally produced foods principally by production and processing principles developed originally in Europe in the late 19th and early 20th centuries, and later in the US. These principles stress production and processing without the use of synthetic chemicals, and soil fertility management using techniques that enhance biological activity in the soil such as composting, green manuring, and rotating crops.

Among the techniques that organic farmers use to protect a crop are bioinsecticides – living organisms that kill insects, and natural 'insect repellants,' toxins produced by a plant that help guard against fungi and insects. The organic industry argues that bioinsecticides have several advantages over

---

[23] *http://www.ncga.com/02profits/insectMgmtPlan/toc.htm*, accessed September 24, 1999.

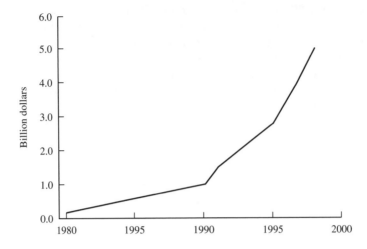

**Figure 7.1**: US organic food sales, 1980–1998

*Source*: Emerich (1996) and Organic Agriculture at FAO website http://www.fao.org/organicag/frame6-e.htm.

synthetic pesticides. It is more difficult for insects to develop resistance because the bioinsecticides can evolve with the insect pests. They are also very specific, sometimes affecting only a single species of insect.[24] Bioinsecticides can be safer than synthetic insecticides if they leave little harmful residue in the environment. They are seen as generally less harmful to people, animals, plants, and beneficial insects. In recent years, researchers have developed several kinds of bioinsecticides based on viruses, fungi, and bacteria.

*Bt* is the most widely used bioinsecticide. It is considered very selective because a single strain may affect only a few insect species and the toxin does not harm humans, birds, fish, or beneficial insects. EPA classifies *Bt* pesticides as toxicity class III – slightly toxic, the same rating as glyphosate. The organic industry relies heavily on *Bt* because of its low toxicity and target specificity, and because it is not synthetic.

For most corn and cotton growers, the development of resistance to specific *Bt* toxins would simply mean a return to synthetic insecticides, or to previous practices. For the organic industry, the development of resistance to currently used *Bt* toxins could have two potential consequences. First, ECB feeds not just on corn but on some 200 other commercial crops, including barley, beans, millet, oats, potatoes, and sorghum. Second, secondary target lepidopteran pests of corn are also pests of other food crops (see Table 7.3). The Federal Insecticide, Fungicide, and Rodenticide Act (FIFRA) advisory panel felt that these pests are 'directly or potentially affected by' *Bt* toxins in corn. Development of resistance by target species would mean organic growers would have to find substitute treatment techniques.

---

[24] However, they can also affect nontarget species. A *Bt* bioinsecticide that affects ECB is also likely to affect the monarch butterfly. We are indebted to Kevin Steffey, Professor and Extension Specialist in Entomology, Department of Crop Sciences, UIUC, for this point.

**Table 7.3**: Secondary target lepidopteran pests of corn and their other host crops

| Nontarget pests | Other host crops |
| --- | --- |
| Stalk borer (*Papaipema nebris* (Guenée)) | Cotton, potato, tomato, alfalfa, rye, barley, pepper, spinach, beet, sugar beet, southern peas, beans |
| Corn earworm (*Helicoverpa zea* (Boddie)) | Cotton, tobacco, strawberry, tomato, beans, peas, lettuce |
| Fall armyworm (*Spodoptera frugiperda* (J. E. Smith)) | Cotton, rye, sorghum, alfalfa, bean, peanut, potato, sweet potato, turnip, spinach, tomato, cabbage, cucumber, cotton, tobacco |
| Potato stem borer (*Hydraecia immanis* (Esper)) | Tomato, sugar beet, canola, hop, celery, beans, wheat, barley, oats, strawberry |
| Webworms (*Crambus* spp.) | Tobacco |
| Armyworm (*Pseudaletia unipuncta* (Haworth)) | Millet, bluegrass, crabgrass, fall panicum, and small grains; occasionally reported to infest various vegetables, fruits, legumes |

*Source*: FIFRA Scientific Advisory Panel, Subpanel on *Bacillus thuringiensis* (*Bt*) Plant-Pesticides and Resistance Management, *http://www.epa.gov/pesticides/SAP/1998/february/finalfeb.pdf*, and Hill (1997).

*Note*: While the European and southwestern corn borers are the primary targets of *Bt* corn, the labels on several of the *Bt* corn varieties mention protection against some of the pests above, including stalk borer, corn earworm, and fall armyworm. The label for at least one type of *Bt* corn also suggests suppression of black cutworms. (Personal communication from Kevin Steffey, Professor and Extension Specialist in Entomology, Department of Crop Sciences, UIUC.)

## Potential substitutes for *Bt* in the organic industry[25]

Among the most likely substitutes for *Bt* are formulations with pyrethrins[26] and rotenone[27] as active ingredients. However, these two insecticides are not pest-specific. Beneficial insects as well as target insects are killed. Fungal pathogens and viruses could also be used in place of *Bt* formulations. An example of a fungal pathogen under development is *Beauvaria bassiana*, which controls more

---

[25] Sources for this section are Burges (1998), Sparks (1998). *http://insects.tamu.edu/extension/bulletins/b-1305.html*; and Bondurant J. Pesticide Chart at Golden Harvest web site: *http://www.ghorganics.com/Page44.html*.

[26] Pyrethrins are produced by certain species of chrysanthemum. The flowers are harvested shortly after blooming and are either dried and powdered or the oils within the flowers are extracted with solvents. Pyrethrins have little residual effect. In stored grain, 50% or more of the applied pyrethrins disappears during the first three or four months of storage. At least 80% of what remains is removed by handling, processing, and cooking. Pyrethrins alone provide only limited crop protection because they are not stable. As a result, they are often combined with small amounts of antioxidants to prolong their effectiveness. Pyrethrin compounds are broken down in water to nontoxic products and are inactivated and decomposed by exposure to light and air (Tomlin, 1994).

[27] Rotenone is an extract obtained from roots, seeds, and leaves of members of the Leguminosae family such as barbasco, cub, haiari, nekoe, and timbo. Rotenone is a nonspecific botanical insecticide used for lice and tick control and for fish eradication as part of water body management. Rotenone is rapidly broken down in soil and in water and with exposure to sunlight. Nearly all of its toxicity is lost in 5 to 6 days of spring sunlight or two to three days of summer sunlight (Tomlin, 1994).

pests than do most microbial insecticides. However, this fungus is not very selective and kills beneficial organisms as well. Baculoviruses, a group of viruses that cause disease in caterpillars, could also be used in place of *Bt* or with it. Unfortunately, there is no single formulation appropriate for different field-spraying equipment and different crops.

## Reduction in pesticide use

Although conventional wisdom is that the use of *Bt* corn reduces pesticide applications, recent findings suggest that this effect is small. As discussed in Chapter 4, for most growers of non-*Bt* corn, the cost of scouting for ECB and the expense of the pesticides were greater than the expected value of the increased yields. Infestation levels are uncertain. Treatment is profitable only in years when infestation levels are high. Scouting costs must be borne every year to identify years when treatment is profitable. Hence, it makes economic sense to scout every year, and treat in high-infestation years only if the average additional profit offsets the cost of scouting. If it does not, then it is economically rational for a grower not to scout.

Giannessi and Carpenter (1999) report that about 5% of US cornfields were treated for corn borers in 1995, before the introduction of *Bt* corn. Insecticides commonly used were chlorpyrifos, permethrin, *Bt*, and methyl parathion. In 1998, when about 20% of the US corn crop was *Bt* corn, the researchers estimate the number of acres using insecticides to treat ECB fell by 2 million acres. In other words, about half the acres previously treated with insecticides for ECB used *Bt* corn instead.

## Food safety and *Bt* corn

Researchers have found no food safety concerns for current varieties of *Bt* corn. The Cry proteins are toxic only to lepidopteran insects and have no effect on mammals. Furthermore, they are destroyed in forms of processing that involve heat. Allergic reactions to the first Cry proteins used, Cry1Ab and Cry1Ac, are unlikely because they break down in human gastric juices. The latest protein to be used, Cry9C.[28] The only commercially available corn variety that incorporates the Cry9C *Bt* protein was sold by Aventis under the name StarLink. The US EPA approved its use only for animal feed and industrial purposes because of concerns about potential allergic reactions. Allergenicity cannot be directly predicted. Novel proteins are subjected to a battery of tests to determine if they match with any of the known properties or features of food allergens. The majority of tests done on the Cry9C protein indicated that the protein was not likely to be an allergen. It does not originate from an allergenic source and does not share amino acid similarity or structural similarity to proteins known to be allergens or mammalian toxins in

---

[28] See *Federal Register* May 22, 1998 (volume 63, number 99), *Bacillus thuringiensis* Subspecies *tolworthi* Cry9C Protein and the Genetic Material Necessary for Its Production in Corn, Final Rule, *http://www.epa.gov/fedrgstr/EPA-PEST/1998/May/Day-22/p13604.htm*. Also of interest is FDA's 'Cry9C Food Allergenicity Assessment Background Document', available at *http://www.epa.gov/oppbppd1/biopesticides/cry9c/cry9c-peer_review.htm*, accessed August 27, 2000.

available protein databases. However, unlike the other Cry proteins, the Cry9C protein is not readily digestible in simulated gastric conditions and has been shown to be stable at 90 degrees celsius.

The approval required Aventis to ensure that StarLink corn was not used for human food. In mid September 2000, a coalition of green groups announced the discovery of StarLink corn in taco shells and later in other products. These products were recalled from the market and Aventis agreed to buy back StarLink corn from farmers. In mid-October, Aventis announced measures to cease all production of StarLink corn and cancelled the animal feed EPA registration.

## Nonmarket Effects of GR Soybeans

### Weed resistance to glyphosate

The development of weed resistance to glyphosate is of concern because glyphosate plays an important role in the set of possible herbicides used in low- and no-till agriculture. Until recently, development of resistance to glyphosate in weeds was thought highly unlikely (Jasieniuk, 1995) because of three factors – genetic constraints to the evolution of a resistance mechanism (i.e., no way to evolve a replacement to the EPSP synthase enzyme that glyphosate inhibits), the rapid inactivation of glyphosate in the environment, and the mode of use in agriculture.[29] Recently, three reports of glyphosate-resistant weeds have been made. Australian researchers at the International Weed Control Congress in June 1996 stated that a GR annual ryegrass (rigid ryegrass, *Lolium rigidum*) population has appeared. According to scientists at Charles Stuart University, the resistance developed in a field in southeast Australia as a result of glyphosate applications made repeatedly since the early 1980s. The resistant ryegrass survived seven times the herbicide concentration that killed susceptible plants (Pratley *et al.*, 1996). Two more cases of resistance have been reported – goosegrass (*Eleusine indica*) in Malaysia in 1997 and rigid ryegrass in the US in 1998 (International Survey of Herbicide-Resistant Weeds, http://www.weedscience.com/).

### Genetic flow potential for soybeans[30]

In commercial operation, fresh soybean seed is produced annually for each new season (TeKrony *et al.*, 1987). However, most remaining seed from one crop is capable of germinating the following season.

Soybean crosses readily only with other members of the species and with other members in *Glycine* subgenus *Soja*. Soybean crosses with members of

---

[29] However, weeds have developed resistance to herbicides such as paraquat that are even less environmentally persistent. It is the persistence relative to the way a weed germinates through a season that is important (Wrubel and Gressel, 1994).

[30] Based on personal communication from Theodore Hymowitz, Professor of Plant Genetics, Department of Crop Sciences, UIUC and *http://www.aphis.usda.gov/biotech/soybean.html*.

**Table 7.4**: Herbicide acre-treatments for soybeans, 1997

|  | Herbicide-tolerant soybeans | All other | Decline in use (%) |
| --- | --- | --- | --- |
| Heartland | 1.80* | 2.34 | 23 |
| Mississippi Portal | 2.09* | 2.62 | 20 |
| Northern Crescent | 2.22 | 2.15 | −3 |
| Prairie Gateway | 2.02 | 2.01 | 0 |
| Southern Seaboard | 1.04* | 2.14 | 51 |

*Source: http://www.ers.usda.gov/whatsnew/issues/biotech/Table3.pdf accessed July 21, 2000.*

*Note*: An acre-treatment is the number of different active ingredients applied per acre times the number of repeat applications. A single treatment containing two ingredients is counted as two acre-treatments as is two treatments containing a single ingredient.

*Significantly different from All other at 5% level.

*Glycine* subgenus *Glycine* have been made, and fertile derived lines are available. However, the amount of research conducted in this area is negligible, and crosses in the field are unlikely because soybeans are almost completely self-pollinated (Carlson and Lersten, 1987; McGregor, 1976). No one has reported successful crosses with any extrageneric relatives (Hymowitz and Singh, 1987).

No members of the genus *Glycine* grow wild in the US. The annual and perennial wild relatives of the genus are found in Asia and the Pacific Basin. Hence, genetic flow is a potential problem in those areas only. Furthermore, because soybeans are almost completely self-pollinated, crosses are unlikely even where wild relatives exist. Caviness (1970) showed that honeybees are responsible for occasional cross-pollination, and that thrips are ineffective pollination vectors.

## Reduction in pesticide use

GR-resistant crops allow farmers to substitute glyphosate for other herbicides. Table 7.4 reproduces results from a 1999 study by the Economic Research Service of USDA. In almost all parts of the country, the use of herbicide-tolerant soybeans (all types included, not just GR soybeans) resulted in fewer herbicide acre-treatments and for three of the five regions the declines were statistically significant. The regions with declines are those where most of US soybeans are grown. Hence, it is reasonable to assume that there was a decline in volume of herbicide applications as a result of the widespread use of herbicide-resistant soybeans. In addition, glyphosate is a relatively benign herbicide from the perspective of mammalian toxicity.

---

[31] For a critique of the studies reported by Monsanto in its submission to the Australian regulatory authorities, see *http://www.biotech-info.net/problem_with_soy2.html*, accessed September 26, 1999 and a response from the regulatory authorities at *http://www.anzfa.gov.au/documents/res01_99.asp*.

## Food safety and GR soybeans

FDA staff reviewed the scientific studies conducted by Monsanto into potential food safety concerns with GR soybeans. They found nothing that would cause them to consider GR soybeans to differ significantly from non-GM soybeans.[31] The US Occupational Safety and Health Administration (OSHA), which requires toxicity reviews of chemicals which workers might be exposed to, reports that glyphosate is not a carcinogen. The oral $LD_{50}$ rate for rats (the quantity consumed that is a lethal dose for 50% of the rats) is 4613 milligrams per kilogram of body weight. By way of comparison, a chemical must be called 'highly toxic' if the $LD_{50}$ rate for rats is 50 milligrams.

## Summary of Nonmarket Effects of *Bt* Corn and GR Soybeans

Table 7.5 summarizes briefly our assessment of the nonmarket effects of *Bt* corn and GR soybeans. The positive benefits are in bold type. A quick glance at the table suggests that the nonmarket benefits of these crops are not very significant. *Bt* corn has only a small effect on pesticide use because farmers in most parts of the country have not been applying pesticides to control ECB. GR soybeans result in increased use of glyphosate and less of other herbicides. The net effect on volume and toxicity may have been a decline, but there is no definitive research on this topic.

**Table 7.5**: Summary of nonmarket effects of *Bt* corn and GR soybeans

| Nonmarket effect | Bt corn | GR soybeans |
|---|---|---|
| Food safety | | |
|   Human toxicity | None from current *Bt* toxins **Potential reduction in aflatoxin, fumonisin** | None known or likely |
|   Allergenicity | Unlikely for current *Bt* toxins; slightly more likely for Cry9C | None known or likely |
| Weediness | No | Small potential from lost seed |
| Genetic flow | In region of origin, crosses with relatives likely because of open pollination | In region of origin, but crosses unlikely because of nature of pollination |
| Resistance | Yes; both in target and nontarget economic pests | Yes for some weeds; slow to develop |
| Changed pesticide use | **Minimal** | Substitution of glyphosate for other herbicides; **probable reduced total volume** |
| Nontarget effects | Other Lepidoptera (e.g., monarch butterfly); species that feed on target pests | None from GMO; potential from increase in glyphosate use, **reduction from decline in other herbicides** |

Based on our review of the existing literature, development of resistance to *Bt* toxins is the most important potential environmental threat from *Bt* corn. Much attention is being paid to resistance in the target pests European and south-western corn borer; we found a lack of attention to the potential for development of resistance in secondary target pests that pose a threat to other crops. The loss of effectiveness of *Bt* toxins would be particularly serious for the organic indus-try, which relies heavily on *Bt*-based bioinsecticides. The agriculture sector is conducting a large-scale experiment to test resistance management strategies for *Bt* crops. Elements of this experiment include optimal size and location of refuges and viability of voluntary compliance approaches. The outcome could have serious consequences for agriculture as a whole and for potential accept-ability of other new technologies.

## Nonmarket Effects of *Bt* Cotton

While this chapter focuses on GR soybeans and *Bt* corn, it is useful to present selected evidence about the nonmarket effects of *Bt* cotton. Since the *Bt* toxins introduced to *Bt* cotton are the same as those in *Bt* corn we expect to see the same (minimal) food safety effects for those products of cotton that might enter the food supply (cotton seed oil and feed from cotton byproducts). In the environmental area, *Bt* cotton has the most important nonmarket effects, a sub-stantial reduction in insecticide applications and the potential for development of resistance in pests that are also important to other crops.

**Table 7.6**: Cotton bollworm/budworm insecticide use in 1995

| Insecticide | Percent of acres treated |
|---|---|
| Amitraz | 4 |
| Bifenthrin[a] | 6 |
| Bt | 9 |
| Cyfluthrin[a] | 12 |
| Cypermethrin[a] | 12 |
| Esfenvalerate[a] | 7 |
| Lambda-cyhalothrin[a] | 21 |
| Methomyl | 9 |
| Profenofos | 13 |
| Sulprofos | 2 |
| Thiodicarb | 12 |
| Tralomethrin[a] | 7 |

[a]Pyrethroid.

*Source*: USDA, Agricultural Chemical Usage: Field Crops Summary, 1995–1998 (separate volumes) as reported in Gianessi and Carpenter (1999).

As discussed in Chapter 6, the important pests of cotton include the tobacco budworm/cotton bollworm complex. The bollworm is also an important pest of corn, where it is known as the corn earworm. The pesticides in use before the introduction of *Bt* cotton are reported in Table 7.6. A large share of these pesti-cides are pyrethroids, and significant pest populations had developed resistance to them. Table 7.7 shows the dramatic decline in insecticide applications after the introduction of *Bt* cotton in 1996.

**Table 7.7**: Average number of insecticide applications to control cotton budworm–bollworm complex, selected states

| State | 1993 | 1994 | 1995 | 1996 | 1997 | 1998 |
|-------|------|------|------|------|------|------|
| Alabama Central | 4.00 | 8.00 | 9.60 | 0.02 | 0.16 | 1.18 |
| Alabama North | 4.00 | 2.70 | 6.70 | 0.18 | 0.56 | 0.52 |
| Alabama South | 7.00 | 5.00 | 5.90 | 0.13 | 1.00 | 2.32 |
| Florida | 5.30 | 5.30 | 5.70 | 1.08 | 0.95 | 2.00 |
| Georgia | 2.70 | 4.30 | 3.40 | 1.67 | 2.53 | 1.54 |
| Louisiana | 4.70 | 4.80 | 4.70 | 3.85 | 3.23 | 3.48 |
| Mississippi Delta | 4.50 | 4.80 | 4.50 | 2.50 | 3.21 | 3.13 |
| Mississippi Hills | 4.00 | 3.10 | 8.20 | 1.51 | 1.33 | 1.55 |
| North Carolina | 2.50 | 3.60 | 2.60 | 3.07 | 1.98 | 3.04 |
| South Carolina | 4.90 | 4.40 | 4.70 | 4.19 | 3.32 | 3.44 |
| Texas South Central | 4.17 | 3.66 | 2.60 | 2.33 | 1.69 | 0.64 |

*Source*: Cotton Insect Losses, Proceedings Beltwide Cotton Conferences, 1993–1998, as presented in Falck-Zapeda, *et al.* (1999).

There is concern about the development of resistance in pests common to both cotton and corn to the *Bt* toxins incorporated in both *Bt* corn (Cry1Ab and Cry1Ac) and cotton (Cry1Ac). EPA has mandated special refuge requirements in regions of the US where cotton and corn are likely to be grown in close proximity. There is little evidence about the extent to which farmers follow these requirements.

# 8

# The Stakeholders and the Struggle for Public Opinion, Regulatory Control and Market Development

**Julie Babinard**
*International Food Policy Research Institute, Washington DC, USA*

**Timothy Josling**
*Institute for International Studies, Stanford University, Stanford, California, USA*

If that useful imaginary figure the 'visitor from Mars' were to drop in on a debate about the health and environmental safety of genetically modified foods, the visitor might be puzzled by the fact that the opponents see little merit in the products of biotechnology in the food area and the proponents see few faults. As most technical advances bring a complex mix of costs and benefits, one might expect to see debates about the relative size of the costs and benefits and whether the one outweighs the other.

One would have to explain to the visitor that the debate is only partly about the scientific properties of foodstuffs produced in a particular way. It is just as much about who owns the rights to the technology, who is benefiting from the sale of these products, and who should regulate the production and sale of these foods. The debate engages a spectrum of interests, from those who believe that the use of biotechnology in food production is a vital part of the equation for feeding the world in the 21st century to those who fear that the technology will harm people through unknown impacts on health and environment and undesirable shifts in income toward corporations at the expense of the poor.

Genetically Modified Organisms in Agriculture
ISBN 0-12-515422-4

The debate is intense because there are significant economic stakes for private firms that market this technology and significant political stakes for non-governmental organizations (NGOs) that oppose its use. Government agencies and scientific researchers also have considerable interest in the outcome of the debate, and one would not expect them to be neutral observers. To them the issue is credibility and the continuation of an implicit pact with the public that allows 'experts' to run much of the regulatory machinery of a modern state. To understand the issues underlying the debate one must step back from the rhetoric and identify the underlying interests of the various groups and how they are expressed.

The debate is being played out in different arenas. These include the 'court of public opinion', where the struggle for public sympathy and support by both the supporters and opponents of GM foods has included major public relations campaigns. A second arena is the regulatory process, where the issues have to do with who regulates GM products and in what way are they regulated. A third location for the debate is the marketplace, where producers have to convince consumers of the desirability of products and compete with other producers for market share and control of valuable assets. To follow the game, it is useful to have a program with at least a little background about the players. This chapter attempts to provide some of the details of the positions of the various groups and their arguments in the three public arenas. Then, Part 2 of this book provides individuals from the different camps with an opportunity to express their views.

The main protagonists in the debate include the firms who are developing and introducing the new technology, the corporations that market and process the crops, and those that sell the food at retail. They are joined by a variety of consumer and environmental groups, who often pool ideas and collaborate to slow down the biotech juggernaut, and by individuals who believe their own roles are to warn or reassure the public. Government departments and food and environmental agencies are also players in the game, as are international bureaucrats. Somewhere caught in the middle are farmers, cautious about growing crops that some consumers mistrust but eager to adopt cost-saving technology, particularly if it is available to competitors; consumers bewildered by the variety of opinions emanating from public and private agencies; and developing country governments unsure as to whether the rich countries are yet again about to stack the deck against them either through over-regulation or oligopolistic control of the tools for a useful technology. All of these actors play on a stage where communication can occur instantly through the internet, and new research findings change the backdrop almost daily.

## The Private Sector

The entities with the greatest financial stake in the GMO debate are private-sector companies that produce and market the products of biotechnology. The private sector can be assumed broadly to have profit-maximizing goals, or at least wish to keep shareholders happy. But this does not mean that companies are focused only on short-run gains. Many companies have longer-term strategic goals in the area of agricultural biotech, including positioning themselves to control certain aspects of the technology and gain a larger share of the rewards. Indeed, one interesting feature of the private-sector role in promoting GMOs is the way the technology has been accompanied by major structural changes in the industry. The research investment needs to be complemented by input supply systems, farmer acceptance, distribution networks, and retail outlets. Each part of the chain stands to gain from the adoption of the new technology. Thus, structural changes in the private sector are in part a reflection of the new technology. But structural changes are also about the distribution of those gains and the extent of control over the use of the technology. There is no guarantee that every element of the private sector will gain from biotech developments. Even without the opposition of consumer and environmental groups, the spread of biotech in agriculture and food might well have been contentious because of its impact on the structure of the food industry. The competition for the consumer dollar and the share that goes to raw material suppliers as opposed to processors and marketers has long been contentious in food marketing. The advent of biotech in the food chain brings this debate to the fore.

## The life science companies (see also Chapter 25)

In the driver's seat of the GMO juggernaut are a handful of companies formed from the marriage of chemical and pharmaceutical firms with biological and agronomic entities. More recently they have acquired (or been acquired by) firms that provide inputs into agriculture, such as seed distributors. These conglomerate enterprises are often called 'life science' companies. Most are based in the United States or in Europe. Among the most prominent are Monsanto and DuPont, with its newly acquired Pioneer Hi-Bred seed company, based in the United States, and Novartis, AgrEvo, Zeneca, and Aventis, based in Europe.[1] The major new products of the life science firms have been biotech medicines, with about 80 new products on the market. GM foods are not so important in the product line but are considered a promising area for future growth, if consumer resistance does not harden against their use.[2]

The companies involved in the creation and distribution of GM crops tend to have confident positions on the benefits of the new technology and few concerns about the threats posed by GMOs. A typical position is that GM crops and biotechnology are safe for both consumers and the environment and offer great benefits for agriculture and food production, and are key factors in the fight against famine (Mack, 1998). Companies often point out that improved food supplies in the future can be achieved only by taking advantage of technological advances. Steven Briggs, head of the Novartis Agricultural Discovery

---

[1] Novartis was formed by a merger between Ciba-Geigy and Sandoz in 1997 and is based in Switzerland. Aventis-Crop Science is the name for a merger of Hoechst and Rhône-Poulenc, bringing the German and French firms into an alliance to develop biotech products. AgrEvo is a German company. Zeneca is based in the United Kingdom and is best known as the producer of Calgene's Flavr Savr tomatoes.

[2] One projection by Zeneca is for a market of $75 billion for GM crops by 2020, compared to the current $33 billion market for agro-chemicals. DuPont's estimate of the biotech market as a whole is $500 billion by the year 2020 (OECD, 1999).

Institute in San Diego, which sequences plant genomes, explains that innovations such as disease-resistant crops or crops containing more calories are the most likely technologies to help feed the world (Mack, 1998). Eventually, as Ganesh Kishore, the head of nutrition at Monsanto, explains, GM crops could be made into a complete balanced meal for the elimination of the nutritional deficiencies that have plagued humankind (Mack, 1998).

The companies also stress the positive benefits to current farmers as potential adopters. They seek to convince farmers that new herbicide- and insect-tolerant crops will allow for more efficient eradication of weeds and pests. Farmers can spray herbicides anytime during the growing season, killing weeds without killing their crops, and control insects without the need for expensive pesticides. Biotech companies argue that this is particularly important in view of the fact that up to 40% of the world's food production is currently lost to weed growth, pests, and diseases (Genetic ID, 1999). In addition, biotech companies argue that genetic engineering will help developing countries feed their population in ways that sustain the environment.

With regard to consumer health and food safety, companies such as Monsanto claim that it is standard for them to thoroughly analyze new transgenic plants to see that they are 'substantially equivalent' to commercial plants (Cohen, 1998).[3] They explain that transgenic crops undergo a series of biochemical checks in which nutrient and protein levels are monitored, and potential poisons and allergens are tested for. In some cases, the crops are fed to livestock to check for normal weight gain and health (Cohen, 1998). Environmental concerns are also met with sangfroid: scares about mutant weeds and insects are the product of the overactive imaginations of environmental activists rather than sober risks to be factored into product marketing. In one respect, however, the commercial interest is impinged by the prospect of health, safety and environmental complications: product liability would be a chilling prospect if the courts were in the future to link human illness or environmental damage to GM foods. Companies no doubt will have learned from the travails of the tobacco industry that it is best not to hide those memos that discuss the discovery of deleterious side effects.

In response to the outbreak of consumer hostility in Europe, biotechnology companies have begun to change their approach to introducing GMOs into the food system. Jeff Stein of Novartis in Greensboro, North Carolina, states, 'there is no clinical concern that the genes pose a threat, but we do worry about public perception' (Cohen, 1998, pp. 43–44). The companies involved in the research and development of GMOs realize that consumer concerns and resistance constitute a major impediment to expanding their markets globally. They believe that the biotechnology industry may have underestimated the strength of feelings about GM food in Europe and that it will have to do better to persuade consumers that its products are safe. Monsanto's UK-based biotechnology development manager, Colin Merrit, declared that '[Monsanto] didn't foresee

---

[3] The phrase 'substantially equivalent' has acquired considerable significance in the regulation of GM foods. If the companies convince the FDA that the modified food is substantially equivalent to the traditionally produced food then no separate testing and labeling is required. See Chapter 9 for a discussion of substantial equivalence and its use in food safety regulations.

the levels of concern which have grown in Europe and [that it] made some mistakes' (*Farmers Guardian*, 1998). In an attempt to regain public confidence, Monsanto launched in June 1998 a multimillion-dollar advertising campaign to promote genetically modified foods and change its image in Europe. By most accounts, this campaign was a public relations disaster that provoked a backlash from consumers who figured that there must be something wrong with the GM foods if Monsanto was trying so hard to convince them otherwise! Another major public relations campaign to sway US public opinion began in early 2000.

The life science firms, as mentioned above, have taken care of their need for seed distribution networks by buying into the seed distribution business, which has led to fears about undue concentration and market power. In the seed sector, a few firms such as Pioneer, Monsanto, and Novartis dominate the US market.[4] This trend is reinforced by the new technology. Life science companies such as Dow, DuPont, Novartis, and Monsanto have in effect been buying their way into the seed business as a way of augmenting the return on scientific research conducted in their laboratories. Such concentration may not be bad in itself; the companies at the moment still have to compete with retained seed, landrace crosses, and seed produced locally. But the potential exists for the life science companies to exercise market power, extracting considerable rents for genetic material in plants.

Not all the agricultural supply industries will benefit from the success of biotech. The more traditional part of the agricultural supply industry that provides chemical and mechanical inputs to the farm sector has a different economic interest than do the life science firms. Chemical suppliers face a remote but real challenge to their position as the anchor of modern agriculture. If plants manufacture their own insecticides, then there is less need to spray chemicals. With respect to herbicides, the issue is more complex. If plants can be made resistant to herbicides, then one might expect the use of that chemical to increase. However, the seed developer can control which herbicides the plant tolerates. Thus there is likely to be some shakeout in the range of herbicides on the market. A highly anticipated event in the industry was the expiration of Monsanto's patent on the recipe for Roundup (glyphosate) in 2000. The herbicide market is gearing up for healthy competition as other manufacturers attempt to compete with the Roundup/GM crop combination.

## Food manufacturers and retailers (see also Chapters 22 and 23)

Public acceptance is the primary concern of food retailers. Until the late 1990s, most European food manufacturers and retailers were not opposed to selling GM crops and GM-derived products. These companies are now rapidly implementing programs to eliminate GM products following consumer and

---

[4] On March 15, 1999, DuPont announced its purchase of Pioneer Hi-Bred, the world's largest seed company. This gave the combined firm over 40% of US corn seed sales and 20% of the soybean seed market. Monsanto holds a commanding 80% share of the cottonseed market (see *The Economist*, March 22, 1999). There are of course still many small firms in the seed industry. There is no current evidence that prices have been forced up by lack of competition.

pressure group backlash against the technology. The companies support the use of genetic modification provided it is legal, safe, environmentally responsible, and has clear consumer benefits. Despite government and scientific reassurances, however, their strategy is to align with consumers' response, thereby avoiding the introduction of new GM products without full customer consultation.

The UK market has shown the lead in this regard. Sainsbury's, a major food manufacturer and retailer in the UK, is keen on accommodating the wishes of its customers. In response to overwhelming customer concern, the company introduced a policy of eliminating GM ingredients from all of its own brand products in 1999. In addition, Sainsbury's has alerted the manufacturers of proprietary branded foods on sale in their stores about its consumers' concerns (see Chapter 23, written by the architect of Sainsbury's plans to go non-GMO).

Among other food retailers in the UK, the views regarding the safety and use of GM crops are divided. In March 1999, Marks & Spencer (M&S) ordered a total ban on GM foods in its stores. The ban, to be implemented across all 268 outlets, will make M&S the first company to offer totally GM-free foods (Poulter, 1999). Recipes on 1000 products will be changed to remove such GM crops as soybeans and corn, and all their derivatives. An M&S spokesperson declared the firm believes 'there is a role for GM. But . . . it is being forced through at such a pace that people are uneasy' (Poulter, 1999). Other suppliers such as Tesco and Safeway have promised to reduce reliance on GM ingredients and to improve labeling. Tesco PLC, Britain's biggest supermarket chain, will remove GM ingredients from foods whenever possible and tighten up on labeling. The policy of Tesco is to deliver a straightforward choice by labeling any products that might contain highly processed, undetectable GM derivatives (Genetic ID, 1999). In a public statement, Safeway also made clear the importance of understanding 'the transparent way in which academia, suppliers, and retailers alike have all been candid with the public.' Safeway also started a successful partnership with the supplier Zeneca by introducing GM tomato purée in 200 trial stores. A Safeway spokesperson declared that the tomato purée was like any other approved food – 'safe to eat' – but that to earn its place on the shelf, it also had to offer benefits to shoppers.

Most retailers and food firms in the United States have not taken action to avoid placing GM products on their shelves. However, Gerber (ironically a part of the Novartis group) and Heinz announced in summer 1999 that they would use only non-GM products in their baby foods.[5] Iams, a high-end pet food company, followed suit. Frito Lay announced that it will only purchase non-GM corn for its snack foods and McDonalds has requested its suppliers to provide non-GM potatoes. A growing number of companies outside the United States have announced non-GMO policies. As of June 2000, the list included two Japanese breweries and a soy products company, and the largest producer of corn tortillas in Mexico.

---

[5] The companies involved have been careful not to indicate that there is any health-related reason to make these changes. One commentator argues that the substitution of non-GM varieties may subject babies to other hazards, such as insect contamination, bacterial infection, or the mycotoxin fumonisin (Miller, 1999).

## Farmers (see also Chapters 13 and 14)

Farmers also are vitally involved in the production of GM foods and ingredients. Most farmers understand and agree with the notion that GM crops can provide them with great benefits. Clearly the farmer has a strong interest in the degree of public acceptance. Whether the consumer is willing to buy GM foods determines whether growing these crops is advantageous. In the United States, the rapid adoption of GM crops is due to their economic advantages and the lack of consumer knowledge. If either or both of these conditions were to change, there could be a general move away from growing transgenic varieties.

In Europe, where consumer acceptance is much less, and there has been outright hostility to GMOs, farmers have often been unwilling to plant genetically modified crops. Farmers argue that biotechnology firms have ignored consumer worries and tried to bring their products to market in Europe without the necessary scientific proof required to show that they are safe.

One body, the International Federation of Agricultural Producers, with widespread membership from farm groups around the world, has taken a cautious position on the adoption of GMOs. Traditional exporters like members of the CAIRNS Group and the US tend to see the GM varieties as yet another way to lower costs. Farmers in European countries who generally need export subsidies to move goods onto world markets are less entranced with GMOs. They see no benefit unless the government supports prices to guarantee a return on costs, an outcome almost impossible after the last round of world trade negotiations. They see the main consequence as being a reduction in prices, further threatening their viability. Farmers in importing countries also are more cautious about the technology, though once it is established, one can imagine a renewed enthusiasm for adopting it in local agriculture.

Farmers are also worried that the new technology will change the relationship between the producer and the input companies in ways that may not prove beneficial to farmers. The trend in the US agriculture and food industries toward concentration is apparent in the major role played by fast-food outlets in setting standards for poultry, beef, and potatoes. But the cereal and oilseed markets are also characterized by a relatively small number of players in the markets into which farmers sell. The prospect of linkups between the input supplier and the output purchaser is of concern to farmers. Not only might the farmer lose freedom of choice among suppliers and buyers, and hence be likely to see margins squeezed, but the role of the farmer would become more that of a manager and less of an independent business owner. In this respect, the GM revolution could be one of a series of developments that change the agricultural sector permanently.

## Public Interest Groups

The main challenge to the private-sector interests described above has come from public interest groups broadly representing consumer and environmental concerns. In addition, some of the non-governmental groups that promote Third World development objectives have become involved in the debate. One would generally associate these groups with promoting the public interest, in that they are usually sensitive to abuses of market power and focus on non-economic quality of life issues that may get overlooked in the commercial world. But it should be remembered that these groups also have agendas, corporate structures, and financial constraints and use the media to persuade the public and the politicians of a particular course of action. This section describes some of the main public interest actors.

## Consumer groups (see Chapters 21 and 24)

Consumer groups, especially in Europe, have generally argued against the introduction of GM foods, at least until much more is known about their long-run impacts. European consumer groups have coalesced behind this issue (and that of food safety more generally) in an impressive way, projecting organizations such as the European Bureau of Consumer Unions to prominence and giving a new lease of life to consumer watchdog groups. Such common cause cannot, however, mask a split between the greater concern in northern Europe and the relative lack of interest in the south. Much of this difference can be put down to political and structural factors, along with chance events, that have exacerbated national sensitivities. The rise of consumerism has clearly fed on concerns about the reliability of regulatory bodies, the honesty of politicians, and the objectivity of scientists.

Consumer activists in the European Union have been enormously successful in their campaign against so-called 'Frankenstein foods.' In a recent survey conducted through the EU, 86% of those questioned believed food that contains GMOs should always be labeled as such. More than 50% trusted consumer associations to tell the truth about the food supply, and only 25% put their faith in national governments or EU authorities (Anonymous, 1999b). Various groups have voiced disagreement with the introduction of GM crops in the food system and have criticized governments for being negligent in assessing the impact on health and the environment.

Political parties have sensed the public mood, and some have sprung into the lead. The Green Group in the European Parliament has a long tradition of fighting for consumer protection and has consistently criticized the GMO and novel food legislation of the EU as being weak. The Greens want the promotion of environmentally friendly production methods and seek an EU position in the upcoming WTO negotiations to argue for a food code.

While consumer groups in Europe have thus far been the most vocal about their reluctance concerning the introduction of GMOs in their diet, there are signs even in the United States that consumer concerns about bio-engineered crops may be growing. In the past, US consumers have been relatively complacent about the increasing proportion of foods made with GM crops. The tide may have begun to turn, as more organizations are voicing an opinion and as discussion heads in the direction of requiring labels on GM foods in the United States. In June 1999, a petition signed by almost half a million US citizens was delivered to Congress. The petition, coordinated by the Natural Law Party, called for the labeling of foods containing GM crops, including soybeans and corn. So far, the pressure has not been to remove such food from the shelves, but the request for labeling could be seen as a precursor to such a demand. After all, it is widely accepted that US consumers do not realize the ubiquity of GM foods in the supermarkets. This level of ignorance is being reduced rapidly. For example, *Consumer Reports*, with 4.6 million subscribers, carried a major article on GMOs and GMO content in US food products in the September 1999 issue. Recognition of the current situation may be necessary before mobilizing support for further action. According to the delegation presenting the Natural Law Party petition, it was not aimed at 'calling for a boycott of genetically engineered foods [but to] simply [promote the labeling of them] so consumers can make a choice about what they buy. [In addition, it is also intended to prompt] the government to conduct an investigation into the long-term effect of [such] food' (*http://www.natural-law.org/*, accessed October 11, 1999).

## Environmental groups (see also Chapters 19 and 20)

Consumer groups in Europe have found natural allies in the environmental groups. If GM foods are potentially bad for both consumer health and the environment, it makes sense to coordinate the campaign. Moreover, one environmental group with a worldwide network stood ready to take up the challenge of leading the opposition to the biotech companies' activities. Greenpeace possessed the organizational skills and name recognition for the task and was prepared to engage in political theater to make its point. Other, less activist groups, such as the World Wildlife Fund, joined the campaign. Unlike the consumer groups, who argued primarily on grounds of possible health hazards, many of the environmental groups focused on the corporate structure of the life science industry. The introduction of GM foods was a symptom of a broader malaise attributable to globalization and the elevation of the profit motive over social objectives. Both consumer and environmental groups have targeted the byzantine regulatory process of the EU and have called for more participation in the approval decisions. Greenpeace, for example, has argued for major improvements in the regulatory process, including a maximum time limit of ten years on market approvals of new GMOs and the inclusion of ethical considerations in the approval process.

Environmental groups worldwide generally argue that little is known about the long-term environmental consequences of GM crops, despite assurances from the biotech companies. Some, like Friends of the Earth in the United Kingdom (a consortium of 61 environmental, developmental, health, consumer, food production, religious, and political groups), call for a freeze on the commercial growing and importing of GM crops, claiming that the health implications are serious.[6] This view was reinforced by the controversial experiments carried out by Arpad Pusztai at Aberdeen on GM potatoes and rats.[7] They also argue that the potential environmental consequences should be taken seriously because biotechnology can lead to the creation of superweeds (see the discussion of genetic flow on pages 62, 67, and 75) as well as to the destruction of natural habitats and flora. Some environmental groups, such as Genetic Concern in the United Kingdom, accuse biotech companies of distorting the

---

[6] See *http://www.foe.org/safefood/index.html*, accessed October 11, 1999.

[7] See 'Leaving a Bad Taste,' *Scientific American*, May 1999, *http://www.biotech-info.net/bad_taste.html*, for more information. This research was published in *Lancet* in October 1999 accompanied by considerable controversy.

reality which many US farmers are experiencing with GM crops (O'Sullivan, 1999). They dispute the claims that GM crops lead to increased yields and less pesticide use, claiming that in the majority of crops and regions surveyed, no statistically significant differences were found in pesticide use or yield between engineered and non-GM crops. They also argue that GM crop tests have failed to investigate the long-term impact of replacing specific herbicides with a broad-spectrum herbicide such as glyphosate (O'Sullivan, 1999).

## The Public Sector

Besides the private sector and the public interest groups, the third major set of stakeholders in the debate is the public sector. This category contains national governments and regulatory agencies, politicians and political party structures, and the plethora of research agencies and committees that surround any modern regulatory activity. As discussed in the next chapter, no single body exists for dealing exclusively with GMOs; the task falls to a variety of agencies in each country. Widespread public interest in the issue has ensured that politicians and others with their finger on the public pulse have voiced their opinions on the matter. One could attribute altruistic motives to these various public sector groups, but each group also sees the issues from its own perspective and may be no more impartial than the lobbyists that knock on their respective doors.

### Government agencies (see also Chapters 15 and 16)

The positions of US government agencies toward GMOs focus primarily on food and environmental safety and competition issues. Part of this reflects the position of the US Congress on the development of the technology. In the United States, there is a bipartisan effort to give top priority to expanding the world trade in biotechnology because, as former Senator John Ashcroft of Missouri explained, 'America is the world leader in biotechnology' and because 'success in world markets will bring major, long-lasting gains to family farmers, consumers, and the environment.' ('United States Official', 1999). But it also reflects a general satisfaction (not shared by all environmental groups) that the regulatory process is broadly doing its job. There is therefore a reluctance to reopen issues such as the way in which the health and environmental impacts of GMOs are tested.

The EU suffers from a less integrated decision structure, in this as in other areas (see discussion on page 109). Moreover, the prospects for technological advances and trade opportunities have not been prioritized at the EU level, though some individual governments such as that in the United Kingdom have taken a strong pro-biotech stand. In fact, environmental ministers have called for imposing stricter standards as well as monitoring GM-derived foods. As a result of pressure from the Council of (Environmental) Ministers, the European Commission declared that it would delay approval of all pending GMOs for marketing in the EU.

The US administration takes a different view of Europe's problems. It sees Europe's fear of genetically modified foods and agricultural products as the 'single greatest trade threat' faced by the United States to agricultural exports (Reuters, 1999c). While US farmers have rapidly adopted genetically modified crops, the United States feels that the EU is trying to slow down the imports of such crops. As Stuart Eizenstat, Undersecretary at the United States Treasury Department, explained, this is particularly problematic for the United States because 'almost 100% of [United States agricultural exports] in the next five years will be genetically modified or combined with bulk commodities that are genetically modified' (Vorman, 1999). The US government has threatened to launch a WTO action against the EU for approval delays that it says cost US corn farmers some $200 million in annual lost sales (Vorman, 1999). US government officials urged the EU to help bolster consumer confidence by creating a food safety regulatory structure with rigorous scientific standards and procedures modeled after the US Food and Drug Administration ('United States Official', 1999). The creation of a European FDA hasd in fact been under consideration for some time, and incoming EU Commission President Romano Prodi has made it a priority.[8] Regulatory authorities in other countries have also moved to slow the approval process for new GMOs and to restrict unfettered use in the food supply.

## Scientists and the scientific establishment (see also Chapters 17 and 18)

Scientists work in a variety of locations and possess the range of views and biases of their nonspecialist counterparts. Most scientists consider genetic engineering an accurate and fast way of producing new crop varieties by changing a piece of genetic code, but some question the safety of releasing GMOs into the environment. Although some of the expected ecological and agronomic changes are being tested experimentally, some scientists believe that the enormous variety of outdoor environments cannot be simulated (Fincham and Ravetz, 1990).

For many research centers and international organizations that focus on finding solutions to world hunger and improving food security, GM crops are seen as having the potential to address nutritional deficiencies and ease Third World hunger. Several of these institutions believe that development of GM crops should focus on their potential beneficial use in developing countries. Improvements they advocate include crops with insect and disease resistance, increased levels of beta-carotene (which turns into vitamin A in humans,) and iron (which prevents anemia). Gordon Conway, President of the Rockefeller Foundation, explains that these benefits could reach the 180 million children in developing countries who suffer from vitamin A deficiency, help save some of the two million who die each year from it, and bring relief to some of the two billion people who suffer from anemia worldwide.

For the international agricultural research centers, the promising technological advances that may help improve the food supply of poor people in developing countries are far from being realized. There is fear that the mounting controversies in Europe and the developing world over the use of biotechnology will delay research, stop field trials, and impose irrational restrictions on the use of foods produced (Rockefeller Foundation, 1999).

---

[8] A European Food Authority was created on November 8, 2000 but its role is limited to coordinating scientific advice rather than taking decisions on the safety of foods.

The unknown risks of genetically engineered crops have led scientists to agree that the long-term effects on food safety and the environment need to be assessed. Some also argue that while there are hidden risks in any form of plant breeding, more research into the fundamental nature of gene action is required. The monarch butterfly controversy (see Chapter 28) has heightened concerns about the prospects of unexpected safety issues. Some ecologists are unsure of the impacts of bypassing natural species boundaries, and some argue that the risks of releasing GM crops into the biosphere are similar to those encountered in introducing exotic organisms in North America (Rifkin, 1999).

## The Struggle for Public Opinion

The first arena in which these actors face each other is the struggle for public opinion. Both the firms and the NGOs know the importance of favorable public sentiment and sympathetic press coverage. This is the key to both market acceptance and political support. The struggle for domestic acceptance is being waged on several fronts, particularly in Europe. The contest for public opinion has made widespread use of press campaigns as well as more modern channels of information such as the internet.[9] The main protagonists are companies with a direct interest, either as life science firms with patents on the new technology or as marketers of food who bear the brunt of consumer concerns, and the consumer and environmental groups concerned about the potential dangers of the spread of GMOs. Participants in the debate include notable public figures outside corporate or government agencies, such as Prince Charles in the United Kingdom and former president Jimmy Carter in the United States.[10]

Despite the philanthropic intentions of most supporters of NGOs and the dedication of the persons who work in these organizations, the tactics employed are not much different from those of the life science companies. Many of these organizations have long histories of skirmishes with corporate opponents and well-known long-term agendas within which the GMO issue conveniently fits. Their survival depends on picking such issues carefully and targeting their resources. Alliances are useful, often through the linking of issues to other agendas. Scientific evidence is marshaled to support positions, as is done by corporate entities. None of this is improper; these bodies have a vital role as a watchdog for the public interest. To perform this role, they must remain financially sound and programmatically credible. They are not infallible or omniscient or even particularly reliable as interpreters of scientific knowledge.

---

[9] GMO information spreads quickly on the web, making it possible for the rush of instant information and misinformation to overwhelm the traditional decision process on technical standards based on limited access to such information by those with a strong interest in the outcome.

[10] Prince Charles is a large producer of organic produce on his farms and therefore has a significant interest in the debate. He has come down strongly against the introduction of transgenic crops. President Carter has spoken in favor of biotech foods. His background as captain of a nuclear submarine before entering politics presumably makes him more positive toward new technology.

By the same token, companies have clear objectives in terms of the acceptance of biotech foods and good reasons to try to persuade consumers of their safety and environmental acceptability. This does not mean that they are always wrong. For one thing, corporations have to face the prospect of financial liability as well as loss of prestige if they market unsafe products. They have an incentive to avoid punitive damage settlements from juries who empathize with the plight of aggrieved consumers. And they need to cultivate an image of reliability and consumer concern.[11]

The battle for public opinion clearly seems to be going against GMOs in Europe. Though one does not know when the media will lose interest in the story, the public perception of the health and safety attributes of GM foods is unlikely to change in the near future. This is not to imply that it could never change. Foods with a clear consumer benefit could become acceptable fairly quickly. Products that fall somewhere between foods and medical remedies (nutraceuticals) could also find a niche market. A significant price difference between GM and non-GM foods could also swing the pendulum back in the direction of acceptability.

In the United States, the initial skirmishes in the quest for favorable public opinion seem to have been won by the companies, though the struggle may be far from over. There has been pressure from consumer groups for a revision of the oversight on GMO approval processes. Environmental groups have made some headway in making the case that the impact on wildlife and on insect resistance may have been underplayed by commercial firms. Until recently though, there has been little or no interest shown by the general public or mainstream press in the United States. However, this does not mean there is no discussion of these topics among interested parties. In particular, the debate over labeling is likely to continue for some time. It is possible that the US administration may decide labeling is a way to appease public concerns without admitting health or environmental problems. Firms, on the other hand, worry that any label may become the focus of a campaign to persuade the consumer to avoid GM products.

## The Struggle for Regulatory Control

The second main battlefield is in the arena of national and international regulation. The protagonists are less numerous in this contest. They include government agencies, both those whose job it is to regulate the domestic market and international agencies that have some coordinating responsibilities. Nongovernmental organizations and companies have limited influence in this arena, though each thinks that the other is having too much input into the deliberation. The struggle for regulatory control is however as intense as the one for gaining public approval in the marketplace. National regulatory bodies

---

[11] In this respect, the fate of the tobacco industry in the United States gives pause to any corporate tendencies to hide research findings that might later be used to show that they knew of possible health or environmental consequences. On the other hand, those who mistrust corporate morality will see the tobacco lawsuits as confirming their worst fears of corporate secrecy and duplicity.

have their own procedures and competencies, and in general dislike the internationalization of regulations. Trade agencies see more merit in harmonization of standards, which tends to promote trade and avoid conflicts. Trade agencies in exporting countries are wary of labeling, seeing the possibility of misuse for protectionist ends.

Until recently, the battle for regulatory control seemed to be going in favor of the steady introduction of GMOs without undue constraints and subject to individual country approval procedures. The struggle continues, however, in the EU to define the regulatory regime and the level at which approval will be granted. Other countries have taken more of a wait and see approach. The problem in the EU is not so much with the EU institutions themselves as with some of the member governments. Northern Europe, in particular, seems to be moving reluctantly on the issue. The French and the Austrian governments have ignored EU rules by delaying acceptance of several GMO products and seem prepared to be taken to the European Court (if the Commission wishes to risk public outcry) for their obstruction. If the EU had not stopped approval of new GMOs, it was quite likely that one of the member governments would have announced a GMO moratorium in contravention of EU directives and regulations that authorize use and importation of these products. Though the fiercest struggle of the turn of the century was in the United Kingdom, where the government tried vainly to calm consumer fears and proceed with the significant investment and research activities that have been encouraged by Whitehall, the challenge to Brussels is more likely to come from France or Germany.

The domestic regulatory struggle is part of a wider set of international forces involving food safety and environmental issues. On the one hand, global agribusiness companies are trying to market agricultural supplies, raw materials, and food products on a worldwide basis; for these companies, the plethora of national regulations is an unnecessary impediment to trade, raising costs and adding uncertainty. On the other hand, environmental and consumer NGOs who would like to see stricter regulation and more accountability for multinational corporate decisions while the development community is concerned that the biotech revolution constitutes a threat to poor countries. To these groups, international regulation is necessary to contain the power of the multinationals.

The regulatory process at the international level is still in the early stages, resulting in confusion and some attempts to control agendas. The tension between the WTO and the guardians of the Convention on Biological Diversity (there is no full-fledged secretariat for the CBD) is an example. Many developing countries, as well as the EU, would prefer that the CBD take the lead through the recently agreed Biosafety Protocol (see Chapter 10) and that the WTO fall into line, allowing countries to choose whether to accept GMOs in their markets. But the United States (not a signatory to the CBD), along with other exporters, insists that the WTO must prevail in matters of trade policy and that the established requirements for customs procedures and market access be respected even in the case of GMOs. The role of CODEX in setting standards and labeling regulations is another question that divides countries and often ministries within national governments.

## The Marketplace

Public sentiment about the safety and environmental risks together with regulatory control at the national and international level will go some way to condition the future for products of agricultural biotechnology. But for the private sector the ultimate test is that of the marketplace, and the ability to sell products that people will buy. The market test has proved to be somewhat inconclusive so far, at least as far as the major GM products are concerned. In the US market, resistance to the introduction of foods including GM corn and soybeans has been minimal. Those groups that have attempted to bring the widespread use of such products in the food supply to the attention of the public have had little response, at least until the StarLink incident, and retailers have continued to market the products with no fear of consumer disapproval. On the other hand, sales of organic foods have increased – in part because those concerned with health and environmental effects of biotechnology have switched their purchasing habits. Feed sales have not been significantly hit by any concern that GM corn fed to animals has any health and environmental impact: at least in this respect scientific opinion, that altered genetic material in feed does not have any detectable impact on the genetic composition of milk or meat, is apparently accepted. Where companies have declared their intention to reduce or eliminate the use of GM ingredients, as in baby food, there has been little impact on the market.

In Europe the situation is very different, with the retail sector judging that it is better to be aligned with the consumer in the struggle against potentially dangerous (imported) foodstuffs, for fear of a negative reaction. This has caused a significant shift in market demand away from GM ingredients, when their use has been approved, and a reluctance of companies to go ahead with marketing plans when approval is still needed. Thus the EU market is shrinking for US corn and soybeans and their products, and other countries are taking up the slack. Retailers are making a virtue out of necessity, and competing to give the consumers the strongest assurance that they sell no GM foods. By staying ahead of the regulators, the retailers are turning a potentially disastrous drop in consumer confidence in the safety of the food system into a prominent marketing strategy. In those countries where concentration levels are high, such as the UK, this has probably strengthened their market position. Their purchasing decisions set standards for the wholesaling of foodstuffs and their development of supply chains with sophisticated tracking capability is leading to structural changes as far back as the farm level.

The market of course takes such differences among countries in its stride. The differentiation of the product by means of tracking of supplies and certification of individual suppliers is a market reaction. (See Chapter 11 for more information about market segmentation.) Consumers that want a particular product express their preference through the market and send the appropriate signals. The market differentiation can also be profitable for the producer, who will charge higher prices for the products that have the 'desirable' attribute (in this case those that are GM-free). The cost of market differentiation, together with a general unease in the market as consumers reconsider their buying habits, will act to reduce profits. It is doubtful whether market differentiation will make the grain and soybean markets in Europe more profitable for US exporters, but suppliers as a whole (including those from Canada and Latin America) may well be able to increase revenues from the EU market if consumer choice or government regulations continue to restrict the use of GM ingredients.

9

# The Domestic and Regional Regulatory Environment

**Gerald C. Nelson**
University of Illinois, Urbana, Illinois, USA

**Julie Babinard**
International Food Policy Research Institute, Washington DC, USA

**Timothy Josling**
Institute for International Studies, Stanford University, Stanford, California, USA

 National governments set up regulatory frameworks to protect society from unintended, harmful consequences of economic enterprise and to encourage activities, such as the development of intellectual property, that the unfettered workings of the market would not. Because commercial production of GMOs has been in existence only for about ten years, the regulatory environment for embodied intellectual property of the technologies and its products is still in flux. Typically, countries have used existing regulatory agencies and authority and added explicit responsibility for GMO crops. Usually, the regulatory process is more complicated than the one followed when introducing conventionally bred crops. In addition, the process is often subject to independent approvals from more than one regulatory agency, with each agency being responsible for different uses or aspects of the product (US Congress, Office of Technology Assessment, 1988).

Table 9.1 gives an overview of the scope of domestic, regional and international regulation of the various stages of introduction of GMOs into the food chain. Regulation typically occurs at four levels of activity – technology patenting, field-testing of the crop, animal and human consumption trials, and approval for commercial use as a crop and for human and animal consumption. Regulations differ greatly in scale and implementation, varying from extremely restrictive in some industrialized countries to nonexistent in some developing countries. Dealing with such disparities will become increasingly important as

Genetically Modified Organisms in Agriculture
ISBN 0-12-515422-4

international trade in GMOs, and the intellectual property on which they are based, grows. This chapter reviews the domestic and regional regulatory environments in selected countries. The next chapter reviews the regulations in place at the international level, in particular the current and potential roles of the World Trade Organization (WTO) and the Biosafety Protocol of the Biodiversity Convention concluded in 2000.

A few industrialized countries and the private sector in those countries dominate GMO development and commercialization. Between 1986 and 1995, over 90% of the trials took place in industrialized countries (Table 9.2). The bulk of the trials, as shown in Table 9.2, was conducted in the United States (6937) and Canada (780). North America accounted for 71% of the trials, the EU for 14%, and the rest of the world for only 15%. Only a few trials were conducted in developing countries. With the exception of a few transgenic crops tested in Cuba, China, and Mexico, all the transgenic crop material tested in developing countries had been developed externally and imported by the developers of the technology (James and Krattiger, 1996). Countries in Latin America and the Caribbean region represented 70% of all trials in developing countries, representing almost 6% of global trials. A total of 14 trials were reported between 1986 and 1995 for three African countries – South Africa, Egypt, and Zimbabwe – with 90% conducted in South Africa.

**Table 9.1**: Regulatory framework for stages of GMO introduction

| Stage | National | Regional | Multilateral |
|---|---|---|---|
| Licensing of GMO technology | National patent laws | EU licensing regulations | WTO/TRIPS |
| Testing of GMOs | National agencies such as APHIS, in conjunction with firms | EU Directives 90/219 and 90/220 | Biosafety protocol |
| Approval for domestic use | National agencies such as the EPA and FDA | EU Novel Foods Regulation and Directive 90/220 | CODEX conventions on labeling |
| Access to market for GMO products | | EU internal market regulations; EU import regulations; and Directive 90/220 | |
| Resolution of trade conflicts | | NAFTA dispute settlement process; Transatlantic Dialog | WTO/DSU |
| Indirect structural and distributional issues | | | Biosafety protocol |

Over 70 crop species have been transformed and at least 54 have been field tested, with most of the tests conducted on eight crop species: corn (maize), canola (rape), potato, tomato, soybean, cotton, tobacco, and melon (James and Krattiger, 1996). Thus far, marketing approvals for GMOs have been sought almost exclusively for crops whose primary use is in processed foods (the Flavr Savr tomato and *Bt* sweet corn and potatoes are exceptions). These genetically modified ingredients tend to become part of the general food supply rather than being discrete food items consumers can either choose or refuse (Nottingham, 1998). Once approval has been granted for a transgenic crop, the crop can enter the diet in a number of different ways. Soybeans, for example, are extremely versatile and are used in a large number of processed foods, including foods as diverse as bread, baby foods, sausages, meat substitutes, ice cream, and chocolate and other candies (Nottingham, 1998). It is estimated that 60% of all processed foods in industrialized countries contains genetically modified soybeans (Nottingham, 1998; *Consumer Reports*, 1999).

**Table 9.2**: Total number of transgenic crop field trials worldwide (1986–1999)

| Country | Trials | Country | Trials | Country | Trials | Country | Trials | Country | Trials |
|---------|--------|---------|--------|---------|--------|---------|--------|---------|--------|
| Argentina | 90 | Chile | 60* | Germany | 100 | New Zealand | 43 | Netherlands | 113 |
| Australia | 57 | China | 60* | Greece | 19 | Norway | 1* | Thailand | 2* |
| Austria | 3 | Costa Rica | 17* | Guatemala | 3* | Portugal | 4 | UK | 178 |
| Belgium | 100 | Cuba | 18* | Hungary | 22* | Russian Federation | 4 | United States | 6937 |
| Bolivia | 5 | Denmark | 34 | Ireland | 4 | South Africa | 11 | Zimbabwe | 1* |
| Brazil | 64 | Egypt | 2* | Italy | 248 | Spain | 153 | | |
| Bulgaria | 3 | Finland | 16 | Japan | 124 | Sweden | 52 | | |
| Canada | 780 | France | 454 | Mexico | 38* | Switzerland | 2 | Total | 9822 |

*Sources*: Information System for Biotechnology, *http://gophisb.biochem.vt.edu/*; OECD Database of Field Trials, *http://www.olis.oecd.org/biotrack.nsf/by+country*, except numbers followed by an asterisk, which are from James and Krattiger (1996).

# National Regulations on Use and Marketing

## United States – FDA, EPA, and USDA[1]

The US regulatory process for GMOs is guided by a coordinated 'framework' established in 1986 (James and Krattiger, 1996). Three federal agencies – the Food and Drug Administration (FDA), the Environmental Protection Agency (EPA), and the Animal and Plant Health Inspection Service (APHIS) of the Department of Agriculture (USDA) – regulate different aspects of GMOs. APHIS issues permits for field trials and commercial release for production. EPA regulates pesticides used in or on foods for safety and sets tolerances or establishes exemptions from them. FDA has authority under the Federal Food, Drug, and Cosmetic Act (FFDC Act) to ensure the safety of domestic and imported foods, except meat and poultry, which are regulated by USDA. FDA monitors foods to enforce the pesticide tolerances set by EPA. In practice, this means that while APHIS issues permits for field trials and also for commercial release, any crop containing a pesticide also requires approval from EPA. If the product from a transgenic crop is designed for food or feed use, FDA is involved in the application process (James and Krattiger, 1996).

In addition, before any commercialization, genetically engineered plants must also conform to standards set by state and federal marketing statutes such as state seed certification laws, the FFDC Act, the Federal Insecticide, Fungicide,

---

[1] A good source of information on US federal regulation of GMOs is *http://www.aphis.usda.gov/biotech/OECD/usregs.htm*.

and Rodenticide Act (FIFRA), the Toxic Substances Control Act (TSCA), and the Federal Plant Pest Act.

## APHIS[2]

The Animal and Plant Health Inspection Service (APHIS) of USDA is responsible for protecting US agriculture from pests and diseases. Under authority of the Federal Plant Pest Act, APHIS regulations provide procedures for introducing a 'regulated article' into the United States. Regulated articles are organisms and products altered or produced through genetic engineering and which might become plant pests. The regulations also provide a petition process to move a product to nonregulated status, in which the product and its offspring no longer require APHIS review.

APHIS exercises its regulatory authority by issuing permits. A company, academic research institution, or public-sector scientist wishing to move or field test a genetically engineered plant must obtain the necessary permit(s) before proceeding. There are two basic types of APHIS GMO permits – for movement and for release into the environment.

APHIS requires a permit and concurrence of individual state departments of agriculture to move any genetically engineered organism that is a potential plant pest into the United States or between states. Permit applicants must provide APHIS with details about the nature of the organism, its origin, and its intended use. After the application is reviewed, APHIS makes a preliminary pest risk analysis and sends a letter to the appropriate state department of agriculture asking it to review the proposed movement. APHIS and state officials inspect the facility that will receive the organism to ensure that the organism will not accidentally be released into the environment.

APHIS also oversees field testing (also called environmental release) of genetically engineered crops. Applicants must provide complete information about the plant, including all new genes and new gene products, their origin, the purpose of the test, the experimental design (how the test will be conducted), and precautions to be taken to prevent escape of pollen, plants, or plant parts from the field test site. As part of the review process APHIS prepares an environmental assessment (EA), a document that analyzes any possible environmental impacts the field test could have. Field trials must be conducted so that the regulated article will not persist in the environment, no offspring can be produced that could persist in the environment, and no viable material remains that is likely to volunteer in subsequent seasons.

Before a genetically engineered crop can be sold commercially, companies must file a petition for USDA exemption. This petition requires more information than a field-test permit, including environmental product safety information.

To simplify the approval process, APHIS has developed a 'Notification System' for those transgenic crops with which the agency has had extensive

---

[2] This section is based on information from the APHIS web page, *http://www.aphis.usda.gov/biotech/*.

experience in processing and monitoring applications. With this system, applicants are not required to obtain a permit to conduct trials as long as APHIS has reviewed a notification and determined that the review needs no further consultation. Effective April 30, 1993, provided they meet eligibility criteria and performance standards, six crops qualify for the notification procedure: tomato, corn, tobacco, soybean, cotton, and potato.[3] The agency does not prepare an EA for these field tests.

## FDA[4]

FDA regulates genetically engineered foods and food ingredients under the same provisions of the FFDC Act used for other food products. This means that a food or food ingredient developed by genetic engineering must meet the same safety standards as other food products. FDA has authority to take legal action against a substance that poses a hazard to the public, including removing it from the market. Foods derived from new plant varieties developed through genetic engineering are regulated under this authority as well.

FDA relies primarily on two sections of the FFDC Act to ensure the safety of foods and food ingredients – section 402 (adulteration) and 409 (additives). Section 402 is the basis for most of FDA's food safety regulations, but GMO-based products are most likely to be regulated under section 409.

---

[3] The notification procedure requires that
- the introduced genetic material is stably integrated in the plant genome.
- the function of the introduced genetic material is known and its expression in the regulated article does not result in plant disease.
- the introduced genetic material does not
  - cause the production of an infectious entity.
  - encode substances that are known or likely to be toxic to nontarget organisms known or likely to feed or live on the plant species.
  - encode products intended for pharmaceutical use.

To ensure the introduced genetic sequences do not pose a significant risk of the creation of any new plant virus, plant virus-derived sequences must be
- noncoding regulatory sequences of known function.
- sense or antisense genetic constructs derived from viral genes from plant viruses that are prevalent and endemic in the area where the introduction will occur and that infect plants of the same host species, and that do not encode a functional non-capsid gene product responsible for cell-to-cell movement of the virus.

The notification process also requires that the plant has not been modified to contain the following genetic material from animal or human pathogens:
- any nucleic acid sequence derived from an animal or human virus.
- coding sequences whose products are known or likely causal agents of disease in animals or humans.

*Source*: Regulation 7CFR340: Introduction of Organisms and Products Altered or Produced Through Genetic Engineering Which Are Plant Pests or Which There Is Reason to Believe Are Plant Pests, *http://www.aphis.usda.gov/bbep/bp/7cfr340.html*.

[4] This section draws heavily on two sources – FDA's 1992 Statement of Policy: Foods Derived from New Plant Varieties, *Federal Register*, May 29, 1992 (reproduced at *http://vm.cfsan.fda.gov/~lrd/fr92529b.html*) and Maryanski (1995).

The first authorization of a GMO-based food ingredient, fermentation-derived chymosin (rennet), now widely used in cheese making, was issued in 1990. The agency foresaw a growing availability of GM foods. An official framework for regulating GMOs under the FFDC Act was published in 1992. Under this framework, GM foods are regulated using an approach identical in principle to that applied to foods developed by traditional plant breeding. The regulatory status of a food, irrespective of the method by which it is developed, depends on specific characteristics of the food and the intended use of the food.

### FDA guiding concepts: GRAS and substantial equivalence

To reduce the regulatory burden, FDA exercises minimal oversight of products that are Generally Recognized as Safe (GRAS). Because foods in use today are GRAS, they are the standard by which the safety of GMO foods is assessed. The safety assessment approach outlined in the 1992 policy focuses on the intended genetic modification and the overall composition of important nutrients and toxicants in the food. This approach starts from two premises. First, GM foods are variants of existing, well-accepted foods. Second, many foods contain components that would present safety concerns if they were present in large concentrations. In addition, some individuals in the population are allergic or intolerant to certain foods. Thus, absolute safety for a food cannot be achieved or expected. If zero tolerance of potentially hazardous compounds were the rule, few foods of any kind would be available for consumption.

An important feature of the 1992 framework is that ingredients (proteins or other added substances such as fatty acids and carbohydrates) produced by introduced genes must receive pre-market approval as food additives if they differ substantially from ingredients already in foods. However, if the ingredients are 'substantially equivalent' to ingredients already in foods, pre-market review generally is not required.[5] For example, if a food derived from a new plant variety contains a novel protein sweetener, that sweetener would likely require submission of a food additive petition and approval by FDA before marketing.

An important early test of the 1992 framework was the development of the Flavr Savr tomato, the first commercial GM food crop. Whole foods, such as fruits, vegetables, and grains, generally are not subject to pre-market approval. However, in the case of the Flavr Savr tomato, FDA chose to conduct an extensive review. The rationale was the potential that the Flavr Savr tomato might not be substantially equivalent to existing tomatoes. After an extensive review, FDA ruled that nutritional characteristics of the Flavr Savr tomato were within the range of existing varieties of tomatoes already in the food supply and substantially equivalent to existing tomatoes.

---

[5] The concept of substantial equivalence as applied to GM foods was first used in an OECD publication, 'Safety Evaluation of Foods Derived by Modern Biotechnology: Concepts and Principles,' 1993, available at *http://www.oecd.org//dsti/sti/s_t/biotech/prod/MODERN.pdf*, accessed October 10, 1999. It is a report of a group of 60 experts from 19 countries, nominated by their governments. An article in *Nature* (Hefle and Taylor, 1999) challenging the validity of using substantial equivalence to evaluate the safety of GM foods has drawn quick rebuttals. See *http://www.biotechknowledge.com/showlib_us.php3?2167*, accessed October 10, 1999.

The possibility that novel proteins could cause allergic reactions in some individuals has received particular scrutiny. Attention has been given to proteins derived from foods to which individuals in the US population are commonly allergic, such as milk, eggs, wheat, fish, tree nuts, and legumes. A developer of GMOs that incorporates genetic material from these sources is expected to demonstrate that the allergenic substance is not present in the new food; otherwise, FDA would require some form of labeling to alert sensitive consumers. In 1995, Pioneer developed a soybean that included a gene from the Brazil nut. Subsequent testing showed that the resulting protein was responsible for allergic reactions to Brazil nuts. The soybean was however not approved for commercial applications because preliminary results suggested that the Brazil nut gene transferred a potential for allergenicity to the soybeans. While it was known at the time that Brazil nuts were allergenic to some consumers, no one had ever identified which Brazil nut gene was the responsible allergen. The preliminary results suggested that the culprit was the specific Brazil nut protein that had been created in these transgenic soybeans (Nordlee *et al.*, 1996).

## Labeling

FDA's 1992 policy also addressed labeling of foods derived from new plant varieties, including plants developed by genetic engineering. The FFDC Act defines the information that must be disclosed in labeling (including on the food label). The Act requires also that all labeling be truthful and not misleading, but it does not require disclosure in labeling of information solely on the basis of consumers' desire to know. The Act does require that a food be given a common or usual name.

FDA requires special labeling if the composition of the GM food differs significantly from its conventional counterpart.[6] For example, if a food contains a major new sweetener as a result of genetic modification, a new common name or other labeling may be required.[7] Similarly, if a new food contains a protein derived from a food that commonly causes allergic reactions (and the developer cannot demonstrate that the protein is not an allergen), labeling would be necessary to alert sensitive consumers. However, if a protein commonly produces very serious allergic reactions (e.g., peanut protein) and is transferred to another food, FDA would need to evaluate whether it would be practical to label the

---

[6] The notion of identity, and therefore the need for labeling, is not clear-cut. For example, if a new soybean variety had a dramatic change in protein/oil ratio – say 90% oil to 10% protein – it would not be identical to existing varieties but neither would it be a new food additive because FDA would advise that the proteins and oils were no different than those found in conventional soybeans. Would it be labeled? Not if it were fractionated into ingredients, but it probably would be if it were sold as a whole food because it is substantially different in composition (based on personal communication from Bruce Chassy, Professor of Food Microbiology, University of Illinois, Urbana-Champaign).

[7] FDA did not require special labeling for the Flavr Savr tomato because the new tomato was not significantly different from the range of commercial varieties referred to by that name. However, Calgene (the developer of the Flavr Savr tomato) decided to provide special labeling, including point-of-sale information, to inform consumers that the new tomato has been developed through genetic engineering.

food throughout its distribution. Circumstances could exist for which labeling would not provide sufficient consumer protection, and FDA would take appropriate steps to ensure that the food would not be marketed (such as occurred with the GM soybean containing a Brazil nut gene).

FDA's current position is that it has no basis to distinguish genetically engineered foods as a class from foods developed through other methods of plant breeding. Thus, it does not require foods produced from crops developed with rDNA (or other) technologies to be specially labeled.[8] As discussed earlier, FDA undertook an extensive review of the food safety aspects of the Flavr Savr tomato, cotton, and canola in the early 1990s. After this evaluation and approval, FDA decided that it was not necessary to conduct either full-scale scientific reviews of new GMOs or comprehensive scientific review of data generated by the developer. FDA implemented a voluntary consultation procedure for new GMOs: 'FDA considers, based on agency scientists' evaluation of the available information, whether any unresolved issues exist regarding the food derived from the new plant variety that would necessitate legal action by the agency if the product were introduced into commerce.'[9] The list of completed consultations can be found at *http://vm.cfsan.fda.gov/~lrd/biocon.html*. Because the objective of most modifications is to effect some kind of change in composition no matter how small, one could argue that a GMO cannot be equivalent. In practice, there is an iterative consultation process between developers and FDA through which it is decided by mutual consensus whether pre-market approval should or should not be required. FDA provides guidance to the developer on a case-by-case basis. Typically, the developer can bring evidence on toxicity, allergenicity, and composition to FDA and ask for an opinion.[10] After holding a series of public hearings on its oversight role, FDA made the consultative process mandatory in 2000.

Unlike a food in the human diet, an animal feed derived from a single plant may constitute a significant portion of the animal diet. For instance, 50 to 75% of the diet of livestock on US farms consists of field corn. Therefore, a change in nutrient or toxicant composition that is considered insignificant for human consumption may be a very significant change in the animal diet. Furthermore, animals consume plants, plant parts, and plant by-products that are not consumed by humans. For instance, animals consume whole cottonseed meal, whereas humans consume only cottonseed oil. Gossypol, a plant toxicant, is concentrated in the cottonseed meal during the production of cottonseed oil.

FDA's 1992 policy statement on GMOs also deals with animal feeds derived from new plant varieties. Most of the attention focused on antibiotic resistance or gene flow to microorganisms that might be caused by use of marker genes. FDA concluded that animal feed production methods will likely denature the enzyme, thereby rendering it inactive against antibiotics.

---

[8] The agency has not required labeling for other methods of plant breeding such as chemical- or radiation-induced mutagenesis, somaclonal variation, or cell culture. For example, there is no requirement to label hybrid sweet corn because it was developed through cross-hybridization.

[9] *http://vm.cfsan.fda.gov/~lrd/consulpr.html*, accessed September 23, 1999.

[10] Based on personal communication from Bruce Chassy, Professor of Food Microbiology, University of Illinois, Urbana-Champaign.

## EPA

The EPA is involved in GMOs primarily through its role in pesticide regulation. The development of pesticide resistance and pesticide resistance management has played an important role in its regulatory decisions (see reviews in Matten *et al.*, 1996, updated in Matten, 1997). The rationale behind EPA approval of pest-resistant GMOs is that pesticide resistance management is likely to benefit the American public by reducing the total pesticide burden on the environment and by reducing the overall human and environmental exposure to pesticides.

Although EPA does not yet have a published policy or standard data requirements in place for pesticide resistance management, it has required the submission of such data on a case-by-case basis. Producers of most pest-resistant GMOs have been given temporary permission to sell their products. Typically the registration is valid for only five years.

EPA has given special attention to *Bt* GMOs. Effective resistance management strategies for *Bt* GMOs are essential because they produce the toxin throughout the growing season. For all pesticides, an effective resistance management plan is likely to include predictive tactics – scouting, sampling, and monitoring for changes in pest susceptibility – and evaluation measures to determine the success of the plan. An example for *Bt* plant pesticides is the use of a high-dose expression strategy coupled with the use of an effective refuge.

EPA believed that during the first five years following commercialization (the approximate time limit of the conditional registrations for *Bt* corn), there would not be enough *Bt* corn acreage to provide substantial *Bt* selection pressure for the development of ECB resistance. Consequently, it did not mandate specific refuge requirements for *Bt* corn. However, as of crop year 1999, refuges were mandated (see discussion of refuges and resistance management beginning on page 68).

EPA mandated specific resistance management data requirements and mitigation measures with resistance management strategy for all *Bt* corn and *Bt* cotton registrations. Typically, studies were conducted on target and secondary pest biology and behavior, population dynamics, cross-resistance potential, refuge strategies, dose deployment adequacy, and discriminating concentration.

For glyphosate-resistant (GR) crops in general and GR soybeans in particular, EPA's responsibility is for the use of the herbicide itself (e.g., Roundup) rather than the seed. In contrast to *Bt* crops, GR crops have no direct toxic effect. When EPA reregistered glyphosate in 1993, the same tolerances for residues of glyphosate were applied to GR and non-GR crops. EPA could in principle monitor weediness and genetic flow; but, these are not likely to be serious problems with soybeans, at least outside of Asia. EPA has apparently not considered it necessary to monitor weed resistance to glyphosate. It was felt that such resistance is likely to develop slowly, if at all.

## US protection for intellectual property in agriculture[11]

US legal protection for intellectual property in the agricultural arena takes three forms – plant breeder's rights, patent law, and trade secrets. Many countries are

---

[11] This section draws heavily on Roth and Shear (1999).

moving towards legal protection for intellectual property in agriculture similar to that in the United States, especially after the ratification of the TRIPS (Trade-Related Aspects of Intellectual Property Rights) Agreement in the Uruguay Round of world trade negotiations.

The Plant Protection Act, adopted in 1930, provided the first protection to intellectual property in US agriculture. It limited protection to asexually reproduced plants such as fruit trees and flowers that can be grown from cuttings. In 1970, the Plant Variety Protection Act was passed. This Act, and subsequent amendments, brought US law into conformity with the 1961 International Convention for the Protection of New Varieties of Plants and provided legal protection to sexually reproduced and tuber-propagated varieties. The term of plant variety protection is 20 years for most crops. The owner has exclusive rights to grow and market the variety and its products. However, a purchaser is allowed to save seed for his or her own use (but not for sale).

The conditions for obtaining a certificate are relatively easy, but protection is provided to identical varieties only. Minor changes to color, plant height, or other characteristics result in a new, unprotected variety.[12]

The first US patent law was passed in 1790, with the current version coming into effect in 1953. The term of a patent is 20 years from date of earliest application.

Patent protection for agricultural technology was not recognized as available until a 1980 Supreme Court decision (Diamond *v.* Chakrabarty). Although patent protection cannot be extended to naturally occurring living beings, the ruling stated that such protection was available if the organism had been altered by human intervention.[13] A patent provides the right to exclude others from making, using, selling, or importing the invention. The invention must be 'new' as defined in patent law. If it has been described in a printed publication anywhere, or has been in public use or on sale before the date of the application, a patent cannot be granted. However, experimental use does not preclude a patent.

A trade secret is information used in the operation of a business that is both secret and sufficiently valuable to provide an economic advantage. Trade secrets generally are protected under state law, although the federal Economic Espionage Act of 1996 protects certain aspects of trade secrets. Trade-secret legislation differs in several key aspects from patent protection and plant breeder's rights. Trade-secret protection has no time limit. Furthermore, if a party obtained the information lawfully and/or independently, the trade-secret owner cannot prevent its use.

---

[12] It is interesting to note that Canada's 1991 Plant Breeders' Rights legislation includes a compulsory licensing provision with reasonable remuneration to be paid to the original rights holder. As of mid-1999, 630 rights had been granted, and no compulsory licenses had been sought or granted. See Horbulyk (1999).

[13] However, well before the ruling, the Patent Office had issued a patent in 1975 to the University of Illinois on a seed product.

# Japan

Research and development of biotech products in Japan is undertaken by the Ministry for International Trade and Industry and by the Biotechnology Research Development Foundation, a conglomeration of 14 top firms involved in rDNA applications, bio-reactor development, and mass cell-culture technologies. Since 1989, the Ministry of Agriculture, Forestry, and Fisheries have issued regulations for field-testing genetically modified crops (Office of Technology Assessment, 1993). In addition, the Japanese Ministry of Health and Welfare has imposed a strict regulatory regime specific to foods and food additives manufactured with rDNA techniques (Miller, 1997).[14] As of March, 1999, Japan had approved 23 GM crops for commercial use. These crops were either herbicide resistant or contained *Bt* toxins.

GMO food safety issues are the responsibility of the Ministry of Health and Welfare. As of April, 1992, the ministry had food safety guidelines for GM products that are not consumed directly. The first product to be approved using these guidelines was GM chymosin (a milk coagulating enzyme used in manufacturing cheese and already approved in the United States and Europe) in August, 1994.[15] In early 1996, the Ministry began the process of evaluating herbicide-resistant food products. These were approved for domestic use.

Following the US news stories about the effect of pollen from GM corn on Monarch butterfly larvae, Japan declared that it would tighten its safety regulations on GM crops (Saegusa, 1999). The Japanese Ministry of Agriculture, Forestry, and Fisheries (MAFF) announced that approval of *Bt* crops would be suspended until its committee on GMOs had established criteria for evaluating the safety of such crops. The committee planned to release revised safety evaluation protocols, basing its final decision on safety studies carried out by Japanese institutions (Saegusa, 1999). MAFF has announced plans to require labeling of 30 types of food for GM content, beginning April 2001 (Tolbert, 2000).

# Australia and New Zealand

Monitoring and regulation of GMOs in Australia are undertaken by the Genetic Manipulation Advisory Committee (GMAC), a nonstatutory body responsible for overseeing the development and use of novel genetic manipulation techniques in Australia. The federal government plans to establish an Office of the Gene Technology Regulator and associated legislation to regulate gene technology in Australia that would be operational by 2001. In the meantime, an Interim Office has been established to work with GMAC.[16]

In New Zealand, regulation of GMO production is the responsibility of the Environmental Risk Management Authority (ERMA). Any research, development, field-testing, or release of GMOs is prohibited unless approved by ERMA.

---

[14] See the Ministry of Agriculture, Forestry, and Fisheries website on biotechnology at *http://ss.s.affrc.go.jp/docs/sentan/index.htm* for more details.

[15] See *http://www.mhw.go.jp/search/doce/white_p/book1/p2_c4/c4_sect3.html*, accessed September 29, 1999.

[16] See *http://www.health.gov.au/tga/genetech.htm* for more details.

The assessment processes as defined under the Hazardous Substances and New Organisms Act (HSNO) are considered to be among the most rigorous in the world. The Ministry of Agriculture and Forestry is responsible for enforcing the conditions for GM organisms imposed by ERMA on approved field trials, and also for taking all reasonable steps to ensure that importers comply with the HSNO Act. There were no approvals to release any GMOs in New Zealand as of late 1999, although there have been some approvals for field trials in containment. In May, 1999, the New Zealand government announced the establishment of the Independent Biotechnology Advisory Council whose role is to stimulate dialogue and enhance public understanding of biotechnology and the broader issues surrounding its use. It also provides independent advice to government on the environmental, economic, ethical, social, and health aspects of developments in biotechnology.[17]

Australia and New Zealand regulate food safety collectively. The agencies in charge are the Australia New Zealand Food Authority and the Australia New Zealand Food Standards Council. The Authority completed a study of a proposal to regulate foods produced using gene technology in 1998. The proposal was accepted by the Council and became law in both countries shortly thereafter. The regulations require an assessment of the safety of the GM food, taking into account nutritional quality, composition, allergenicity, and end use. Labeling was originally required only when the nature of the food was significantly altered. However, on August 3, 1999, the Council extended the labeling requirement to all foods made from GMOs. The labeling is a new requirement that takes effect from December 2001.[18] Labeling is required when novel DNA and/or protein is present in the final food in concentration greater than 1%.

## Developing countries

Most developing countries are only at an early stage of the process of developing regulations (James, 1998). With some exceptions, developing countries lack operational field-testing regulations (James, 1998). Some developing countries are pursuing efforts to align their regulations to that of developed countries. In South Africa, for example, the government is working to harmonize its regulations with the United States so that applicants in the country will no longer need to seek a permit for using a product from a transgenic crop. This would be possible as long as the crop has already been approved for use in the United States and as long as applications for field trials have been submitted (James, 1998).[19]

---

[17] See *http://www.maf.govt.nz/MAFnet/index.htm* for more information on GMO environmental regulation in New Zealand.

[18] See *http://www.anzfa.gov.au/* for more information on food safety issues and joint regulation of GMOs.

[19] See *http://www.isb.vt.edu/cfdocs/globalfieldtests.cfm* for links to international field tests of GMOs.

# Regional Regulations

Regional trade agreements (free trade areas and customs unions) once dealt exclusively with trade policy as manifest in tariff and non-tariff actions at the border. Domestic regulations were neither discussed nor regulated. As tariffs were reduced and the more blatant non-tariff barriers such as quotas and import licenses were removed, attention turned to the less obvious barriers to trade within regional agreements. These hidden barriers included government regulations and industry standards. A number of techniques evolved for dealing with these regulatory differences, including most notably the harmonization of standards and the mutual recognition of each other's regulations. Harmonization of GMO regulations within regional agreements is possible as a way of preventing intra-regional trade disputes, but mutual recognition of each partner's domestic GMO regulations seems possible in only the closest of regional integration schemes, where both mutual regulatory trust and the cost of conflict is high. For this reason, the regional approach has been most fully developed in the European Union.[20]

# The European Union[21]

The process that regulates transgenic crops in the European Union is complex and detailed in terms of application procedures and requirements.[22] There are at least two underlying sources of this complexity. First, some of the Union's regulations were enacted after EU member countries had already established their own procedures, and the full regulatory integration of EU member states is still in progress (James and Krattiger, 1996). Second, some member countries, especially Germany and Denmark, have developed particularly strict regulations and require a case-by-case governmental review of any use of biotechnology in agriculture (Miller, 1997). Denmark had established strict laws related specifically to the environmental applications of biotechnology-derived products

---

[20] As a regional rather than a national policy, the EU biotech policy is important in two respects. First, it recognizes the differences which still exist within the EU, among its member governments, on the issue of food regulation. Second, it demonstrates the similarity between the problems faced by the EU Commission in reconciling different national laws, traditions and institutions with the demands of a single market, and the ones faced by the WTO in reconciling different approaches among its members.

[21] The European Union is the most recent name for the organization known previously as the European (Economic) Community. The European Commission, a body of appointed officials, initiates proposals for legislation and executes Union policies and programs. The Council of Ministers approves legislation, sets political objectives, and coordinates national policies. The Council is made up of national ministers, and the members depend on the topic under discussion. For example, Union environmental discussions are held by a Council of ministers of the environment from national governments. The European Parliament is an elected body with some oversight and veto power over Union decisions. See the EU website, *http://europa.eu.int/*, for more information.

[22] For a full discussion of the regulation of biotechnology in the EU see Patterson (2000).

(Wiegele, 1991).[23] Similarly, under an Order for Genetically Modified Organisms, the Netherlands established a national regulatory framework requiring a permit from the Dutch Ministry of the Environment to use GMOs in an experimental or commercial setting.

At the Union level, three main regulations are in place – Council Directive 90/219, Council Directive 90/220, and Regulation No. 258/97 (the Novel Foods Regulation). Directive 90/219 provides EU-wide rules for the use of GMOs in both research laboratories and industrial facilities and specifies measures to protect human health and the environment. Field trials and marketing of GMOs are regulated under the 90/220 Directive (IPC 16). The Novel Food Regulation makes labeling of any product containing GMOs, or that may otherwise be considered a 'novel' food, mandatory throughout the European Union.[24]

### Directive 90/219

Directive 90/219 provides for a system of risk management during the GMO development process. The microorganisms and the activities in which they are used are each classified into one of two categories according to their potential risk.[25]

---

[23] The 1986 Environment and Gene Technology Act in Denmark referred all cases involving the free releases of GMOs to the jurisdiction of the Danish Ministry of the Environment and established a set of mandatory product-approval mechanisms for commercial biotechnology innovations, including a backup enforcement system (Gibbs *et al.*, 1987).

[24] A directive is a framework agreement at the EU level that must be implemented by legislation at the national level. By contrast, regulations have the full force of law in all member countries and do not need to be implemented through national legislation. Key health and safety laws are regulations. GMO legislation has generally been in the form of directives.

[25] Criteria for classifying GM microorganisms in Group I (lower risk) include
- Recipient or parental organism is nonpathogenic; not adventitious agents; and there is an extended history of safe use or built-in biological barriers that confer limited survivability and replicability, without adverse consequences in the environment.
- Vector/insert is well characterized and free from known harmful sequences; limited in size as much as possible to the genetic sequences required to perform the intended function; should not increase the stability of the construct in the environment (unless that is a requirement of intended function); should be poorly mobilizable; should not transfer any resistance markers to microorganisms not known to acquire them naturally (if such acquisition could compromise use of drug to control disease agents).
- Genetically modified microorganisms are nonpathogenic; as safe in the reactor or fermentor as recipient or parental organism, but with limited survivability and/or replicability without adverse consequences in the environment.
- Other genetically modified microorganisms that could be included in Group I if they meet the conditions in the point above include those constructed entirely from a single prokaryotic recipient (including its indigenous plasmids and viruses) or from a single eukaryotic recipient (including its chloroplasts, mitochondria, plasmids, but excluding viruses); and those that consist entirely of genetic sequences from different species that exchange these sequences by known physiological processes.

Microorganisms of Group II are all those microorganisms other than Group I (*http://europa.eu.int/eur-lex/en/lif/dat/1990/en_390L0219.html*).

Containment and control measures are specified, including procedures for record keeping, emergency planning, and notification of authorities. Member countries are expected to take steps to implement the directive, including designating a national regulatory authority that receives notifications and organizes inspections and other control measures.

Since 1990, Directive 90/219 has been amended several times. The most recent substantial changes were made by Directive 98/81 to bring Directive 90/219 in line with scientific knowledge and experience acquired since the original directive was issued. The main changes were simplification of administrative procedures, introduction of a link between notification requirements and the risks posed by contained use, and addition of a list of genetically modified microorganisms posing no risk to human health or the environment.

### Directive 90/220

Directive 90/220 covers the deliberate release of GMOs into the environment for research and development purposes and for sale. Examples include microorganisms to aid in creating artificial snow or to act as pesticides or fertilizers; viruses modified for use as vaccines; crops that are resistant to pests or herbicides; and tomatoes with a longer shelf life. The directive is based on the view that all deliberate releases should be reviewed on a step-by-step and case-by-case basis before being approved.

The directive's main elements are as follows:

- An environmental risk assessment must always be carried out before any experimental or commercial release of GMOs into the environment.
- No releases may be carried out without the consent of a designated national authority.
- An approval procedure must be developed by a national authority in the country of release for experimental releases, lasting a maximum of 90 days.
- A Union approval procedure for commercial releases must be developed. (Theoretically, while a product that has been approved in a member state is automatically approved in any other member state, most companies usually prefer to register the product in each of the EU member countries for marketing purposes.)

This directive requires the release of selected information on the nature of the GMOs while providing protection for confidential information. Publicly available information includes the description of the GMO, name and address of the notifier, purpose of the release, location of the release, and plans for monitoring the GMOs and for emergency response.

For a transgenic crop to be tested in the field, approval from a national regulatory body is required. For commercial production, two steps must be completed. First, a growing permit must be issued under Directive 90/220. Second, a variety registration must be obtained for all new varieties, irrespective of whether the crop is transgenic (James, 1998).[26]

---

[26] See Chapter 31 for a list of GMOs approved for field-testing in the EU.

To obtain marketing authorization within the EU, a request must be made to the member state in which the product will be sold first. A copy of the application is then sent to all member states so that objections can be made within 60 days to the EU Commission, the main regulatory body. If a member state opposes the introduction, the Commission must resolve the dispute.

Shortly after allowing *Bt* corn onto European Union markets, the Commission amended Directive 90/220 with Directive 97/25 to require the labeling of products that contain or may contain GMOs. This amendment was generally accepted throughout the member states and was agreeable to members of the European Parliament who had been and still are wary of biotechnology.

On June 25, 1999, the Council of the European Union met to recommend (to the European Commission) an amendment to Directive 90/220 on releasing GMOs. The Council reached an agreement, with France, Ireland, and Italy abstaining. The provisions of the Council's recommendation were

- to take a thoroughly 'precautionary' approach in dealing with notifications and authorizations for placing GMOs on the market,
- not to authorize placing any GMOs on the market until it is demonstrated that there is no adverse effect on the environment and human health,
- to the extent legally possible, to apply immediately the principles, especially regarding traceability and labeling.

The most important consequence of the recommendation to amend Directive 90/220 is that all countries have agreed to follow its provisions. All further approvals of GMOs were halted, including several that were in the final stages.

### The Novel Foods Regulation

The Novel Foods Regulation (Regulation No 258/97) was adopted in January, 1997. The law requires not only GMOs in the form of raw materials to be labeled, but all foods, processed and nonprocessed, that may contain GMOs to be labeled as such. When EU officials began to develop this regulation, several member states, such as France, Denmark, and the Netherlands, had already instituted such regulations. The law was written primarily as a response to the need for harmonized legislation on novel foods throughout the European Union to facilitate the functioning of the internal market.

As in the original Directive 90/220, the Novel Foods Regulation required that there must be prior notification to the individual member states when a GMO seed is marketed, including specific details on the method of labeling for resulting products. In addition to the 'may contain GMOs' labeling required under Directive 97/25, labels must indicate whether characteristics of a food make it no longer equivalent to an existing food. The decision whether a food is not equivalent to an existing food, and thus novel, is determined by a scientific assessment process that is (at least superficially) very similar to that used by the FDA.

The directive did not specify what level of GMO content required the use of a label, an omission that created uncertainty in the marketing chain. On October 21, 1999, the EU's Standing Committee on Foodstuffs approved a Commission proposal that the threshold be 1%; that is, all foodstuffs, additives, and flavors containing at least 1% of GM corn or soybeans will have to be labeled as GM products. Products under the threshold do not have to carry a label.

## Summary of EU agricultural GMO releases

In the period from 1992 to January 10, 2000, 1531 notifications of the experimental releases of genetically modified organisms in the environment (see Chapter 31) and 29 applications to market GMOs have been submitted. Ten GMO products were approved for commercial use before the June 1999 moratorium.[27] They are[28]

- Three rabies vaccines with virus-recombinant DNA (initial approval: 1 Belgium, 2 Germany)
- T102 test for *Streptococcus thermophilus*. The product is a test kit for the detection of antibiotic residues in milk (initial approval: Finland)
- Three canola (rape) varieties (initial approval: 2 Great Britain, 1 France)
- One soybean variety (initial approval: Great Britain)
- One tobacco variety (initial approval: France)
- Four corn varieties (initial approval: 3 France, 1 Great Britain)
- One radicchio rosso variety (initial approval: The Netherlands)
- One carnation variety (initial approval: The Netherlands)

## EU versus national policies

As a pioneer of modern regional integration, it is not surprising that the EU has progressed furthest in the direction of regional regulations for GMOs. Individual member states of the EU are governed by regulations adopted collectively. But national regulations still exist, and as a result of individual national regulatory concerns, the EU appears reluctant to adopt and approve GMOs.

The European Commission, which initiates proposals for legislation and implements decisions made by the Council of Ministers, has an interest in scientific and transparent processes of testing and approval. It would prefer a permissive trade regime and consumer information through labeling.[29] Member states, however, do not always agree. Certain countries, notably Austria, France, and Germany, remain very cautious about potential environmental effects. This reflects in part the influence of the Green Parties in Austria and Germany and in part consumer preferences in France and the United Kingdom, coupled in some cases with a latent anti-multinational sentiment.[30] For example, although the Commission approved the use of one *Bt* corn variety, Austria and Luxembourg prohibited the use and sale of the product. The Commission is trying to take

---

[27] A Commission Decision is the official EU document that allows a GMO to be marketed. Approvals are sometimes called Commission Decisions.

[28] Source: Various issues of *Official Journal of the European Communities, http://europa.eu.int/eur-lex/en/oj/index.html.*

[29] Some of the other EU institutions are not so strong in their support of liberal solutions. In 1999, the EU environmental ministers supported a moratorium on the approval of new GMOs. The European Parliament, with a dose of populism and a strong Green Party representation, has always been skeptical of GMOs and their place in the food chain. This is a natural progression from their views on the issue of beef treated with hormones (see Chapter 29).

[30] Belgium joined the list of countries in which the Greens share power as a result of the election in summer 1999.

action against these two countries for refusing to abide by its decision. France also expressed discontent with the decision, but later decided to allow domestic sales with strict labeling requirements. The French delay in implementing a decision to authorize the import of some varieties of *Bt* corn was the subject of a major trade dispute between France and the United States. The United States claimed that the delay cost corn exporters some $300 million in lost exports to the EU. Other disputes arose with Canadian and US canola exporters, who also grow GMOs.

If regional rules prevailed, the EU policy situation would be more clear-cut. But national legislatures still essentially make the laws governing GMO adoption, albeit within the framework of EU directives. Thus, the inherent ambivalence of the EU position on such matters as GMOs is exacerbated by the complexity of multiple levels of decision-making, creating confusion for the rest of the world.

The EU also has regional rules on intellectual property. In this respect, it is interesting to note that the EU Parliament recently approved new rules regarding the patenting of new biotechnology processes.[31]

## NAFTA

At the time of the North American Free Trade Agreement (NAFTA Agreement), in 1992, few people were aware of the potential issues related to GMOs and how they might influence trade. The agreement therefore makes no mention of the rules for adoption of transgenic plants. The NAFTA Agreement does, however, potentially touch the issue in two respects. First, the NAFTA Agreement did address the issue of sanitary and phytosanitary (SPS) regulations, anticipating the eventual SPS Agreement in the WTO by installing 'science' as the legitimate basis for national policies. It envisaged a progressive reduction in differences in technical regulations that might lead to trade conflicts but did not set up distinct mechanisms for harmonization. The NAFTA Agreement does not explicitly support mutual recognition. However, a form of harmonization does pervade the North American market, as the regulations of Mexico and Canada have often been brought into line with those of the United States. This could also be regarded as a form of *de facto* asymmetric mutual recognition, as US standards become recognized as adequate for domestic purposes in the other two countries.

The other area in which the NAFTA Agreement touches the GMO debate is with respect to environmental regulations. The NAFTA Agreement does not require a common approach to environmental safety. Each country maintains its own regulatory system for the environment, including the testing and adoption of pesticides and insecticides. But a side agreement ensures that each country has in place some form of environmental protection and makes it possible for individuals and groups in other NAFTA countries to challenge the way in which the regulations are or are not being implemented.

---

[31] See 'Europe Approves Move on Biotechnology Patent Laws,' *Financial Times*, Thursday 13 May, 1998, page 12.

So far, Canada and Mexico seem to be in step with the United States regarding GMOs and their incorporation into the food chain. No internal disagreement has threatened to escalate into a trade conflict. NAFTA partners have maintained cohesion in international fora, such as the Convention on Biological Diversity. Canada and Mexico have adopted GMOs, if at a somewhat slower pace, and have not required labeling of GMO-based foods.

## APEC

The Asia-Pacific Economic Cooperation (APEC) process is an arena for discussion of trade policy initiatives among a growing number of Asian and American states.[32] It has no regulatory or administrative function. In terms of the GMO debate, three aspects of APEC are significant. First, the APEC process has been focused on ways to facilitate trade at the sectoral level, driven in large part by the APEC Business Advisory Group. This piecemeal approach has included agricultural and food sectors, and some limited steps have been taken to improve market access. Countries have discussed an APEC 'Open Food System' that would include an agreed approach to the introduction of GMOs. The potential disruption for Pacific Rim trade of a major conflict over GMOs is clearly understood. However, countries within APEC do not yet have a common position on GMOs. Japan and South Korea have recently been slowing the process of GMO commercialization, moving toward market segmentation and considering labeling. China, on the other hand, is a major innovator, and environmental and consumer movements in the country are relatively restrained.

The second feature of the APEC process is that until now it has been successful in persuading non-APEC members (notably the EU) to go along with major liberalization initiatives, such as the agreements on information technology and on telecommunication services, so as to avoid the emergence of APEC preferences. The question is whether this would still be possible in an area such as GMOs, in which the EU might feel obliged to remain apart from an agreement among APEC members. The third feature of APEC is its support for the multilateral process, initially encouraging the EU to conclude the Uruguay Round and subsequently agreeing to speed up its implementation. But it also poses a threat to the multilateral system by providing a trade forum from which the EU is excluded. Thus, the APEC process could be used to coordinate views on GMOs in the context of the next WTO round, and hence may be seen to coerce the EU into eventually supporting such a view.

## Summary

One constant of GMO regulation around the world is that it is a patchwork. Existing regulatory authorities have been given new responsibilities for GMO

---

[32] As of late 1999, APEC has 21 members including Australia, China, Canada, the Association of Southeast Asian Nations (ASEAN) countries, Mexico, the United States, New Zealand, Japan, Korea, Taiwan, and Chile. With the admission of Russia, the APEC initiative is encroaching on the European continent.

products. Because the safety of the food supply is a universal concern, food safety aspects of GMOs generally have been grafted onto the responsibility of ministries of health. These agencies have started from existing food safety regulations and modified them somewhat to deal with specific GMO issues. Environmental regulatory authority has more often been split between ministries of agriculture and ministries of the environment and natural resources. Agriculture ministries traditionally regulate development and release of new varieties, so it was a natural fit to add release of GM varieties. Environment ministry roles have been less clear. For example, EPA has paid a great deal of attention to *Bt* corn but much less to GR soybeans. Because commercial varieties of GMOs have been grown only since the mid-1990s, this patchwork pattern is not surprising. However, the potential for duplicative regulation is obvious, as is the potential for nonmarket effects to escape regulation by falling between the cracks. Perhaps the most important point to emphasize is that regulatory efforts are rapidly changing, both in response to consumer and environmental groups and to keep up with technology. The balance between under-regulation and over-regulation is precarious.

# 10

# International Institutions, World Trade Rules, and GMOs

**Timothy Josling**
*Institute for International Studies, Stanford University, Stanford, California, USA*

The focus of this chapter is international institutions, the roles they currently play in the GMO controversies and what might be their future contributions. What we mean by international institutions are those organizations to which some degree of national sovereignty has been ceded in the interest of greater national and international gains. An example, much in the limelight, is the Dispute Settlement Understanding (discussed in detail below) of the World Trade Organization (WTO). Countries have voluntarily agreed to accept and implement decisions arrived at by this process, even to the point of changing national laws.

International institutions currently play a limited role in the regulation of biotech products. In general, countries can and do implement their own domestic rules governing the testing, production, and marketing of GMOs. In this regard, the situation is little different from the situation for most other safety and environmental regulations, which are administered at the national level. Trade issues arise mainly as a by-product of these national regulations. At the international level, few specific legal instruments for regulating GMOs exist. But the trade issues and the international institutions that govern trade are becoming increasingly important, and have attracted a significant share of the attention in the discussion of the regulation of GMOs. Reconciling diverse national regulations is necessary if trade conflicts are to be avoided. This chapter discusses the most important of these international institutions and agreements for the regulation of international trade in the products of agricultural biotechnology.

Genetically Modified Organisms in Agriculture
ISBN 0-12-515422-4

# The WTO

Since its establishment in 1994, the WTO has become the preeminent multilateral institution dealing with international trade issues. It has no role in the testing of GMOs or with the rules governing the adoption of GMOs in farming or the food industry. The right of countries to institute policies that protect human, plant, and animal health, together with public morals, is clearly stated in WTO Article XX.[1] The WTO is, however, involved in at least two aspects of the use of GMOs in the food chain. One of these is restrictions on national import regulations for GMOs embodied in the GATT (94), the Agreement on Technical Barriers to Trade (TBT Agreement), and the Agreement on the Application of Sanitary and Phytosanitary Measures (SPS Agreement).[2] The second aspect is the operation of the Agreement on Trade-Related Aspects of Intellectual Property Rights (TRIPS Agreement), governing the patenting of biotechnical processes and certain resultant products. In addition, the provision of a mechanism for the settlement of trade disputes through the Understanding on Rules and Procedures Governing the Settlement of Disputes (the Dispute Settlement Understanding, or DSU) makes the rules of the WTO more important. The DSU was seen as a major step towards a rule-based trade system, though it has come under criticism from those who complain that it has obliged countries to change domestic policies when they contradict WTO provisions. The existence of a strong international dispute settlement mechanism will have a significant impact on the way governments frame their domestic and trade policies toward GMOs.

## Restrictions on national import policies

The core of the WTO, and the General Agreement on Tariffs and Trade (GATT) before it, is a code of commercial conduct governing the types of trade barriers that a country can use and their operation in practice. The main principles are nondiscrimination among WTO members (unless in the context of a regional trade agreement); 'national treatment' of imports to avoid discrimination against them once they have entered the domestic market; and transparency of customs procedures.[3] Each of these principles applies to the regulation of GMOs entering into world trade.

How might these principles operate in practice when GMO-related issues are involved? And how might they restrict what individual countries can do? At

---

[1] Article XX specifies several 'general exceptions' to the normal rules constraining what countries can do in the area of trade policy. For the issue of biotechnology, the two most important provisions of this clause cover 'measures necessary to protect human, animal or plant life or health' and 'measures relating to the conservation of exhaustible natural resources.' The interpretation of these exceptions is discussed below.

[2] The provisions of the original General Agreement on Tariffs and Trade (GATT) were incorporated into the WTO articles in a slightly amended form and are collectively referred to as GATT (94). The WTO articles cover a wider area than the GATT, including trade in services as well as trade-related intellectual property issues.

[3] The nondiscrimination clause is often referred to as the most-favored nation principle, or MFN. Each member is entitled to the conditions of access enjoyed by the most-favored nation.

one level, the nondiscrimination principle seems clear cut. It would not be WTO-consistent, for instance, for the EU to ban imports of a good from one country if the same good were being imported from other WTO members without hindrance. It is unlikely that the EU, or any country, would blatantly discriminate in this way. But discrimination could occur in more subtle ways. Suppose, for instance, that the EU were to impose a ban on imports of GM foodstuffs based on the different production methods used. The EU could merely impose a ban on imports of *Bt* corn (from all suppliers) and not violate the nondiscrimination clause.[4] But the countries producing *Bt* corn would have a good case for arguing that this was discriminating against their exports. The nondiscrimination clause therefore probably does inhibit the EU from targeting imported *Bt* corn as a matter of trade policy. The alternative approach, involving domestic regulations that apply to trade incidentally, is more likely to be defensible in the WTO than an import ban unrelated to domestic regulations.

The principle of national treatment might seem also to have implications for national import regulations regarding GMOs. The principle states that no additional tests or hurdles can be set up for imported goods, once border taxes have been paid and the goods have gone through entry inspections. It would not be possible under this principle to require additional tests for the presence of GMOs on imported food ingredients if those same requirements were not imposed in some way on domestic products. On the face of it, the principle of national treatment seems to pose no threat to domestic regulations. It is unlikely that countries would ever be tempted to test imported GM foods while allowing domestic production to go untested. For this reason, countries may not find the 'national treatment' provision too restricting.

Though the WTO injunction is restricted to regulations that 'afford protection to domestic production' (Article III:1), it cannot always be assumed that discrimination will be against the imported product. Assume that domestic corn is tested for genetic modifications, but the testing is waived on imported corn because it does not pose the same environmental threat locally. This may be entirely logical in terms of the aims of the policy, and legal under the WTO. Considerations of equity may, however, point in a different direction. Domestic producers may well argue that a form of 'national treatment' should apply to favorable as well as unfavorable treatment of imports.[5] In practice, there will be strong pressure to impose similar restrictions on imported and domestically produced goods.

---

[4] In practical terms, such a ban could be implemented by assigning separate customs classification numbers for, say, *Bt* corn and non-*Bt* corn. But such a change might be interpreted as a challenge to the entire customs classification system, which regards goods as having recognizable characteristics unrelated to their country of origin and (in most cases) method of production. This is a 'can of worms' that countries would be most reluctant to open.

[5] Even if a more symmetrical interpretation of 'national treatment' were to be adopted, it is worth noting that the WTO has no provision for private entities to challenge government decisions. The issue would have to relate to a conflict between governments, and foreign governments may not wish to become involved in a dispute between a government and an industry that feels unfairly treated.

The main trade conflicts are more likely to arise as a result of challenge to domestic regulations. Rules on the types of regulations that countries can impose on imports for health and safety reasons have become more clearly defined in the past few years. This has largely been in response to trade tensions, such as the conflict over the ban on hormone-treated beef in the EU. With the growing internationalization of the food industry, trade conflicts over food regulatory issues have become more common. Most of these conflicts arise from differences in food trade regulations imposed for the ostensible reason of protecting plant, animal, or human health from disease or other affliction as a result of trade. The major strengthening in the rules has come with the negotiation of the SPS Agreement and the modification to the TBT Agreement in the GATT Uruguay Round of trade talks. These two agreements have the full weight of the WTO dispute settlement process behind them, and are beginning to have an impact on national behavior.

## The SPS Agreement

The international rules governing allowable regulations on imports have been slow to develop. As mentioned earlier, Article XX (b) of the original GATT agreement in 1947 allowed countries to employ trade barriers 'necessary to protect human, animal, or plant life or health' which would otherwise be illegal so long as 'such measures are not applied in a manner that would constitute a means of arbitrary or unjustifiable discrimination between countries where the same conditions prevail, or as a disguised restriction on international trade.'[6] However, Article XX had no teeth. There was no definition of the criteria by which to judge 'necessity,' and there was no specific procedure for settling disputes on such matters. The attempt in the Tokyo Round to improve this situation through an agreement on technical barriers to trade known as the Standards Code (1979) also failed. Though a dispute settlement mechanism was introduced and countries were encouraged to adopt international standards, relatively few countries signed the code, and a number of basic issues were left unresolved.

Intensive negotiations in the Uruguay Round led eventually to the new SPS Agreement, which tried to repair the faults of the existing rules, and to the TBT Agreement, which modestly improved other aspects of the Standards Code. The SPS Agreement defined new criteria to be met when imposing regulations on imports more onerous than those agreed in international standards. These included scientific assessment of the risks involved, scientific evidence that the measure reduced risk, and recognition of the equivalence of different ways of achieving the same risk level. In addition, the dispute settlement mechanism was considerably strengthened under the WTO to make it easier to obtain an outcome that could not be avoided by the losing party.[7] The force of the SPS

---

[6] For a fuller discussion of the importance of Article XX to agricultural trade see Josling *et al.* (1996), p. 209.

[7] The Decision on the Application and Review of the Understanding on Rules and Procedures Governing the Settlement of Disputes (the Dispute Settlement Understanding, or DSU) provides a framework for better enforcement of panel rulings. To block the adoption of a report from a panel now requires consensus. Any party may appeal the ruling (on issues of law), but the Appellate Body Report is final. The importance of the DSU is discussed below.

Agreement comes in part from the more precise conditions under which standards stricter than international norms can be justified and partly from the strengthened dispute settlement process within the WTO.

A key unresolved question is how far the SPS Agreement covers GMOs. Arguably, if a country imposes a ban on GM imports based on animal, plant, or human health or safety concerns, the ban could be challenged under the provisions of the SPS Agreement. This implies that the ban would have to conform to the risk assessment criteria of the SPS Agreement and show scientific justification if the risk exceeded international standards. If all these conditions were met, the ban could be defended under the SPS rules. But the experience with the beef hormone dispute suggests that the case would not be easily made.[8] Scientific opinion, as discussed above, has not identified significant human health risks for GM foods as such.[9] The question would be what to do when there are unspecific concerns about health by a significant segment of the population (as represented by political bodies and interest groups) that are not easily met by existing scientific knowledge. So far, the interpretation of the SPS Agreement has been that one cannot take such concerns into account. Unless countries choose to modify the SPS Agreement, it is unlikely to provide much cover for GM import regulations.

## The TBT Agreement

The TBT Agreement is directly relevant to certain aspects of the GMO controversy. In particular, it covers such issues as labeling and transparency of regulations. Its provisions also include constraints on national health and environmental regulations, as well as those instituted for national security. Such constraints include that these measures not be more trade-restrictive than necessary and not discriminate against imports in favor of domestic products. As with the SPS Agreement, the TBT Agreement suggests that countries use recognized international standards where possible. Disputes over the effect of national regulations are handled within the WTO dispute settlement process.[10]

The distinction between coverage of the TBT Agreement and the SPS Agreement is important in analyzing food safety disputes. This distinction affects how regulatory authorities in several countries decide, for example, whether to restrict imports of bio-engineered products or how standards for organic foods should be developed. The TBT Agreement covers all technical regulations and conformity assessment procedures, *except* sanitary and phytosanitary measures as defined by the SPS Agreement.[11] In other words, the

---

[8] See Chapter 29 on the beef hormone dispute.

[9] Threats to animal and plant health (which include the health of wild populations) could also justify import restrictions under SPS Agreement rules, but it might be difficult to argue that these represent a great risk in the importing country.

[10] This corrects one weakness with the Standards Code, which had its own dispute resolution process. This limited the usefulness of this procedure.

[11] Article 1.5 of the Uruguay Round TBT Agreement. (General Agreement on Tariffs and Trade, 'The Results of the Uruguay Round of Multilateral Trade Negotiations: The Legal Texts,' Geneva, 1994, p. 139).

applicable regulations for a given measure are determined by the objective of the measure. If a measure that prescribes the use of an additive were to be adopted to safeguard human health, it would come under the SPS Agreement provisions. However, if the measure were adopted to regulate labeling or to ensure the compositional integrity of a product, it would fall within the parameters of the TBT Agreement. This is particularly important because the SPS Agreement holds governments to a higher standard than that required by the TBT Agreement. There is no requirement in the TBT Agreement that domestic standards be based on science. The standards must be administered in a nondiscriminatory way and distort trade as little as possible, but the domestic standard does not have to go through the risk assessment process.

## Environmental restrictions on imports

The most difficult issue for the WTO rules to handle might be a ban on GM food ingredients on the basis of their potential environmental impacts. Environmental issues such as biodiversity, effects on wildlife in the country of origin, and impact on the ecosystem are not covered by the health and safety provisions of the SPS Agreement. The TBT Agreement also does not help greatly in sorting out the possible conflicts between trade rights and domestic regulatory regulations, even though it ostensibly covers environmental regulations.

Two distinct problems emerge in this area. One is the possible conflict between multilateral environmental arrangements and the WTO. This topic was discussed in the Uruguay Round and led to a Ministerial Decision on Trade and Environment (DTE) mandating the creation of a Committee on Trade and the Environment in part to deal with this issue. As trade in GMOs are not yet covered by a multilateral environmental agreement, it is not clear how the DTE will be made operational. But with the coming into force of the Biosafety Protocol (see below) the chance of a conflict is increased.

The other possible problem is when domestic environmental goals are pursued through trade measures. These goals in turn can refer to domestic use regulations that appear to discriminate against imports or to production regulations imposed on other countries indirectly through regulations on imports.[12] The former is a case in which the principle of national treatment could be invoked. The second raises the tricky issue of 'extraterritoriality,' discussed below. Moreover, the case of import regulations that impose domestic environmental regulations in supplying countries raises squarely the issue of regulating production and processing methods for imported goods.[13] A ban on GM foods

---

[12] There is rarely a WTO-type trade problem with respect to excessively stringent domestic production regulations, though the non-enforcement of domestic regulations could lead to complaints by exporters selling into that market.

[13] Standards in food and agricultural markets can be specified either in terms of product standards (PS), defining the composition or characteristics of the good in question, or in relation to production and processing methods (PPM), which define the nature of the process by which the product was made. Health standards typically have emphasized PS, and environmental regulations have used PPMs. The GATT had a preference for PS regulations, but the SPS and TBT Agreements both recognize that PPMs are legitimately used.

could be interpreted as being based on how such foods were produced rather than on their product characteristics. As such it would be setting a precedent that some countries would rather avoid.

The issue of extraterritoriality has been at the heart of many of the tensions between the trade policy establishment and environmental groups. The trade system would seem to run more smoothly if each jurisdiction implemented its own regulations and traded goods that satisfied its own standards. Harmonization of these standards would be desirable but not strictly necessary, though there would be a tendency for countries to orient their standards to the sensibilities of their major markets. Environmentalists, at least in the industrial countries, have taken the view that other countries, in particular some developing countries, have more lax laws and little enforcement. Both protection of the environment of the producing country and preventing unfair (and possibly unhealthy) competition from imports require the export of regulations as a precursor to imports of the product.[14] But to do so calls into question the independence of each WTO member to administer its own standards without interference. This issue will have to be clarified within the next round of world trade talks, or else the credibility of international trade regulations could well suffer. The GMO issue could become the stimulus for such clarification.

## Ethical trade restrictions

It is also possible that countries may choose to enact an import ban on GM food ingredients for reasons other than human, plant or animal health or safety, or the safeguarding of the environment. If this were the case then other grounds for a challenge by the aggrieved exporter would have to be found.[15] Suppose a ban were implemented to keep off the market biotech goods that (some) consumers considered an affront to an ethical principle.[16] An argument could be made that as ethical and moral concerns fall within the parameters of GATT Article XX(a), allowing the use of trade barriers to protect public morals, GMO bans on such non-health bases are covered by this provision. Because this provision of Article XX has not been elaborated and is rarely the basis for trade disputes, it is difficult to predict how a panel in a dispute would rule. But on the face of it, this provision is vague enough for countries to define ethical and moral protection to

---

[14] The tuna-dolphin and shrimp-turtle disputes revolved around the compatibility of such attempts to export US environmental standards with international trade rules. In the former case the question was whether US law regarding the protection of dolphins from being caught in tuna nets could be enforced on other countries through trade sanctions: the latter case involved US law on the use of turtle-excluder devices when catching shrimp. The WTO has generally expressed the view that such actions could be taken in the context of an international environmental agreement but not by unilateral action.

[15] Under GATT/WTO 'lore,' a country is deemed to be in compliance with its obligations until another member successfully challenges. Thus, the complaining country in effect decides on what grounds to make the challenge. The defendant must show its action is in conformity with that rule or provision.

[16] This is more likely to happen in the case of animal biotech products than those from plants.

include banning goods considered offensive by the population. Biotech products could fall under such a heading. In the case of GM foods, it seems unlikely that the major governments would follow this route for fear of where it might lead, but some smaller countries could well decide that this was a safe haven for anti-GM bans.

## Trade-related intellectual property

A major international legal regime exists to manage the patenting process for inventions and technical innovations. This regime is administered through a number of agencies. As a direct application of biotechnology, GMOs are regulated by intellectual property laws aimed at protecting the developers of new crops. Instruments include the Paris Convention, the International Union for the Protection of New Varieties of Plants, the Patent Cooperation Treaty, the European Patent Convention, the Budapest Treaty on the International Recognition of the Deposit of Microorganisms for the Purposes of Patent Procedures, and the World Intellectual Property Organization.

In addition, with the establishment of the WTO, all members of the organization are now bound by the TRIPS Agreement. The concept behind the TRIPS Agreement is to ensure that all WTO members have at least minimal intellectual property protection on traded goods. This was particularly important to the software and pharmaceutical industries and to those that produce audiovisual media. Agricultural interests were involved through provisions on geographical origin (e.g., champagne can be only from a particular region in France). The impact of the TRIPS Agreement on biotech was itself foreseen at the time of the Uruguay Round, with developing countries arguing that the protection of intellectual property in this area could act against their best interests.

Under Article 27 of the TRIPS Agreement, on patentable subject matters, countries are required to grant patents for 'inventions whether products or processes, in all fields of technology, provided that they are new, involve an inventive step and are capable of industrial applications.' Patents on GMOs and the techniques used to produce them are held mostly by the private sector in industrialized countries. The patents give exclusive property rights on organisms, genes, or processes for up to 20 years. As patent owners these companies can license patent rights in exchange for royalty payments or license fees. For example, royalties can be charged for the use of a transgenic crop seed and on all subsequent seed produced from these transgenic plants for as long as the patent lasts (Nottingham, 1998).[17] This provision has many farmers and development experts worried; on the face of it, farmers could not use seed from transgenic

---

[17] Cross-licensing and inter-firm cooperation agreements are also becoming the norm in agricultural biotechnology. With these arrangements, companies that have complementary and similar market interests cooperate on a selective basis and develop networks of alliances and joint ventures. A typical licensing agreement was made between Calgene and Monsanto in 1996, whereby Calgene received a royalty-free license to use Monsanto's Roundup Ready technology with Calgene-owned canola genes. In return, Monsanto was allowed to use Calgene's technology for developing crops free of royalty (Nottingham, 1998).

**Table 10.1:** Comparison of implementation of TRIPS provisions on patents for biotechnological inventions in select WTO members

| Subject | US | JP | EC | SW | AU | CA | NW | KR | SA | TT | AR | BR | CO/CA | CH | SI | MY |
|---|---|---|---|---|---|---|---|---|---|---|---|---|---|---|---|---|
| 1. Product patents on microorganisms, if otherwise patentable | Y | Y | Y | Y | Y | Y | Y | Y | Y | Y | Y | Y | Y | Y | Y | Y |
| 2. Process patents on : | | | | | | | | | | | | | | | | |
| (a) Essentially biological processes | N | Y | N | N | Y | N | N | N | N | N | N | N | N | N | N | N |
| (b) Microbiological processes | Y | Y | Y | Y | Y | Y | N | Y | Y | Y | Y | Y | Y | Y | Y | Y |
| (c) Non-biological processes | Y | Y | Y | Y | Y | Y | N | N | Y | Y | Y | Y | Y | Y | Y | Y |
| 3. Product patents on biological or genetic material as found in nature i.e., discoveries | N | N | N | N | N | N | N | N | ? | N | N | N | N | N | N | N |
| 4. Patents on plants and animals *per se*, if otherwise patentable | Y | Y | Y | Y | Y | N | N | Y | Y | Y | N | N | N | N | Y | N |
| 5. Patents on plant and animal varieties | Y | Y | N | N | Y | N | N | Y | N | Y | N | N | N | N | Y | N |
| 6. Exclusion on grounds of morality or public order | N | Y | Y | Y | Y | N | Y | Y | Y | Y | Y | Y | Y | Y | Y | Y |
| 7. Patents on human body | N | ? | N | N | N | N | N | N | N | ? | N | N | N | N | Y | ? |
| 8. Patents on human genes | Y | Y | Y | Y | Y | Y | N | Y | Y | Y | N | N | N | N | Y | Y |
| 9. Breeders' exemption for patents | N | N | N | N | N | N | N | N | N | N | Y | Y | N | Y | N | N |

*Source:* Watal (1999); compiled by author from *http://www.upov.int.*

US = United States; JP = Japan; EC = European Communities; SW = Switzerland; AU = Australia; CA = Canada; NW = Norway; KR = Korea; SA = South Africa; TT = Trinidad and Tobago; AR = Argentina; BR = Brazil; CO/CA = Colombia and other members of the Cartagena Agreement; CH = Chile; SI = Singapore; MY = Malaysia.

crops grown on their own farm without royalty payments. They would presumably be required to purchase new seed from the supplying firm.

Under Article 27.3(b) of the TRIPS Agreement, member countries are permitted to exclude plant varieties from being patentable, provided other intellectual property rights (IPR) protection is available, such as a plant breeder's rights system (Horbulyk, 1999; Roth and Shear, 1999). Table 10.1 reports the status of TRIPS implementation in selected countries.[18]

## Settlement of trade disputes

The Uruguay Round trade negotiations created a new settlement mechanism for legal disputes. Two significant improvements reduce the inadequacies that had hindered the resolution of such disputes as the US–EU beef hormone conflict (see Chapter 29 for more details on this conflict). First, all members became parties to the WTO's single integrated rules system, converting for instance the Standards Code to the multilateral SPS and TBT Agreements. Second, the DSU no longer allows a single country to block a dispute ruling or request for a panel.

The WTO dispute settlement mechanism is meant to provide credible and effective ways of dealing with disputes. A significant number of disputes have been resolved through the WTO consultation stage process. The DSU has offered a more predictable and transparent process. A prominent expert on trade law points out that a rule-oriented approach to international economic affairs has considerable advantages over negotiation or diplomacy approaches (Jackson, 1998, Chapter 4). Rule-oriented dispute settlement procedures focus disputing parties' attention on the rule, and 'on predicting what an impartial tribunal is likely to conclude about application of a rule. This in turn will lead parties to pay closer attention to the rules of the treaty system, and hence can lead to greater certainty and predictability – essential in international affairs, particularly economic affairs driven by market-oriented principles of decentralized decision-making, with participation by millions of entrepreneurs' (Jackson, 1998, pp. 60–61). The DSU allows for an appeal, on legal grounds, within a specified time period. DSU procedures also ensure that unless the Dispute Settlement Body rejects or reverses the Appellate Body decision, it will come into force as a matter of international law. Thus, a losing party will have no opportunity to block a panel decision. The existence of this stronger dispute resolution system in the WTO makes it necessary for governments to pay much more attention than in earlier years to the issue of the consistency of standards (such as GMO regulation) with international rules.

## The Role of International Standards-setting Bodies in GMO Trade

The SPS Agreement gave a boost to the role of international standards from bodies that had been previously set up to consider norms for domestic food

---

[18] For an excellent review of the TRIPS Agreement and especially the interests of developing countries, see Watal (1999).

safety regulations. Of these bodies, three were mentioned specifically in the agreement – the Codex Alimentarius Commission (CODEX), the International Plant Protection Convention (IPPC), and the International Office of Epizootics (known by its French acronym, OIE). The CODEX is run jointly by the World Health Organization (WHO) and the Food and Agricultural Organization (FAO) and sets standards for such things as pesticide residues and standards of identity. If countries agree to implement CODEX standards, other countries can be assured that certain basic health standards have been met. This usually simplifies the entry procedure for imports. The IPPC deals mainly with plant pests and diseases. The OIE focuses on disease control in animals and tries to coordinate the response to outbreaks of infectious diseases. Neither the IPPC nor the OIE is currently involved with GMO regulation but could be in the future if GMO technology touches on their areas of responsibility.

By contrast, CODEX has become a potential locus for GM regulations. Currently, CODEX has not set standards for GMOs. Indeed it may never do so, because GMO exporters will tenaciously argue that CODEX standards should be based on product characteristics not on methods of production. But this argument turns on whether GM foods are considered different from their non-GM counterparts. If it is accepted that they are 'substantially equivalent' then CODEX may have no role to play in setting standards. If they are agreed to be different products then CODEX could be asked to define standards of identity for GM foods. The focus of the debate may thus be on 'how different is different?'

CODEX is, however, involved in the debate in another capacity. The body has taken up the issue of labeling GM foods. Labeling has been a part of CODEX activity for some time, but labeling in the case of GM foods raises a host of questions. So far the committee considering a labeling standard under the CODEX has not been able to agree. Meanwhile a new CODEX committee has been set up to consider the specific question of GM foods (the Ad Hoc Intergovernmental Taskforce on Biotechnology). This group will report its initial findings by the year 2001.

## Convention on Biological Diversity

GMOs are covered under the Convention on Biological Diversity (CBD) and by the United Nations Industrial Development Organization (UNIDO). The CBD was established during the 1992 UN Conference on Environment and Development (UNCED) held in Rio de Janiero. The United States has not ratified the convention and associated treaty. The CBD is concerned primarily with threats to biodiversity, but an extension, the Biosafety Protocol, was proposed to deal more explicitly with potential environmental problems associated with GMOs.[19] A convention to agree the Protocol, held in Cartagena, Colombia, in February, 1999, was suspended after the United States, Canada, Australia, Chile, Argentina, and Uruguay (known as the Miami Group) objected to key provisions. The aim of the protocol was to define the conditions under which

---

[19] The Biosafety Protocol does not deal with the human health issues related to GMOs.

countries could ban the import of GM materials. The main concept developed for this purpose was that of 'advance informed agreement' (AIA). This agreement would be necessary before any GM product was intentionally introduced into the environment of an importing country. The key sticking point was whether the requirement for advanced approval should apply to viable seeds of GM crops (living modified organisms, or LMOs) meant for feeding, eating or processing, or just those destined for planting. Further discussions on the 'Cartagena Protocol on Biosafety' were held in September, 1999, in Vienna, but again no agreement was reached.[20]

Agreement on the Biosafety Protocol came somewhat unexpectedly at a negotiating session in Montreal, at the end of January 2000.[21] It was agreed that the principle of advance consent be applied to some but not all parts of the trade in GMOs. For the purposes of the Biosafety Protocol, genetically modified material has been separated into three categories: (i) non-living GM material (such as processed products that contained detectable traces of modified corn or soybeans) that cannot reproduce and hence poses no threat to the environment in which it is introduced; (ii) living GM material (LMOs) such as seeds intended for further cultivation; and (iii) LMOs destined for food, feed, and further processing, and hence of more remote threat to environmental systems. The non-living GM material was outside the AIA provisions of the Protocol, and the full advanced consent rules apply only to seeds. LMOs for feed and food are subject to a weaker kind of consent arrangement which requires exporters to agree to share information on approvals for their own use of GM material (e.g., seeds) and importers to notify a Biosafety Clearing House if they wish not to import these products.[22]

Three aspects of the Cartagena Protocol are significant in judging its impact on the regulation of biotech trade. First, the fact that an agreement was reached which has been broadly accepted by the governments concerned and NGOs is itself significant. It has acted to lower tensions between governments and to free up the agenda in other on-going discussions, such as the next round of talks on agricultural trade in the WTO. Second, it is important to note that the US is not a party to the Protocol, as it has never signed the Convention on Biological Diversity to which the Protocol is attached. To exclude the major producer of GM products would seem to weaken the agreement, but it is possible that the US will cooperate in informational activities in such a way as to allow importers to operate under the terms of the Protocol. Third, the agreement itself has some 'constructive ambiguity' necessary to allow countries with different views to sign on to the Protocol. These include the relationship between the Protocol and the WTO (in particular the SPS Agreement), which had been a major source of concern for the Miami Group; the extent to which the withholding of AIA should

---

[20] See *http://www.biodiv.org/biosafe/Inform-cons-vienna.HTML* for more details.

[21] For a discussion of the issue surrounding the negotiations see Gupta (2000).

[22] Three other categories of GM material are mentioned in the Protocol. LMOs destined for research laboratories or greenhouses ('contained use') are excluded from AIA, as are LMOs in transit, though documentation must accompany such shipments. LMOs for pharmaceutical use are excluded on the assumption that they are handled in other international agreements.

be based on scientific evidence; and whether countries could adopt a 'precautionary' approach in the absence of adequate scientific knowledge. On each of these three issues the ambiguity served to facilitate agreement by postponing the day when conflicting interpretations have to be reconciled.[23] The Protocol therefore changes little in the short run as regards trade in corn and soybeans, though it gives some recognition to the concerns and approaches of those countries that have been seeking to control such trade.

## Summary

International regulation of biotech is in its infancy. Some hope that it will never grow up. But as trade becomes a more significant part of marketing plans for companies, and as consumers come to expect a wider range of foods on the table, regulating the new global marketplace will become more important. A basic problem that will have to be addressed is that of overlapping competencies of different organizations. As multilateral environmental agencies develop, and as their reach extends to international transactions, potential conflicts will multiply. Countries have different preferences for how these tensions will be resolved. Currently the EU and the developing countries favor environmental agreements such as the Biosafety Protocol taking precedence over the WTO in areas such as the control of GM imports. The US and many other exporters are against this, as it would essentially weaken the SPS Agreement. But differences within countries, such as between trade and environment ministers can be just as large as inter-governmental disagreements. One could imagine an eventual outcome that would involve an International Environmental Agency with a dispute settlement process of its own which could mandate corrective actions or sanctions that would otherwise be seen as WTO inconsistent.

The apparent success of the Biosafety Protocol in reducing tensions surrounding the issue of GM food trade gives governments a brief 'breathing space' to make progress on some of these issues. The agricultural talks, aimed at completing the reforms started in the Uruguay Round, can continue without the burden of having to tackle the GM issue. The CODEX discussions of GM labeling seem to offer the prospect of an agreement, in particular as the US appears to be softening its strong opposition to certain types of label requirements. The issue of the 'precautionary principle' has been demystified somewhat by the EU Commission's effort to clarify its use, and by the response by the US that its own laws are based on 'precaution' (EU Commission, 2000). As with that other phrase which caused so much alarm among agricultural exporters, the 'multifunctionality' of agriculture, an approach which focuses on the implications of

---

[23] On the relationship with the WTO, the Protocol emphasizes that it does not change the rights and obligations of parties under existing international agreements but then indicates that it is not subordinate to other international agreements. Scientific evaluation (risk assessment) is required for the notification to the Clearing House of the intention not to accept a particular LMO for food, feed or processing, though the inclusion of a precautionary clause allows countries some scope to delay approval for imports until scientific evidence is available.

its application rather than reacting to it as a piece of political rhetoric seems to work best.

The intellectual property issues may well prove to be the most difficult to resolve. Even if rich consumers lose their qualms about eating foodstuffs designed in a laboratory rather than bred over the centuries in the field, the question of who controls the essential components of the technology will still be controversial. One obvious answer would be to incorporate the technology firmly in the public sector, but this would hardly be feasible in political terms even if economically desirable. Another way would be to make the licenses needed for the use of technology developed by companies more readily available to other firms and public agencies. The length of time that patents grant exclusive use could be shortened, or the difficulty in obtaining patents could be increased. Each of these moves could cut the attractiveness of research and investment in the sector, but would also reduce what many consider to be the most problematic aspect of the spread of the technology. Unless the public is convinced that agricultural biotech in general and GM foods in particular are broadly in the public interest, the regulatory and civil society hurdles will continue to be placed higher until the private-sector interest in the technology disappears.

# 11

# Market Responses to Consumer Demand and Regulatory Change

**Laurian Unnevehr**
*University of Illinois, Urbana, Illinois, USA*

**Lowell Hill**
*University of Illinois, Urbana, Illinois, USA*

**Carrie Cunningham**
*University of Missouri, Columbia, Missouri, USA*

## Introduction

Consumer concerns about current commercially grown GMOs will determine their marketability in some cases. For example, several major food firms in different countries have announced that they will accept only non-GM crops. At the same time, GMO adoption has proceeded rapidly in the US, as well as in other major producing countries. As a result, the world market for corn and soybeans is developing a new channel for guaranteed non-GM crops to meet demand in markets where GMOs are not desired. For the remainder of bulk commodity trade, supplies in all likelihood contain at least some GMOs due to the mixing that occurs in the marketing channel.

This chapter explores the determinants of this process of market segmentation and the implications for the agricultural supply chain. We consider four sets of issues, beginning with consumer demand and moving up the supply chain to

Genetically Modified Organisms in Agriculture
ISBN 0-12-515422-4

consider costs. First, we examine the size of emerging non-GMO market segments created by consumer demand or by regulation. How discrete are these segments and what degree of substitution is possible? Second, we examine labeling alternatives. What impact will mandatory or voluntary labeling have on consumer choice and market structure? Third, we examine alternative marketing arrangements for supplying non-GM crops. What are the costs of these marketing arrangements? How will the GM status of supplies be monitored and verified? Finally, we use our examination of the first three issues to speculate on the direction of change for world markets. What are the implications for long-run price and quantity equilibria in world markets?

The focus in this chapter is on emerging negative reactions to GMOs that are creating segments of demand for non-GMO supply. However, it is possible that there will be future 'positive' demand for segmentation to capture value from GMOs with consumer benefits. Thus we also speculate on how the current process of segmentation could evolve to capture value from 'second-generation' GMOs.

## How Are Market and Regulatory Actions Leading to Market Segmentation?

Both industry and regulatory actions are driving market segmentation to separate GM from non-GM products. Table 11.1 shows examples of current market events or regulatory actions that create market segmentation. These actions influence market structure in two ways: first, by determining how exclusively non-GMOs are demanded; and second, by determining the size of the non-GMO market as a share of the total market. In other words, how many products or what proportion of total use will completely exclude GMOs?

**Table 11.1**: Examples of demand and regulatory responses to GMOs classified by market impact

| Degree of exclusion | Specific food product markets | Categories of food products | | Entire country's market |
| --- | --- | --- | --- | --- |
| | | Contain novel genetic material | Derived from, but do not contain novel material | |
| Complete exclusion of GMOs | Baby food | Sainsbury private label US organic products | | Austria |
| Search for substitutes where possible | McDonald's excludes GMO potatoes but no specific exclusion for corn products | Reformulation or substitution in most EU markets | Less substitution necessary (livestock feed in EU) | UK |
| GMOs tolerated | Rennet in most cheeses | Products labeled 'contains' or 'produced with' GMO, but product composition unchanged | | Australia |
| GMOs not an issue | – | Most US foods | | US Canada |

Consumer concerns have led private companies to announce that they will not use GMOs in particular products, including Gerber baby food, IAMS pet food, Frito-Lay corn products, Asahi beer, and McDonald's french fries (Barboza, 2000). Often such actions are preemptive, as firms wish to avoid consumer outcry or loss of specific overseas markets. The global marketing of food brands makes it necessary for multinational firms to be cognizant of consumer perceptions throughout the industrialized world. For example, a decision by Grupo Industrial Maseca in Mexico to exclude GMO corn from their tortillas was motivated by their desire to maintain access to export markets in the EU and the Far East (Bailey, 2000). Another type of private action is the guarantee by major retail chains in the UK, France, and Germany that their private label products have no GMO content (Kalaitzandonakes, 2000). This creates a broader category of products for which exclusion is demanded, sometimes including the use of GMO crops in animal feed for meat products.

The current controversies over corn and soybean GMOs are complicated by their multiple uses in food production, many of which are not readily apparent to consumers. Soy- and corn-derived products are used as ingredients in many different processed products, sometimes in very small amounts. Furthermore, some processed products derived from corn and soybeans contain little or no novel genetic material, such as soybean oil from GM soybeans (NZMAF, 2000). Corn and soybeans are also used in animal feeds to produce meat, milk or eggs, none of which will contain the novel genetic material. Thus, one determinant of the size of the non-GMO market is whether consumers avoid any product derived from GMOs or only products with novel genetic material.

The widespread use of corn and soybeans in different products also means that substitute ingredients are a possibility for many uses. Some food suppliers have undertaken searches for substitute ingredients (see Chapter 23). When substitution is less costly than sourcing guaranteed non-GM corn and soybeans, then such substitutions will reduce demand for these two crops. In other cases, substitution is impractical, but firms make the substitutions that are economically feasible. An example of the latter is McDonald's, which has requested non-GM product from its potato suppliers, but uses corn oil or other corn products that are not specifically non-GM (Barboza, 2000).

Regulation (discussed in Chapter 9) can also create a non-GMO market. Regulatory actions include bans on the production, use and importation of GMOs in a few countries; required separation of products containing GMOs; or required labeling of products containing GMOs. Bans on use create a large non-GMO market, but other types of regulation do not necessarily segment the market, in the absence of accompanying consumer demand for non-GMOs. Labeling that only requires a 'may contain' type of message, such as that used in many EU countries before 1999, effectively tolerates the presence of GMOs, rather than requiring their separate identification or excluding them.

An important subset of regulation enables, rather than requires, market segmentation. This includes regulations regarding tolerance levels of GMO content in products labeled non-GMO or recognition of particular testing methods as certifying no GM content. For example, in the US, the USDA/FGIS (US Department of Agriculture/Federal Grain Inspection Service) will accredit independent laboratories that demonstrate the capability to differentiate between conventional and biotechnology derived grains, on blind samples provided by

FGIS (USDA/FGIS, 2000). In the EU, tolerance for GM content in labeled non-GM foods has been established at 1% (Reg EC 49/2000). Another type of enabling regulation sets standards for niche markets, such as organic foods in the US, which do not allow GMOs. These kinds of regulations enable the segmented market by creating a set of rules that inform both producers and consumers.

Table 11.1 summarizes how both market actions and regulatory actions may create segmentation of different degrees and different extents. The degree of segmentation varies from banning GMOs entirely, to searching for substitutes where possible, to tolerating GMOs, to markets where GMO status is not an issue. Product coverage can vary from niche markets of the most concerned consumers to entire food categories, including indirect consumption through processed products or through the food chain (feed into meat); or it can extend to all food products in one country's market.

We see examples of all of the above but no clear signals yet about the extent of market segmentation and the discreteness of the segments. At one extreme, a few countries such as Austria have banned GMOs entirely. At the other extreme, GMOs are likely ubiquitous in many US food products, but are not labeled as such and most consumers have not reacted negatively. In between are many different market segments with different sizes and degrees of elasticity in substitution. It is possible that the size of EU market demand for guaranteed non-GM food is stabilizing, and will not increase significantly in the future. But new non-GMO demand could emerge if more firms decide to exclude GMOs or as countries such as Australia and Japan implement new and more stringent labeling regulation. The extent of market segmentation is evolving and there is great uncertainty about how large or how discrete future demand for non-GMO products may be.

We can make some predictions about the economic implications of different paths of market evolution. Highly discrete inelastic market segments of limited size create high premiums and incentives for separation in the supply chain. Examples include foods for subpopulations at risk or private labeled products in UK retailers. Given the inelastic nature of this demand, many of the costs of segmentation will be passed along to consumers. This type of segmentation is already apparent in grain markets, as we discuss below.

The question for the future is which path market segmentation will follow beyond the markets that are being established now. If there are less discrete, more elastic segments that encompass less than half of end-use, then there will be little response in the marketing chain. The primary response would be to label food products for disclosure and/or to substitute only where least costly to do so. More elastic demand would follow from the fact that most consumers would not change their demand. More concerned consumers would seek out brands or labels for non-GM products, which would expand the non-GMO market share somewhat. On the other hand, if larger, discrete segments of inelastic demand develop, for example through effective bans in more countries or through greater effective demand for certified non-GM products in more countries, then the ultimate result will be lower adoption of GMOs worldwide. Current and future GMOs would then become the niche market, with end use either where consumers are less concerned (e.g., animal feed in the US) or in very specialized markets that demand characteristics of the GMO (e.g., specific health benefits).

# How Will Labeling Affect Market Structure?

Consumer demand will determine the size of the market segments for GM and non-GM foods. But in order for demand to be effective, consumers will need to know whether a product does or does not contain GMOs. There are many different ways that this information could be provided (Table 11.2), either as a result of regulation to mandate labeling or voluntary industry response.

**Table 11.2**: Different kinds of labeling and their impact on consumer choice and market structure

| Type of labeling | Impact on consumer choice | Potential impact on market structure |
|---|---|---|
| *Mandatory* | | |
| None | Provides no information | None, unless voluntary labeling emerges |
| 'May contain GMOs' | Provides very limited information | Not likely to lead to provision of alternatives or segmented demand |
| 'Does' or 'does not contain GMOs' | Provides clear information about whether or not GMOs are present | May lead to segmented demand |
| 'Contains specific kinds of GMOs' and/or 'in specific amounts/% of product' | Provides more specific information about GMO content; could allow consumers with personal risks or ethical concerns to avoid specific GMO products | May lead to segmented demand |
| 'Contains specific kinds of GMOs with these known impacts on health and the environment' (could be linked to off-label sources for further clarification) | Provides most information about GMO content and what is known about potential risks or lack of risks | May lead to segmented demand |
| *Voluntary* | | |
| Non-GMO content implied by label covering other attributes (e.g., organic, food product or retailer brand) | Provides information through the 'filter' of the label; other label characteristics may or may not be of interest to consumers wishing no GMOs or to consumers wishing to avoid specific kinds of GMOs | Will only lead to expanded demand for label beyond current market if additional consumers are motivated by non-GMO content |
| Any of the above types applied to specific food products with GMOs | Provision of information as noted above | Would be adopted because producer perceives market advantage from disclosure and education; does not create segmented demand |
| Any of above types applied to specific food products on a voluntary basis to indicate no GMOs | Provision of information as noted above | Would be adopted because producer perceives advantage from non-GMO content; creates segmented demand |

Voluntary industry labeling includes signaling in the marketplace through brand name or retail chain, such as when a firm announces a non-GMO policy (e.g., Gerber or Sainsbury's). Non-GM content could be implied through product certification that covers many processes of interest to particular consumers, such as organic or natural food labels. If the demand for non-GM food is primarily among market segments already served by particular suppliers, certified processes, or brands, then consumers will rely on firms and certifiers to distinguish GM from non-GM sources. Firms will do so to satisfy their core constituencies; they may or may not attract new demand from excluding GMOs. Many of these kinds of firms or labels already have highly integrated supply chains. Thus, it has been relatively easy for these firms or for the organic industry to verify and reward non-GM products in much the same way that vertically coordinated channels currently reward other specific quality characteristics.

Alternatively, information about GM content might be placed directly on food labels, and would appear regardless of other brand characteristics. Such information could be mandated, as it is currently in several countries (see Chapter 9), or could be provided voluntarily by food firms. If such information is general and vague, e.g., 'may contain GMOs,' then consumers have received very little information and have no basis for making choices among foods. Labeling that states whether or not GMOs are present provides somewhat more information, but only allows consumers to choose GMO or non-GMO. This would be most useful to consumers who have ethical concerns about all GMOs. Other consumers may react negatively to all GMO-containing foods whenever any GMO receives negative reports in the media. Somewhat more information is provided if the specific GMO is identified (e.g., 'contains GR soybeans'), and possibly the amount of GM content. Then consumers might be able to exercise choice based on environmental concerns or personal health risks. Thus, they could discriminate among GMOs. This might become more important if and when GMOs are introduced with specific consumer benefits. Finally, labels might include information about scientific evidence concerning the potential risks (or lack of risks) from the GM product. Labels might also provide internet sites or hotlines for further information, for consumers who want to know more than can be stated on the label. This kind of information would provide the greatest opportunity for consumer education.

One can imagine different scenarios with respect to the possible interactions between labeling and consumer response. Widespread labeling of food products as containing GMOs may lead to indifference on the part of consumers over time, and eventual acceptance of GMOs. Alternatively, labeling of products may lead to a larger number of consumers exercising choice for non-GM foods and expanding the size of the segment over time. This market segment might demand even more explicit labeling through the marketplace or the political process.

The controversies surrounding GMOs are similar to other debates about production processes and labeling (Golan *et al.*, 2000). The evolution of organic foods and certification in US food markets is an example. Consumer and environmental groups are concerned about pesticide use, and many consumers would like to see reduced pesticide use for a number of reasons. The organic market has developed to respond to concerns about the food safety and environmental consequences of mainstream agricultural practices. Many different private and state-level certifying agencies exist in the organic market. In 1990, Congress mandated the establishment of federal standards for organic products. Proponents argued that a unified national standard would facilitate the growth of the organic market and thus enable consumers who are concerned about pesticide use to exercise their choice in the marketplace more effectively (Golan *et al.*, 2000). This is an example of the use of a 'label' to meet both political and market demand for reduced risks not satisfied through the regulation of pesticides.

When there is no political consensus on how to regulate a particular risk, then labeling has political appeal as a compromise solution (Golan *et al.*, 2000; Magat and Viscusi, 1992). Golan and Kuchler (2000) point out that labeling is a very inefficient way to address any environmental externalities from GMOs. Where such externalities exist, it is much more efficient to address them directly rather than indirectly through consumer demand. Whether labeling should be mandatory depends on the extent to which such labeling would lead to more informed consumption decisions that better conform to social objectives. If health risks from approved GMOs are small, then the principal benefit from mandatory labeling might be to reinforce consumer confidence and to reinforce other kinds of consumer education (Caswell and Padberg, 1992). These benefits might apply to either GM or non-GM products, as labeling would facilitate faster adjustment of consumer demand to these new products and processes.

## What Kind of Alternative Market Structures Will Evolve to Address Segmented Demand?

The current GMO controversies and segmented demand have emerged at a time of rapid change in food marketing throughout the world. There is growing demand for specialty production, for farm to table management of quality, and for the ability to trace supplies to their source (Caswell *et al.*, 1998). So it is likely that segmentation of markets by GM status will be addressed through existing and evolving mechanisms for meeting small, specialized demands. But large, discrete segments requiring large-scale separation might mean that these institutions are inadequate, and new institutional and physical arrangements will be needed. Below we explore the marketing implications of segmentation.

## Marketing costs and price premiums

The marketing costs and price premiums associated with the development of a segmented market include any premium paid to a producer to supply a particular variety, the costs of segregation in storage and handling, and the costs of verifying that the crop is truly non-GM. The complexity of the US grain marketing channel makes it difficult and expensive to segregate crops when variety is the only clear difference. Three alternate marketing strategies exist— test the product at selected points in the market channel, accept producer assurances at the first handler and maintain identity through the market channel, and use third-party supervision and certification from seed to final processing. Each alternative has its advantages and its limitations.

### Product testing
This approach provides assurance that the sample selected for testing meets the standards of the buyer. It focuses on the attributes of the product rather than on the process by which it was produced and delivered. The disadvantages arise from the costs of sampling and testing. Current sampling methods provide a

low level of confidence that a large bulk shipment is adequately represented by the sample analyzed. The standard error for even the best sampling strategy (e.g., automatic diverter samplers in the inbound or outbound grain stream) is large for the current low tolerance levels. Sampling of inbound deliveries by farmers has the same problem of obtaining representative samples, and it has the additional problem of time required for testing and segregating, given the speed with which inbound trucks must be unloaded.

In 2000, the technology to test for the entire range of possible genetic modifications with a single test at the first-handler level was not available. The most sensitive GMO test in use was the polymerase chain reaction (PCR). It is very accurate but takes 3–5 days to complete and costs as much as $250 per sample. Immunoassay tests in the form of a strip test ('dipstick') were available that identified GR soybeans and two of three *Bt* toxins then in corn. These tests were cheaper (about $7.50 per test) and provided immediate results, but did not provide information about the amount of GM content (Bullock *et al.*, 2000). Two newer technologies are being tested for use in rapid GMO detection – immunoassay using the ELISA (enzyme-linked immunoabsorbent assay) and near infrared (NIR) technology. An ELISA test takes 5 to 20 minutes and costs less than $10 per sample. An NIR test takes only 1 to 2 minutes and costs less than $5 per sample.

For labeling purposes, testing sensitivity may be important. It is possible with the PCR to test for either the promoter or marker DNA, which are common to many GMOs, or for the specific genes that confer the desirable traits. The first type of test would identify only whether a crop was genetically modified; the second type would identify what type of modification had taken place (Hurburgh, 1999; Gachet, 1999).

### Producer validation and market segregation

This strategy segregates the non-GM crops at the beginning of the market channel. If the product is shipped in small containers dedicated to non-GM grain, guaranteeing the process will also guarantee the final product. Prior contracts or arrangements are made with the producer, and the first handler must make arrangements for verifying compliance. This strategy also requires the elevator to maintain separate dump pits or separate locations for facilities. It is impossible to clean a dump pit of every kernel of GM grain between loads. Some elevators designate one of their facilities for handling non-GM grain, thus simplifying the problem of identification at the time of delivery. Yet a major obstacle to maintaining purity through the rest of the market channel still exists, because trucks, rail cars, barges, and port equipment must also use dedicated equipment to guarantee that all GM kernels have been removed if the equipment has previously been used for GM grain.

### Third-party certification

This strategy reduces the danger of misinformation, questionable methods of isolation in the field, and incomplete knowledge on the part of the producer. The strategy is well known and frequently used for delivering food-quality corn and soybeans to foreign destinations. Organic, pesticide-free, and variety-specific qualifications are common in international trade, but third-party certification adds significant costs per bushel. Illinois grain handlers currently use third-party certification from seed to river elevator and shipping in containers or small-volume segregated barge loads that are transferred directly from barge to vessel to avoid contamination. This strategy is based on the premise that it is more effective to guarantee the process than to guarantee the product. No shipper can guarantee that a few kernels will not be introduced into a shipment from any of many sources, including the importing elevator in the destination country. Instead of a 100% purity guarantee, the supplier provides assurances that the grain has been handled in such a way as to minimize the possibility of contamination. Some exporters employ a combination of all three strategies.

## Evidence regarding additional marketing costs

There will be a cost for any of the three strategies described above. A 1998 survey of 200 US firms by Bender *et al.* (1999) examined the marketing costs associated with specialty grains. Such specialty grains have particular characteristics, such as oil or protein content, that bring high value in particular end-use markets. The survey reported an average additional handling cost of $0.17 per bushel for corn and $0.48 per bushel for soybeans in 1998, over and above the premium for specialty characteristics. These handling costs are around 10% of the crop price at farm level. These reported costs are similar to the 6 to 10% additional marketing costs estimated by Buckwell *et al.* (1999) in their review of several segmented or identity-preserved markets.

Segregation of non-GMOs will be much like the marketing practices for other specialty grains, at least when demand is small. As demand in this segment grows larger, it might be expected that some costs of handling would decline, but also that new types of market organization might be necessary. Bullock *et al.* (2000) noted that the major cost of non-GMO segregation would come about through the 'reshuffling' of grain handling facilities, as farms are forced to deliver to different elevators, and elevators become more specialized in supplying certain demands. These longer-run costs of market reorganization are more difficult to estimate than the additional costs of short-run handling.

## Market developments in 1999 and 2000

Some US firms announced that they would accept only certain varieties of corn and soybeans in 1999. In April 1999, the Archer Daniels Midland Company (ADM) and A.E. Staley, both based in Decatur, Illinois, announced that they would pay a premium for Synchrony Treated Soybeans (STS), a non-GM soybean variety. In addition, they would reject any genetically modified corn not accepted in EU markets during the 1999 growing year (Grainnet, April 14, 1999; Reuters, 1999a). STS, produced by DuPont, are bred to resist Synchrony herbicide, also produced by DuPont. This variety is not genetically modified, as it was developed through conventional breeding. The STS program ADM offered provided an 18-cent per bushel premium over the Decatur market price for soybeans grown in 1999. Monsanto announced that it would help growers of their products find domestic market outlets for varieties that are not approved by the EU. In August, 1999, ADM requested that its suppliers segregate GM and non-GM crops, in response to growing consumer requests for such segmentation, but later withdrew this request in February 2000 (ADM website). These private-firm actions reflect the rapidly evolving market response to consumer concerns.

We interviewed nine firms in May, 1999 who advertised on the internet to contract with farmers for non-GM corn or soybeans. The means of verification for non-GM products included spot testing, segregated on-farm storage, segregated on-site storage at the elevator, and segregated transportation measures. Some elevators did not do any testing and relied on the word of the farmer regarding the non-GM product. These firms also reported widely varying premiums for non-GM product that was contracted in spring 1999. This premium varied from zero to 20 cents per bushel and had declined from the previous year when the non-GMO market was very small. Price and quantity were not yet in equilibrium between the two segments, as would be expected. Marketing costs differ widely among firms (Good *et al.*, 2000). Furthermore, adjustment to the new equilibrium could be lengthy, due to the costs of shifting current fixed capital in the grain marketing system (Bullock *et al.*, 2000).

The market for non-GM crops seems to have expanded from 1998 to 2000, and the premium paid may have fallen, as would be expected. Good *et al.* (2000) surveyed Illinois grain handlers in 1998 and report that only a very small number of firms were handling guaranteed non-GM crops at that time. In their sample, elevators received a premium of 34.6 cents per bushel for non-GM soybeans. Bullock *et al.* (2000) report that elevators in Illinois received 22 cents per bushel premium for exporting guaranteed non-GM soybeans to Japan in 2000. Golan *et al.* (2000) report a market survey indicating that one out of 10 grain handlers would offer a price premium for non-GM crops in 2000.

The development of assured marketing channels for segmented GM and non-GM crops suffered a major setback in September 2000. StarLink, a GM corn variety approved in the US only for use in animal feeds, was discovered in food products (see Chapter 7 for further discussion). This outcome may be the result of lack of information among producers and first handlers regarding the limited approved uses of this variety. The failure of the marketing system to adequately segment when required to do so by regulation reduces consumer confidence in market guarantees of product content. This is likely to reinforce the use of the most extensive certification measures for non-GM content in markets with inelastic demand.

A survey of agricultural leaders in Spring 2000 was undertaken by Cunningham (2000) to elicit their opinions regarding the future agricultural market structure. Most leaders believed that a segmented market would continue to evolve. Several key issues were identified as factors in this evolution, including: the level of consumer concern in the major crop producing and consuming countries; the need for standardization of tolerance levels and regulatory frameworks to facilitate international trade; and the potential for second generation high value consumer benefits (Cunningham, 2000).

## What is the Future for a Segmented World Market?

It is useful to think about GM and non-GM corn and soybeans as separate products with separate supply and demand curves. There is substitution in supply, but in some markets there is little substitution in demand. GM varieties reduce costs of production and/or increase yields for some US producers, shifting the supply curve out and down compared to non-GM varieties (as discussed in Chapters 3 and 4). Thus, without any change in demand, prices of corn and soybeans would fall (slightly) in the long run following the introduction of GMOs, as the world market simulations in Chapter 5 have shown. However, if EU and Japanese demand, for example, is restricted to non-GMOs, their markets will clear wherever EU plus Japan excess demand intersects with the non-GMO supply curve. The market for the rest of the world will clear at the intersection of the new GMO supply curve with demand in the rest of the world. This process is complicated by the possibility of substitution on the supply side between GMOs and non-GMOs. If GMOs have lower production costs, then producers would need a premium to supply non-GMOs at higher cost. Segmenting introduces additional costs that shift the export supply curve inward for non-GMOs.

The following scenario is one possibility, and may already be taking place. Some portion of EU and other demand will be for guaranteed non-GM crops. At the same time, there will be widespread adoption of GM varieties in major producing countries. The demand for non-GM varieties will be met from segmented market channels that will develop in all exporter countries, and this supply will carry a modest marketing surcharge. In addition to this marketing surcharge, producers will receive a price premium to cover the higher costs of production with non-GM crops. However, this premium will be in relation to the somewhat lower world prices for corn and soybeans brought about by adoption of GMOs. The costs of the producer premium and additional marketing are likely to be passed to buyers in the EU and elsewhere because this demand is relatively inelastic and, in the case of the EU, these products are protected by tariff barriers.

So who will gain and lose from market segmentation? In the short run, the uncertainties in markets create difficulties for both producers and buyers. In the medium term, once segmented markets have evolved, some producers will find the premium for non-GMOs attractive, and others will choose the lower costs of production from adopting GMOs. Producers will choose to enter the GMO or non-GMO market segment depending upon their relative cost savings from adoption or their proximity to particular markets. Grain handlers and producers with the ability to deliver non-GM corn and soybeans will view the segmented market as an opportunity for collecting 'economic rents' in the short run. As other suppliers enter the non-GMO market, profits will decline. Consumers of GM products will have slightly lower food prices. Consumers who are concerned about GMOs will be able to avoid products they do not want, although they may pay slightly higher prices. The long-run effects are more difficult to predict, as they depend upon the eventual size of the GMO and non-GMO market segments.

## Concluding Comments

Disagreements over the risks and benefits from the current set of commercially produced GMOs have led to demand for non-GM food products in some markets. This non-GMO demand has been addressed and facilitated by private voluntary labeling, particularly in the major EU retail chains. Worldwide demand for non-GM food may be further facilitated in the future as mandatory labeling is implemented in more countries. An important area of uncertainty is whether concerned consumers will also want to avoid products derived from or produced with GM crops, in addition to products that contain novel genetic material.

So far, the discrete non-GMO market segments seem to be limited to specific products or brands that have well-established supply chains for guaranteeing product characteristics. The additional handling costs appear to be similar to those for delivering other kinds of specialty crops and utilize the same kind of market institutions for verification. One possibility is that demand for non-GM foods will be met entirely through the expansion of these specialized markets and private labels. Another possibility is that the market for non-GM foods will expand significantly in the future, and will extend beyond specific products and brands. Such expanded demand requires different kinds of market organization to maintain segregated supplies, including new investments in dedicated facilities.

In the long run, there are further possible changes in market structure that might arise from this new technology. Demand for non-GM foods might become large enough to preclude widespread adoption of GMOs, reversing the adoption trend that is well underway in major producing countries. Or, demand for specialized GM foods with particular benefits to consumers might emerge, which would increase the demand for market segmentation. These and other possibilities are discussed further in the next chapter.

Both the supply and the demand side of the market will require some time to find a new equilibrium. On the demand side, the provision of information will be a key determinant of how fast consumers learn about and decide whether to consume GM or non-GM products. Labeling will be an important part of such information and may facilitate more rapid market adjustment. On the supply side, the reorganization of production and marketing to find least-cost channels for segmentation will also require long-run adjustments. Thus, world markets for corn and soybeans are likely to experience increased uncertainties from the introduction of GMOs for some time.

# 12

# Looking to the Future

**Timothy Josling**
*Institute for International Studies, Stanford University, Stanford, California,*
*USA*

**Gerald C. Nelson**
*University of Illinois, Urbana, Illinois, USA*

To hazard a guess about the future of agricultural biotechnology it is useful to consider the following questions. What is so different about GMOs that the technology has raised so much controversy? What has the controversy so far taught us about the public reaction to biotechnology? Can one restore consumer confidence in the new technology so that advances in farming and food production do not get sidetracked? How adequate are the regulatory systems to perform their functions in this new area of consumer and environmental safety? Can one be certain that legitimate environmental objectives are met without resorting to a zero-risk strategy that would kill the technology? Will GM foods continue to be a problem area in international trade regulations? And what factors might emerge in the future to change our current assessment of the value of the technology in the farming and food sectors?

## Why Are GMOs Different?

New agricultural technologies appear regularly, bringing with them structural change and transitional problems. In the last 100 years, world agriculture has seen two previous waves of technical change involving biological processes. The first Green Revolution, in the 1930s, was stimulated by the widespread use of Gregor Mendel's pioneering work on inheritance in plant breeding, the identification of male-sterile corn varieties so that the private sector could exploit

Genetically Modified Organisms in Agriculture
ISBN 0-12-515422-4

hybrid vigor in corn, and inexpensive nitrogen fertilizer. These three innovations together meant rapid increases in yields of corn and other temperate-climate crops. The second Green Revolution, starting in the 1960s, took these same technologies and applied them to crop varieties widely grown in the tropics – rice and wheat. Both revolutions changed the structure of the farm sector and the input supply industries, first in the developed world and then in developing countries. Though generally welcomed, the adoption of high-yielding varieties caused tensions and required new institutions and policies to capture the benefits without imposing unacceptable costs. The recent advances in agriculture involving recombinant DNA can be seen as a third Green Revolution, starting in the 1990s.

The parallels between the biotech and earlier Green Revolutions are strong. Both are potential sources of improved food supply. Both created commercial opportunities for the sale of inputs, making agriculture part of an industrial process rather than a self-contained traditional rural activity. Both have a tendency to reward adopters and those that are already more integrated with other sectors. Some potential problems also arose in earlier decades: concerns with the thinning of the gene pool were expressed at the time of the earlier Green Revolutions, and efforts were made to preserve genetic diversity.

But two key differences between the two technical advances are significant. First, in the case of the earlier Green Revolutions the consumer did not make any distinction between crops from improved and traditional seeds. Obviously there were genetic and quality differences,[1] but they were deemed to still be natural variations as a result of 'ordinary' plant breeding. By contrast, the biotech revolution produces a product that many consumers consider to be different. The fact that they see no benefits from the first set of commercialized GMOs further contributes to the negative perception.

Second, both the first and second Green Revolutions stemmed from public-sector research efforts, and the improved seeds, with the notable exception of hybrid corn, were generally distributed through public agencies. The intellectual property developed was a public good, freely exchangeable. The current biotech revolution, on the other hand, has been driven by the private sector, though there is no intrinsic reason why the technology could not be developed in public laboratories if funding were available. The involvement of the private sector, and in particular the large multinational corporations in the life science area, seems to account for a major part of the concern and opposition. Thus, the combination of a technology that entails a change in consumer perception of the product coupled with the greater involvement of the private sector in the development and marketing of the new varieties set this revolution apart from the earlier ones.

## What Have We Learned So Far?

GM foods have been on the market for less than a decade. The widespread use of transgenic seeds by farmers is even more recent. First-generation lessons are

---

[1] For example, the first generation of Green Revolution rice had poor milling and eating quality.

beginning to be learned, but there are bound to be some surprises in store. The main lesson, which is clearly emerging, is that it is risky to let technology get too far ahead of consumer acceptance. Technology should not, and ultimately will not, drive the market.[2] Consumers have to see a benefit to the technology before they will embrace it. A technical change that is uncontroversial is of little concern to the consumer, but in this case public interest groups and the media have persuaded European consumers that there is a difference between GM and non-GM crops and the foods produced from them. These consumers have demonstrated a fear of unknown, negative environmental consequences that seems immune at present to attempts by governments and the scientific establishment to put the issue in perspective. This does not mean that consumers will never accept GM foods, but it does mean that a marketing effort that ignores possible consumer resistance can prove a costly failure.

## Can Consumer Confidence Be Restored?

The elusive solution to the problem of regulating food safety aspects of biotechnology lies in ensuring that each national regulatory body has the full confidence of consumers and the public and is neither under the influence of self-interested local producers nor captured by political movements with agendas broader than public safety and information. These national bodies should themselves be involved in the dissemination of information reflecting scientific consensus. They should also assist in the construction of international standards that they can recommend to governments. They should work with industry to devise appropriate labeling systems that would give consumers the choice when controversy surrounds the properties and consequences of particular foods. Such actions are likely to happen only if those bodies themselves are free of direct influence from vested interests (on both sides of the issue) and have their independence guaranteed by governments. The hope for the future of GMOs and similar technologies therefore rests with the establishment of bodies that have the confidence of consumers.

## How Adequate Are Regulatory Systems for Safeguarding Human Health?

Regulatory systems for foodstuffs are in a state of flux in many countries. Clearly they have been put under strain by increased attention from consumer groups and the public. One aspect of the question most such agencies face is that of convincing the public that regulations set up before GM crops hit the market are adequate to control any deleterious side effects on health or the environment.

---

[2] Technology did drive the rapid adoption of the GM crops by farmers. When the new crops were released, some farmers saw distinct benefits, as demonstrated in Chapters 4 and 6, and voted with their seed-buying dollars. If consumers in Europe had reacted with the same unconcern as in the US, adoption of the technology would likely have continued throughout the world.

Part of this revolves around the question of whether there is a significant, detectable difference between foods produced as a result of a biotech process and those from more traditional methods. This issue may never be resolved. There clearly is a difference between crops that have been designed with particular characteristics and those without. Supporters of the current regulatory system claim that the 'differences,' in the case of those that have gained approval, do not affect the safety of the product. Opponents counter that we do not have enough information to make that claim. Supporters cite the fact that genetic modifications occur all the time in nature and that biotech is essentially speeding up and controlling such modifications. Opponents reply that not enough is known about the impact of genetic manipulation to be sure that there will be no effect on human health. Agreement on such a fundamental regulatory issue may be elusive.

Whether it is strictly necessary to conduct further widespread tests on the safety of GM foods is therefore debatable. But to address food safety fears, it would seem wise to conduct toxicology and other tests to evaluate more carefully the properties of GM and non-GM foods. Many traditional foods contain ingredients that are harmful at high doses; subjecting GM foods alone to extensive testing would clearly not be adequate. Disinterested parties, rather than organizations with a stake in the outcome, should undertake the research. Eventually, scientific evidence on the issue of food safety should become convincing to all but the committed skeptic.

Communication of scientific evidence is crucial. And it should be communicated at various levels of detail, to provide information to consumers of varying levels of sophistication. The traditional label on the side of a food product is a very limited place to attempt communication. But a standardized approach to information dissemination that includes use of the web could provide a wealth of information for the discerning consumer.

## Can the Environment Be Protected?

Environmental protection is an objective enshrined in the legislation of most countries. International environmental agreements urge governments to coordinate these actions. Of course, there are cases in which such objectives seem to interfere with commercial activities, and they are sometimes based on uncertain scientific evidence. For the food and farming industries to be seen as obstructionist is likely to be self-defeating. It is better to focus on ways in which technical progress in the agricultural and food industry can be made consistent with or actually contribute to legitimate environmental goals.

A much stronger case exists for stepping up research into the environmental impacts of transgenic plants than into their health effects. The ecosystems involved in modern agriculture are complex and poorly understood. But this fact does not argue for moratoria or the destruction of field trials. The challenge for the private sector and regulatory agencies is to persuade environmental groups of the value of allowing relevant evidence to be accumulated over time and adequately monitored.

# Will Biotech Continue to Be a Cause of Trade Friction?

As the food industry becomes more global, trade rules seem destined to come into increasing conflict with the domestic politics of food and the environment. Trade rules help the transparency of the trade system, but the politics of food revolve around consumer confidence in regulatory mechanisms. These two elements come together in the GMO issue. At present there is no open trade conflict involving GMOs, but two fears haunt the trade ministers. One fear is of an explicit ban on GM food ingredients by a major trading country. The second fear is that countries with the view that labeling GM food ingredients is sound policy might try to put labeling on the agenda of international trade negotiations. Countries such as the United States that have opposed labeling would be placed in an awkward position.

The broader question that will challenge the trade negotiators revolves around a simple but fundamental choice: Should one base trade rules on 'hard' scientific evidence and thus judge domestic regulations by that standard? Or should one allow countries to take into account consumer sentiment as well as scientific opinion when setting domestic (and hence import) standards? The SPS Agreement (Agreement on the Application of Sanitary and Phytosanitary Measures) already allows countries to take account of market impacts in the case of animal and plant health. From the point of view of trade policy, a rule-based system is necessary to guard against implementation of domestic policies that react to the headlines of the day and to pressure from groups looking to manipulate consumer opinion for other purposes. The SPS Agreement appeared to put in place a rule-based system built on the principle that scientific evidence is required to justify a stricter standard than those in international use. There will be considerable resistance to tinkering with this principle.

From the point of view of politicians, however, consumer confidence and voter sentiment are strongly connected. Because politicians are servants of the public and not answerable to trade dispute panels, it may not be wise to appear to elevate the niceties of trade policy above that of concerted public opinion. This state of affairs suggests that the attempt to impose the use of scientific evidence exclusively, especially where the science is uncertain, rather than respond to the more complex set of psychological and political factors that governs consumer concerns, may be at best controversial and at worst counterproductive. Indeed, it poses the threat that trade rules may lose the support and credibility necessary for their effectiveness.

Thus, the task of governments is to respond to the concerns, rational or not, of those who think that their governments are paying more attention to pressures from commercial interests in expanding global markets than to the needs for healthy food and environmental sustainability. The need to establish in the public mind a 'human face' of globalization goes far beyond the issue of GMOs, but public perception of the trade system is formed in cases such as this.

# What Is the Future for GM Foods?

No one can predict the future of GM foods and the broader biotech revolution. But three different futures seem possible at this moment. First, the objections to

GM foods could evaporate as consumers in Europe and other countries decide that their fears on the health effects were groundless, as consumer-friendly traits begin to appear in foods, and as the price of GMO-free foods rises to reflect the cost of segregation. Regulatory authorities could recapture the confidence of both consumer and environmental groups and set up an adequate monitoring system for spotting environmental hazards before they became widespread disasters. The press would soon drop the subject from the front page, moving on to the next crusade on behalf of those abused by corporate and bureaucratic power. Environmental groups might come to trust the regulatory checks on environmental side effects and be content to keep a watch on the spread of the technology. This rosy picture seems implausible. It suggests that there will be no surprises and that public interest groups will give up their 'moral high ground' in the debate without a struggle once they see the tide turning.

A second possibility is that the industry could decide that the world is not ready for the biotech revolution and that there is no point in investing further in a process that faces real or imagined obstacles. Investment could dry up and farmers could shift back to using more traditional seeds. Consumer and environmental groups would celebrate their victory and move on to other issues. Regulators would return to dealing with less controversial issues. And world agriculture would have to wait for another decade or two for the promised benefits from designer plants. This scenario also looks implausible. There is too much at stake for the industry to give up at this stage, and there is no indication that governments will allow promising technologies to fold at the first sign of serious opposition.

A third possible outcome is more credible, though less easy to specify. This scenario includes a period of perhaps five to ten years of disorganization as firms and pressure groups continue their play for the ear of public opinion and the pen of the regulator. Governments will take different views, depending on the importance of the industry to their economies and on the strength of public opinion. Then the situation will settle down, with a number of successful GM food products competing with their traditional and organic counterparts. By that time, new GMOs will have been introduced that provide clear benefits to the consumer. 'Green' GMOs will emerge that are beneficial to the environment and will be accepted as safe. Regulations on labeling will have been developed that give consumers meaningful information, and non-GM foods will sell for a premium in certain markets. The countries where GM crops are grown will have developed 'best practice' techniques that minimize environmental impacts. Certain types of GM crops will be restricted to certain parts of a country or regions of the world. Organic foods will enjoy rapid growth in high-income countries but will continue to be too expensive to replace the products of chemical and biochemical farming.

If this outcome is plausible, the current task is to develop the regulations, tests, and procedures that will smooth the passage through the uncertainties ahead. This would suggest that the industry, including food retailers and farmers, join in a dialogue with (moderate) consumer and environmental groups to set up monitoring and evaluation processes that increase the comfort level of all. This outcome argues for cool heads, searching for ways for governments to collectively manage the trade conflicts and avoid projecting commercial and other differences on the GMO debate.

# Part 2

## Perspectives on the
## Controversies

# 13

# Biotechnology Crops – A Producer's Perspective

**Mark W. Jenner**

*American Farm Bureau Federation (AFBF), Park Ridge, Illinois, USA*

U S farmers and ranchers across the country have adopted new biotechnology crops, because they have significant benefits. Commercial farming is very competitive and farmers are very good at producing wholesome, healthy food at a very low cost to consumers. Access to biotechnology-enhanced crops is a great assistance to farmers.

There are questions for which we don't have all the answers, but we have trust and confidence in the scientists and industries that have developed biotechnology. Farmers trust the federal determination that our food is safe. Farmer concerns about biotechnology are less over the science and more about emerging market control and access. Farmers already bear the costs of numerous environmental mandates that are driven by emotion, rather than science.

The food production industry challenge and goal is to have policies that build trust and confidence in food products without putting farmers out of business. Emerging policies directed at agricultural biotechnology will play a significant role in securing our food supply.

## AFBF Policy Related to Biotechnology

The American Farm Bureau policies speak for the nation's farmer members. Policies in place for the year 2000 address the biotechnology-related areas of regulatory process, intellectual property rights, international trade and labeling. The quotes below are taken from the policies adopted for the year 2000 by delegates to the 2000 annual meeting of the AFBF.

AFBF supports, 'Increased efforts through biotechnology to more rapidly develop consumer beneficial traits, to increase the marketability of our products

---

Genetically Modified Organisms in Agriculture
ISBN 0-12-515422-4

to solve environmental concerns, to increase net farm income by decreasing input costs and to improve product quality. We urge state and national political leaders to develop a positive national strategy for biotechnology research, development and consumer education.'

## Regulatory process

AFBF believes, 'The approval of new products should be based on safety and efficacy criteria. US government agencies, particularly the USDA and the Food and Drug Administration (FDA), should continue to serve their respective roles in providing unbiased, scientifically-based evaluations concerning the human and animal safety and wholesomeness, as well as the environmental impacts of biotechnology-enhanced commodities. US government agencies should evaluate whether there are improvements in the regulatory approval process that could be made to further enhance consumer confidence.'

AFBF is, 'Opposed to any law or regulation requiring registration of farmers who use or sell products approved for sale by the Food and Drug Administration. FDA should set acceptable standards for determining what is non-biotech.'

## Intellectual property rights

AFBF favors, 'Strong patent support to encourage these new technologies. Patents should be broad enough to provide reasonable protection of development costs, but should not be so broad as to grant one developer the right to a whole class of future developments. Royalties from patents on transgenic animals must be structured in a manner which allows producers a clear understanding of their obligations and does not disrupt the existing livestock marketing systems.'

## International trade

AFBF supports, 'Increased efforts to educate the public worldwide regarding the safety and benefits of products developed through biotechnology. We believe that our competitive advantage in world markets will be maintained only by the continued support and encouragement of technological advancements.'

AFBF opposes, 'The imposition by foreign countries of any import restrictions, labeling or segregation requirements of any genetically modified organism, once such commodity has been certified by the scientific community as safe and not significantly different from other varieties of that commodity.'

## Labeling

AFBF supports, 'Consumer friendly, science-based labeling of agricultural products that provide consumers with useful information concerning the ingredients, nutritional value and country of origin of all food sold in the United States. Labels should not be required to contain information on production practices that do not affect nutrition or safety of the product. Agricultural products that are produced using approved biotechnology should not be required to designate individual inputs or specific technologies on the product label.'

AFBF supports, 'The science-based labeling policies of FDA, including no special labeling requirement unless a food is significantly different from its traditional counterpart, or where a specific constituent is altered (e.g., nutritionally or when affecting allergenicity); and voluntary labeling using statements that are truthful and not misleading. We also support the voluntary labeling of identity-preserved agricultural and food products that is based on a clear and factual certification process.'

Since AFBF believes in the safety of our food system, we do not believe there is any reason to treat foods produced through biotechnology as anything but safe. We do not believe they necessitate warning labels. 'Warning labels on products should be based on conclusive scientific proof.'

## Economic Benefits

There are significant benefits from growing biotechnology crops. Consumers have the most to gain from agricultural uses of biotechnology and transgenic crops. In 1996, only 10.7 cents of every disposable dollar is spent on food in the US (US Department of Agriculture, 1998). While this statistic precedes the most recent advances from agricultural biotechnology, it does reflect the market incentives that provide rewards for safe food at low cost. US food consumers vote with their dollars to spend less of their scarce resources on food. Like all the preceding advances in agricultural production technology, biotechnology will allow the food consumers' expenditure on safe abundant food to go even lower.

The benefits to the food production industry are only just beginning to be documented. The advent of herbicide-tolerant, biotechnology soybeans has created a decrease in all soybean herbicide prices (Nelson *et al.*, 1999). Companies relying on non-biotechnology seed must remain competitive with the benefits of using biotechnology soybeans. Gianessi and Carpenter (2000) estimated that all US soybean growers spent $220 million less on weed control in 1998, even after compensating for increased technology fees paid for by growing Roundup Ready soybeans. Documentable benefits from using the pest-protecting, *Bt* cotton include a 250 000 gallon reduction in 1996 chemical use and a 30–50% reduction in the number of insecticide applications from 1996 to 1998 (National Research Council, 2000).

Other farm-level benefits from agricultural biotechnology are less easily measurable. Recent in-depth reports, such as that by the National Research Council (2000), provide an excellent chronology for why we protect our crops against insect damage. There are significant producer benefits in finding new and better ways to increase yields and lower costs. Agricultural biotechnology combines the successes of improved breeding and pest protection. The cumulative effect is fewer acres planted and less impact on the environment.

The most difficult benefits to measure are the gains in management and timeliness. The insect-resistant and herbicide-tolerant crops are effective and provide comparable productivity with fewer trips across the field (lower costs and greater flexibility). Added (earlier) planting flexibility can increase yields significantly. Farmers have a high level of confidence in these technologies.

Biotechnology, like any other farm production technology, is a tool. Farmers are generally not impassioned about being 100% biotechnology or non-biotech. Farmers do get upset about not having access to the tools that allow them to excel at growing crops. Farmers don't use chemicals if they can get by without them. They don't till if it isn't necessary. A farmer may be a no-till farmer until the wet spring when he needs a dry field – then the disk is his friend. The same is true for biotechnology.

## Benefits and Costs Determined by Technology

Many of the costs associated with biotechnology are based on perceived risks. We forget that risks are relative. The net value of biotechnology is dependent on the perspective of each individual. Technologies discussed outside of the production environment lose considerable depth. The producer's perspective has merit in discussions concerning human health and the environment, to keep the discussion balanced.

Chemical application practices change radically with the use of herbicide-tolerant and insect-resistant crops. With the insect-resistant crops, insect control chemicals are nearly eliminated. A thousand commercial cultivars have been released for production that have been enhanced genetically through mutagenesis (Gianessi and Carpenter, 2000). This is the bombardment of genetic material with either radiation or strong chemicals to force genetic mutation. The current innovations in genetic biotechnology are far more precise, reducing the risk of unintended effects by magnitudes.

Genetic contamination of plants grown near the biotechnology crops is a concern, but this is already being addressed, and we are learning about the value of buffers. One proven technique that has been reviled by environmental groups is the use of sterilization or a 'terminator' gene. Yet, sterile commercial hybrids have enhanced the corn industry. The development of sterile triploid grass carp has allowed the introduction of alien fish species into the environment, without fear of genetic contamination.

The effects of *Bt* corn on monarch butterflies have provided fodder for lively discussion of the unintended consequences of corn borer controls. Misinformed critics have missed the whole value from the very specific results of using the *Bt* toxin, which is much more species-specific than many of the more traditional pest control alternatives.

And finally insect resistance is a real phenomenon. Nature adapts. It is the speed at which this process occurs that is under debate. The companies involved in these technologies plan to succeed, but a wholesale resistance to a single pest protectant in any given year would destroy the confidence farmers have in the companies and end any chance for commercial success. These sensitive agricultural input market feedback mechanisms are overlooked in current policy debates.

# Benefits and Costs Yet to Be Determined by Policy

Other benefits and costs have yet to be determined. These relate to emerging policies directed at biotechnology and other related issues.

## Market control

The question of who controls the market has yet to be determined. With few competitors for the major herbicide-tolerant and insect-resistant technologies, farmers are concerned with access to competitive prices. Emerging evidence indicates that this is not a problem. There are opportunities for farmers to establish value-chain alliances with all the other industries in the chain. Farmers that enter into attribute-specific or identity preserved crop production contracts want to be confident that they are appropriately compensated for additional costs involved and additional value they add. Farmers are concerned that emotions, politics and incomplete information will limit their access to markets. This is already happening with some of our international trading partners.

Labels can either enhance markets or they can destroy them. There is much discussion about labels for biotechnology products. Labels may have a role, but they must convey usable information to be useful. If consumers value information about products that are non-biotech, labeled products should command a higher price. A higher valued, non-biotechnology product price-premium must cover the extra costs of preserving a desired attribute from the farm to table. Defining a label assumes that standards of purity are also resolved.

An identity preserved (IP) product infrastructure protects the movement of specific attributes through the food production chain. A small portion of our grain moves through this more costly system. Products that have a high value justify the cost of segregation. This system is very different than our traditional bulk grain system. Most of the grain we grow is stored and handled as generic bulk commodities. These homogeneous commodities are handled very effectively and inexpensively. There are two very different and distinct systems.

To date only a few premiums have been paid for the specific attribute of a 'non-biotech' crop. Moving to a higher-valued, identity preserved market carries with it added costs. Non-biotechnology market premiums are not common, but producers targeting that market should consider the following factors. All grain handling equipment must be thoroughly cleaned. Planters and combines must literally be partially disassembled, after working with biotechnology enhanced seed and grain, to ensure that some biotechnology grain has not accumulated inside the machine, before using conventional, non-biotechnology seed. Explicit contracts need to be signed before planting to protect the interests of the farmer and to understand the expectations of the grain buyer. Arrangements need to be made, particularly for non-biotechnology corn, for adequate protection and buffers from surrounding crops.

This market is currently very small, but producers need to begin to understand the added costs involved before a decision can be made about the net benefit of operating as an IP product.

## Biotechnology Policy Must Add Benefits

The policy benefits to input suppliers, farmers, processors, retailers, and consumers must outweigh the costs to input suppliers, farmers, processors, retailers, and consumers. The policies must be transparent (make sense) and contribute to the trust and confidence levels of all involved. The cost of segregated, identity preserved (IP) product markets must add enough value to justify those added costs. Benefits must outweigh the costs. Food labels, such as 'non-biotech,' must convey useful information and market value to be effective. The emerging communication and transportation infrastructure will provide new opportunities for market access for biotechnology producers, and others in the food value-chain.

# Genetically Modified Crops and the American Agricultural Producer

**14**

**Gary Goldberg**
*Former Chief Executive Officer, American Corn Growers Association (ACGA),*
*Tulsa, Oklahoma, USA*

It would be hard to imagine a more complicated, confusing and contentious issue facing American agriculture than the issue of genetically modified (GMO) food products. What was presented as clear-cut technology that would save farmers money and allow for increases in productivity and efficiency has instead become an albatross around their necks. Farmers find themselves caught in the middle of a debate between chemical manufacturers, seed companies, agribusiness concerns, grain exporters, foreign and domestic consumers, and governments around the world.

The rapid acceptance of GMOs by production agriculture shows that farmers may want to have these products as part of their planting options for the future. But the uncertainty over marketability, cross-pollination, segregation, testing, certification, labeling, and liability, is driving farmers away from the technology. In 1999, close to 32% of all corn acres or approximately 25 million acres were planted to GMOs. In 2000, a national survey conducted by the United States Department of Agriculture showed 19.8 million acres actually planted to GMOs or 24.8% of total corn acres. This amounts to a 20.4% drop in 2000 planted acres compared to 1999.

To the American farmer, this debate over genetically modified crops is not an issue of science, environment or health. It is an issue of economics. Simply put, can US farmers afford to grow a crop that they may not have a market for come harvest in the fall? Or can US farmers deal with the concerns over on-farm segregation and the risk of liability caused by pollen contamination? Consumer resistance in Europe, Asia, New Zealand, Australia, Canada, Mexico, South Africa and the growing resistance in the United States makes it unlikely that many market opportunities will remain available for GMO products. Therefore, rather than take the risk, corn growers are making a move back towards conventional, non-GMO seeds.

## The Role of the ACGA

The American Corn Growers Association could very well be the single most important agricultural player in this debate because the ACGA remains the only major, national agricultural organization taking a neutral role in this issue and because corn is the nation's largest cash crop. The ACGA's questioning of genetically modified crops is the first time a major farm organization has made farmers look at the financial effects of GMOs on their individual farming operations. The ACGA realized that farmers had not been getting all the information they needed to make educated decisions about what to plant and why they should plant it. The message they were receiving from the biotechnology industry was that GMOs are safe, environmentally friendly and pose no health risk. They also heard that GMOs were the best thing to hit production agriculture since the invention of the tractor. However, at no time did they hear any concerns over growing foreign market resistance or other concerns over segregation and liability. In fact, they were presented with no risks at all. Even the US Department of Agriculture was advocating the planting of GMOs. On the other side of the spectrum, environmental and food safety groups were totally negative about GMOs and found no advantages to their continued planting. They were even making US farmers feel guilty about what seeds they planted. To make sure that information was available on all sides of the issue, the ACGA developed their Farmer Choice–Customer First program. This program has many different objectives including providing unbiased, even-handed and honest information about GMOs, helping to find marketing alternatives if farmers decide to plant conventional seeds, protecting farmers from liability concerns whether they plant GMOs or non-GMOs, understanding the potential burdens of segregation and recognizing the importance of giving customers whatever products they demand.

The biotechnology companies and many within the traditional, more conservative farm organizations have roundly criticized the ACGA for our Farmer Choice–Customer First program. They believe that either you are totally supportive of genetically modified crops or you must be totally against them. On the other hand, some within the environmental community have also been critical of the ACGA because we have taken a neutral stand and separate the medical benefits of biotechnology from the concerns over the agricultural aspects of biotechnology. These different groups cannot understand how anyone could be neutral with the only objective being to protect their membership within this entire debate. But we have seen that farmers overwhelmingly support having access to the information we have provided and appreciate getting honest analysis about the benefits and risks associated with GMOs.

## Benefits and Risks of GMOs

So just what are the benefits? To US farmers, they view the benefits as increased yield opportunities for corn, up to 15% by some accounts. These products also allow farmers to use fewer chemicals, thus making them more efficient and saving them money. Estimates have shown up to a $25 per acre saving with reduced herbicide and insecticide use. Anything that allows a farmer to use their time more efficiently is going to find many supporters.

On the other side of the issue, the risks include questions about increased yields at a time of record overproduction, the substantially higher cost of GMO seed and technology fees, the cost associated with testing, segregation, and certification of non-GMOs, the risks of liability brought about by cross-pollination and contamination, and the loss of markets, both foreign and domestic.

## A Declining Market for GMOs

The marketing concern is the single major issue surrounding the decision by corn producers to reduce GMO planted acres this growing season. In the 1997–1998 marketing year, the US shipped 2 million tons of corn to Europe. In the marketing year 1998–1999, we shipped only 137 000 tons, resulting in the loss of well over $200 million in sales. Corn sales will continue to drop. Soybean sales to the EU were 398 million bushels in the 1997–1998 marketing year. Estimates for the 1999–2000 marketing year are 221 million bushels. In addition, other countries, especially Brazil, are filling the void left by the US insistence that we advocate GMOs. Carrefour, one of the largest French grocery store chains recently contracted with Brazil to supply 180 000 metric tons of non-GMO soybeans to feed their meat supplier non-GMO fodder. Most major grocery store chains in the UK, including Tesco and Sainsbury, eliminated GMO ingredients from their own-brand products. Other stores in the UK and Germany do not sell any products that contain GMOs.

Japan is the single largest purchaser of corn from the United States. In 1999, Japan purchased 15.891 million metric tons. But now Japan is demanding that their food grade corn be non-GMO, causing segregation of the total shipment. In addition, Japan's plan to require safety screening of genetically modified products is impacting US sales and beginning in April 2001, Japan will require labels on all food products that contain GMOs. It is clear that if the United States does not adhere to the import requirements of the Japanese, American farmers could very well lose this vital market. This would cause severe economic hardship to US producers.

In the US, Gerber and Heinz baby foods, IAMs Pet Foods, and Frito-Lay all announced they will not accept GMOs in the future, and most of the major fast food restaurants and food companies stated they will begin to source non-genetically modified potatoes for their french fries and potato chips. This includes McDonald's, Burger King, and Procter and Gamble. The two largest natural food store chains, Wild Oats and Whole Foods Markets, stated they would no longer use GMOs in their private-label products. In Canada, Seagram's and Azteca, Mexico's largest tortilla maker, also stated they will no longer accept GMO corn. This movement towards non-GMOs products will demand the segregation of GMO from non-GMOs in the US. Segregation will be difficult and will cost hundreds of millions of dollars. Much of this cost will fall on the backs of American farmers, as grain elevators are not physically equipped to keep the products separated. This segregation will result in two different channels for grain, one for GMO markets and the other for non-GMO markets. The Environmental Protection Agency has called for the planting of refuge zones to prevent insects from becoming resistant to certain insecticides. The Biosafety Protocol agreed to in Montreal by 133 countries, including the United States, will require the labeling of grains. The precautionary principle allows countries to refuse to accept a GMO product if they believe it poses a health or environmental risk to the population. All these events are encouraging American farmers to stay away from genetically modified crops.

A recent, scientific, randomly selected, statistically valid survey of 500 corn producers by the American Corn Growers Association found that 66% of those surveyed expressed concerns that the burden for segregating the GMO corn from traditional, non-GMO corn will rest with them. Sixty-four percent of the farmers say that their decision to plant more or fewer acres of GMO corn in the future will be influenced if grain elevators, grain processors and grain exporters require segregation, and 76% say they will plant fewer acres of GMO corn if the grain industry requires segregation. These survey results are a clear indication that farmers are not wed to genetically modified crops and if the burdens become overbearing, they will revert to planting, traditional, non-GMO corn seeds.

As concerns occur with overseas customers, US governmental officials and American grain exporters are discouraging our European customers from purchasing the products they desire and insisting they buy GMOs. Instead of honoring the long business tradition that says the customer is always right, US trade representatives are arrogantly dictating what the customer should buy and threatening sanctions if they refuse. This is a careless position that will continue to cost American farmers valuable foreign customers. The American Corn Growers Association believes in the precautionary principle, and believes that every individual country has the right to keep out whatever products they deem to be questionable. Most American farmers feel the same way. The ACGA farmer survey found that 82% of the nation's corn growers felt that foreign export customers should have the right to choose between buying traditional, non-GMO corn and GMO corn. We compliment the consumer and environmental community in Europe for having the courage to stand up for what they believe to be right and in the best interests of their citizens. After all, the concerns about health and safety, while substantiated or not, rest with individual countries to determine.

## Feeding a Hungry World

To many people of the world, biotechnology and genetically altered food products are an ethical issue, as is the ethical responsibility to feed a hungry world. American farmers have a desire to provide food to those who need it. After all, each US farmer grows enough food to feed 128 people, 96 in the United States and 32 overseas. However, even with current record world grain production, we are still not getting food to the world's hungry.

Is biotechnology the salvation for feeding a hungry world? Answering this question involves a detailed discussion of food distribution, infrastructure improvements, population control measures and the fate of over 3 billion indigenous people who depend upon agriculture for their survival. If one were to listen to the biotechnology industry, the only reason for the proliferation of genetically modified foods is their desire to provide enough food for everyone who wants it. Unfortunately, we all understand that the real reason is profit, clear and simple. While there is nothing wrong with a profit motive, let us admit the obvious. Let us also understand that there is no ethical responsibility in making a profit on the backs of hungry children.

At a time of record worldwide food production, is biotechnology needed to feed a growing population? Will continued poverty allow poor countries to purchase food even though it is available? Will the food reach those who really need it or will corrupt governments and the lack of a local transportation system still prevent the food reaching those who need it the most? And will the advent of the 'terminator' technology bring about increased hardships upon the poor? Holding back part of a harvested crop to be used as seed for the next growing season has been a tradition for hundreds of years. Yet the 'terminator' technology, which renders second-generation seed sterile, makes this tradition no longer available. With the purchase of Monsanto by Pharmacia, any pledge by the CEO of Monsanto to discontinue the research and development of the 'terminator' technology is null and void; therefore future development is moving forward. However, if the biotechnology industry is serious about the development of technology to provide more food to a growing population, it must stop the development of the 'terminator' immediately.

## Opposition to GMOs in Europe

European countries have the right and authority to restrict the importation, use, and consumption of genetically modified foods. Many within the European Union will not allow the United States to dictate what should or should not be bought and sold. Recent events only add to the decisions by many European countries to restrict any access by GMO crops. In the United Kingdom, Germany, France and Sweden, GMO contamination of oilseed rape caused the accidental and unauthorized planting of thousands of hectares when the seed was sold as non-GMO. Not only did this cause deep financial problems for those farmers who planted the rape, but it added to the mistrust that many in Europe have with international biotechnology companies. Another event that will have long-term implications on the proliferation of GMOs in Europe is a recent three-year study conducted by the University of Jena in Germany that showed that genes from GMO crops could spread from plants into other forms of life.

These events will make it increasingly difficult for American grain exporters to sell GMO products into Europe. This message is important for US farmers because if they believe the opposition to GMOs in Europe will disappear in the short term, they are mistaken. Suspicion and mistrust of GMO products, the United States and grain exporters will continue for the foreseeable future. American farmers cannot count on regaining the European market for years to come.

# US Farmers Must Make Their Own Decisions

The questions over the uncertainty of agricultural biotechnology will be with us into the foreseeable future. As worldwide opposition grows to the use of genetically modified crops, and as American consumers begin asking many of the same questions that have been asked by consumers in Europe and Asia, the demand for GMOs will likely diminish. Will the United States be prepared to provide whatever the world's consumers demand or will the seed supply become so contaminated with GMOs that guaranteeing non-GMO becomes impossible? If the desire of the customer continues to demand non-GMOs, will American farmers be prepared to grow it and segregate it from GMOs, or will the corn fields of GMO so contaminate non-GMO fields that there is no longer any distinction between the two? If we are not prepared, we will continue to lose our customer base and market share, opening up new markets for our traditional trading competitors. That message of growing for the customer has not sunk into the mindset of the biotechnology and grain exporting companies. It needs to sink into the mindset of American agricultural producers as well.

American farmers have no control over market prices, domestic agricultural policy, weather, and international events, but they do have control over the planting of genetically modified crops. After all, it is the farmer who determines what seed to plant. If every farmer in the United States were to plant GMOs, there would be no reason to address the issues of segregation and liability and the marketing of non-GMO to Europe and Asia because farmers would intentionally be giving away the market. On the other hand, if every farmer only planted conventional, non-GMO seeds, there would be no worry over cross-pollination and providing our overseas customers with the products they desire. Therefore, the long-term battle over genetically modified crops will be waged in rural coffee shops, local radio and television stations, agricultural publications and individual farmsteads. For the first time since the American Civil War, a battle will be fought on American soil. The future of US agriculture, as we know it today, could well be at stake. This battle pits multibillion dollar biotechnology companies with limitless advertising and public relations budgets and their funded agricultural allies against environmental, social and farm activists with their limited dollars but a wealth of determination. How US farmers receive the message from these competing interests will determine the outcome. The American Corn Growers Association will be in the middle, interpreting and dispensing the rhetoric from both sides and encouraging farmers to make their own decisions based on what is best for their own individual farming operations. After all, when all is said and done and the seeds have been planted and harvest approaches, neither the biotechnology industry nor the activists will be standing by the American farmer. The farmer will have to live with his or her final decision and live with the consequences of that decision. It is the goal of the ACGA to make sure that farmers have the information to make these important decisions based on knowledge, understanding, and facts.

# 15

# Toward Common Ground: Roles of Markets and Policy[1]

**Nicole Ballenger**
*Economic Research Service, USDA, Washington DC, USA*

**Mary Bohman**
*Economic Research Service, USDA, Washington DC, USA*

In the earlier renderings of the biotech debate the issue was cast as a trade dispute between the United States and European Union – as a case of the US wanting to sell it and the EU not wanting to buy it. The focus on the differences emerging in the US and EU was understandable as the two appeared to be headed for a standoff that could have consequences for the entire world. Indeed, countries all over the world are looking to the US and the EU for guidance and even technical support in developing and adapting regulatory principles and processes to biotech demands. Many countries are still waiting to see how the US and EU resolve questions about the biotech approval process, risk assessment and environmental monitoring, and consumer information. Fortunately, considerable progress is being made on these fronts.

At the core of the agricultural biotech debate is a more complicated story about the intersection of technological change in agriculture and an increasingly consumer-driven food system. That intersection is reshaping public-sector roles in the agricultural and food system in the US, the EU, and elsewhere. The US and EU systems are sure to evolve differently because the historical, legal, political, and cultural contexts are different (Jasanoff, 2000). The different ways

---

[1] The views expressed in this chapter do not necessarily represent those of the US Department of Agriculture or the US government.

Genetically Modified Organisms in Agriculture
ISBN 0-12-515422-4

they evolve are a potential source of tensions in an increasingly globalized food system, with implications beyond bilateral relations and the biotech case. But because the food system is increasingly global, there is great impetus to find common ground. It is also important to recognize that there is no single EU viewpoint or single US viewpoint but rather a diverse array of consumer and producer preferences and attitudes in both places; indeed, regulatory systems everywhere are confronted with balancing a wide array of economic and social interests. In the case of biotech, multinational efforts are underway to build a shared knowledge base that improves the scientific foundation for making difficult social policy decisions and that can enhance the potential for agreeing upon 'equivalencies' and 'crosswalks' between different regulatory frameworks.

## The Biotech Issue as an EU–US Trade Dispute

The biotech debate has been cited as a factor in the performance of US exports. The EU approval process for bioengineered seed varieties has in particular been a source of consternation for US exporters. A number of corn varieties approved and planted in the US have not received approval in the EU and a *de facto* moratorium currently exists on EU approvals. Only a small fraction of US acreage has been planted to these non-EU approved corn varieties, but fears of having shipments delayed or halted if unapproved varieties are commingled with approved varieties prompted some US corn exporters to forgo the EU market altogether. The volume of corn exports to the EU fell more than 90% in 1998, due largely to these regulatory delays. (This is the one documented loss of US biotech export due to the actions – or lack thereof – of a trade partner.) Moreover, the EU market represents an import quota allocated to trading partners to compensate them for the loss of market when Spain and Portugal joined the EU, so it is important in a symbolic sense (Ballenger *et al.*, 2000).

Despite the troublesome nature of the situation, the loss of sales to the EU makes only a small dent in total US corn sales because the EU is a minor source

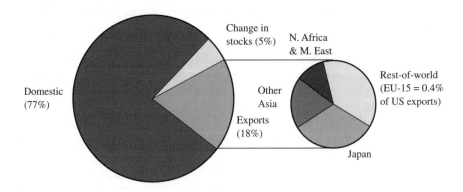

**Figure 15.1**: Use of US corn supply: 1998/99

*Sources:* Total use for marketing year from November issue of *Agricultural Outlook*. Trade shares from unpublished, unreconciled United Nations annual trade data for 1998.

of demand for US corn. In the 1998/99 marketing year the US domestic market claimed around 80% of total corn use (Figure 15.1). The large domestic component, coupled with the fact that most US corn (61% of domestic use, for example) goes into animal feed rather than directly into food production, has insulated the US corn market significantly from biotech concerns. Most of the 18% of US corn that is exported is shipped in a nearly even distribution to Latin America, Japan, 'other Asia,' and 'Africa and the Middle East,' and none of these regions has restricted imports of biotech corn to date. At its peak in the early 1990s, exports to the EU represented less than 5% of US corn exports. In fact, even before the biotech issue arose, the EU was relatively self-sufficient in corn, as shown by the large volume of trade among member countries relative to imports from non-members. This relatively high self-sufficiency may have made it relatively costless (in terms of potentially higher feed costs for its livestock producers) for the EU to accommodate its own regulatory delays and to forgo or seek alternative supplies.

The soybean situation is different. The one biotech soybean variety commercially grown in the US is approved in the EU so regulatory delays have not been a bilateral issue. But bilateral trade ties are much more significant than in the case of corn, with potentially larger implications for US and global markets should consumer preferences shift significantly or EU regulations evolve to pose barriers to US exports. A larger share of US soybeans and products is shipped to foreign markets (42%) than is the case for corn (18%). In addition, the EU share of US exports is quite large (26%). If soybean exports to the EU were to fall suddenly, there would indeed be a significant impact on the US soybean market unless US exporters were able to quickly find alternative buyers. However, less often noted is the likelihood that in order to switch to alternative suppliers, EU buyers would have to pay higher prices to obtain enough soybeans to replace US supplies (Ballenger *et al.*, 2000). EU buyers have been as reliant on US suppliers as US producers have been on EU buyers, which may have something to do with the fact that the EU regulatory system has not to date posed serious problems for US soybeans.

Broad shifts in corn and soybean markets have not yet occurred due to differences over biotechnology because in fact the demand for non-biotech commodities has been small.[2] There are signs, however, of new markets for non-biotech foods and that some food companies are gearing toward supplying them (Barboza, 2000). What may be more relevant over the long run than whether the EU will accept biotech commodities or not is how the US agricultural system (and other countries') will adjust to the growing diversity of consumer preferences, both within the US and internationally, for how food is produced. This shift toward a differentiated product system is the focus of the next section.

---

[2] Japan and some other countries are in the process of implementing mandatory labeling regulations that could act as trade barriers, although it is too soon to assess their impacts. There is concern that tolerance levels for biotech content may be set unduly low in relation to what is technically feasible to provide or to test for, or in relation to what consumers actually want. Divergent standards among countries could also significantly increase the costs of market segregation.

## The Biotech Issue as an Extension of the Shift from Bulk Commodities to Differentiated Products

Many studies have pointed out significant differences between EU and US consumer attitudes regarding biotech foods (e.g., Gaskell *et al.*, 1999). The reality, however, is that in any country consumer demand may and is indeed likely to coexist for both biotech and non-biotech products. In response, agricultural marketing systems are developing to offer consumers products differentiated according to their biotech status (USDA Economic Research Service, 2000). Our view is that such product differentiation is merely an extension of an important trend already established for wine, specialty crops, organic foods, specialized cuts of meat, and high-value products in grain and oilseed markets. Other differentiated grain and oilseeds such as high-oil corn, hard endosperm corn, white corn, waxy corn, nutritionally dense corn, high oleic soybeans, and improved food-quality soybeans are already fixtures in the marketplace (Riley, 1999).

The Japanese soybean market is one example of how US agriculture has already tapped into opportunities presented by potential demand for non-biotech commodities, and how new marketing channels emerge in response to 'differentiated' demands. Although Japan has continued to import biotech soybeans for use in animal feed, the US has also been exporting both organic and non-biotech soybeans to the Japanese food-use market at a considerable price premium (Ballenger *et al.*, 2000). Clearly, some US producers and companies have seen the potential to profit from the niche overseas demand for non-biotech foods. If entrepreneurs perceive there are profits to be earned for non-biotech commodities (or for any varieties with specific traits of value to processors or consumers), then marketing services that help producers meet these specific demands are likely to emerge.

The transition from an agricultural system characterized largely by homogeneous (or bulk) commodities toward one in which differentiated products are the norm comes with some new and challenging policy questions for USDA and other federal agencies. Biotechnology is itself one of the most important sources of product differentiation, which adds another dimension of policy issues. We characterize those policy questions as falling into two groups: those related to facilitating markets for differentiated products and those related to managing consumer and environmental risks.

### Policy issues related to markets for differentiated products

Policy makers are now considering possible roles for the public sector in facilitating efficient markets for differentiated agricultural and food products. The issue is how producers are able to deliver the product attributes sought by consumers (and provide assurances regarding those attributes), and how consumers are able to assess the attributes of products (particularly where the attributes of interest are not visible) such that their purchases can match their preferences. Product differentiation is far from a new modern phenomenon (it is observed throughout the economy), but its proliferation in agricultural commodity markets is new and the private and public institutions to enhance the functioning of these markets are still developing.

There has been some concern that it will be extremely costly for US agriculture, which has been so efficient at delivering bulk commodities, to make a wholesale shift to a differentiated product system. Lin *et al.*, for example, describe and quantify the expenses through the marketing chain to form supply chains on a larger scale that keep the non-biotech product separate from undifferentiated grain (Lin *et al.*, 2000). Whether there is a public sector role in facilitating the development of such supply chains, and to what extent government should share in or redistribute the costs, is still an open issue (USDA Economic Research Service, 2000).

The ever-larger array of products entering the agricultural and food system multiplies the demands on the system for product information. The government is increasingly asked to think about its role, vis-a-vis that of private firms and also non-governmental 'third parties,' in verifying, standardizing, and providing consumer information regarding agricultural and food product characteristics. Firms have incentives to provide information about some food characteristics (particularly those they think will enable them to sell more products), which is known as 'voluntary labeling.' In other words, firms who see a market demand for non-biotech products will want to label their products as such in order to capture that market. Even with these incentives, there is a wide range of potential roles for 'third parties' (a category that includes but is not limited to government) in providing services such as standard setting, testing, certification, and enforcement in order to bolster the value of voluntary labeling (Golan *et al.*, 2000).

The US government has recently announced several steps designed to enhance the efficiency of the US marketing system in delivering both biotech and non-biotech products. In May 2000, USDA announced its intent to assess the need for a 'quality assurance program' for the production, handling, and processing of non-biotech crops. The program's purpose would be to facilitate marketing by providing an independent, third-party verification and certification process for differentiating and segregating conventional and biotech crops, thereby strengthening consumer confidence in food industry claims. USDA has also moved ahead to establish a reference laboratory that will evaluate and verify analytical procedures for detecting and quantifying bioengineered traits; evaluate the performance of detection methods; evaluate and accredit non-USDA testing labs; and establish sampling procedures for use in testing biotech grains and soybeans (USDA Grain Inspection Packers and Stockyards Administration, 2000). In these ways, the US government has decided it can enhance the functioning of voluntary labeling of non-biotech products. However, the appropriate roles for the government vis-a-vis the private sector in the provision of product information is likely to be a cause for international dialogue and debate for many years to come.

## Policy issues related to potential risks of biotechnology

The US has extensive regulatory infrastructure to assess and protect against the potential risks of introducing new crops into the fields and new foods into the food system.[3] Biotechnology poses special challenges to these systems, however,

---

[3] USDA's Animal and Plant Health Inspection Service (APHIS) regulates field testing of genetically engineered plants and certain microorganisms to determine potential risks to other plants and animals. The Food and Drug Administration (FDA) has responsibility for the safety and labeling of food and feed, excluding meat and poultry. And the Environmental Protection Agency (EPA) examines any products that can be classified as pesticides.

in part because it makes the potential number and variety of new agricultural and food products virtually limitless. The pressures this suggests, coupled with some public concerns, have spawned a number of government actions to ensure the adequacy of the regulatory systems. Are the current regulatory processes appropriate for biotech products and processes; are there important gaps in the science base for biotech risk assessment and if so how to fill them; how does the regulatory system incorporate inevitable uncertainties about long-run environmental and safety effects?

The US Food and Drug Administration (FDA) has found that substances introduced into foods through bioengineering are, in general, 'substantially the same' as other substances found in the diet and that they have therefore not needed pre-market approval. Nonetheless, a Presidential initiative announced in May 2000 would require companies to notify the FDA in advance of their intention to market new biotech-based products (a step that was previously voluntary) (USFDA, 2000). Advance notification is intended to give FDA sufficient time and means to identify any safety issues. Also because of the 'substantially the same' finding, FDA does not require labeling of bio-engineered food ingredients.[4] For example, if corn is listed as an ingredient on a food product, there is no requirement to list biotech corn separately. However, to facilitate some consumers' desire for information about the biotech status of the product, FDA will now develop guidance for voluntary labeling by food companies. The focus of the guidelines will be on whether genetic engineering was or was not used to develop a food or whether biotech ingredients were or were not added to a food product – there are no health or safety implications.

There are also new initiatives to ensure that the US regulatory programs are keeping abreast of the latest developments in science. A standing committee under the auspices of the National Academy of Sciences has been tasked with considering how the scientific evaluation of transgenic plants is working. It has also been asked to make recommendations on environmental risk assessment processes, gaps in the science base for risk assessment, and a system for environmental monitoring. Two White House offices (the Council on Environmental Quality and the Office of Science and Technology Policy) have also been asked to coordinate an interagency assessment designed to identify opportunities to strengthen the existing framework for regulation of biotechnology products. The US administration has also called for significant funding increases for competitively awarded university research to examine safety issues surrounding biotechnology and to support US regulatory decisions (Glickman, 2000).

---

[4] The FDA requires that food products carry the following information on their labels: common or usual name/statement of identity; quantity of contents; name and place of business of manufacturer, packer, or distributor; ingredient information; and nutrition information (Rodriguez, 1997). The FDA also requires labeling of food, including bio-engineered food, whose composition has been significantly altered, whose nutritional values or intended use is different from conventional food, or which has allergenic properties.

# Reconciling International Discords

Both markets and government policies contribute to reconciling international discords over the introduction of biotechnology. While the magnitudes differ, consumers exist in all countries who accept biotech crops along with those who are willing to pay higher prices for non-biotech crops. As farmers and food companies across countries respond to these demands with products differentiated by biotech status, the debate will no longer be cast with US interests aligned with biotech crops and EU interests focused on preventing biotech crops. To some extent this has already happened with US firms exporting non-biotech crops, a market for non-biotech crops developing in the US, and other countries (e.g., Argentina, Canada) producing and trading biotech crops. The new US regulatory initiatives described above work towards reconciling differences with Europe and other trading partners by supporting development of differentiated markets and reviewing scientific assessments of biotechnology.

The US and other countries have ongoing initiatives to reconcile international differences through development of international standards and sharing information on scientific processes. An example of an initiative to facilitate development of differentiated markets is participation in the CODEX Committee on Food Labeling's Biotechnology Working Group that has been meeting to develop recommendations.

The US participates in bilateral activities to improve mutual understanding of scientific aspects of the biotechnology approval processes. In areas where science provides clear guidelines, countries can move towards harmonization of regulations. For example, the molecular genetic characterization components of the review process have been harmonized, where possible, by USDA's Animal Plant Health Inspection Service (APHIS) and the counterpart Canadian agencies. Part of the process involved simultaneous reviews of transgenic plants prior to their commercialization. In addition, the regulatory agencies meet on an ongoing basis to exchange information and prioritize future areas of cooperation (USDA/APHIS, 2000). A similar process of exchange of scientific information and simultaneous review of applications is underway with the EU under the Trans-Atlantic Economic Partnership (TEP).

In areas of less consensus, international cooperation can develop a common base of facts on scientific information and existing regulations to support development of national policies. For example, the US participates with other countries at OECD on two committees that bring together national technical experts to synthesize the scientific information about food safety and risk assessment: the Working Group on Harmonization of Regulatory Oversight of Biotechnology and the Task Force for the Safety of Novel Foods and Feeds (OECD, 2000b). The Working Group has published eight science-based consensus documents on specific topics that can be used in the environmental risk assessment of genetically modified organisms. Other work at OECD to develop a common information base includes development of a database, BioTrak Online, with information on biotechnology regulations.

## Conclusions

As a result of limited global demand, trade impacts of the biotechnology debate have been small, but international tensions have nonetheless been high. Our chapter has identified two broad approaches to reduce international tensions, including those between the US and EU, over biotechnology. First, support development of markets so consumers with different preferences for biotechnology can make choices. Consumer choice can partially solve problems of scientific uncertainty where individuals can bring their individual beliefs to bear via choices in the supermarket. Second, develop information so that policy makers, NGOs, consumers, and producers have access to factual information on scientific issues and actual policies across countries. Where a common understanding of science exists, move towards mutual recognition or harmonization of policies.

In our view, a positive aspect of the biotech controversy is that the initiatives discussed above have the potential to bring new and fresh perspectives to the broader issue of the interface between science and regulatory processes and the strengths and weaknesses of that interface in today's world. This will, we hope, have positive spillovers for a number of other issues that have and are likely to plague agricultural and food trade relations. It is simply impossible for science to bring certainty to every regulatory decision, and the lack of certainty inevitably leads to the consideration of other factors, which may be based in cultural, ethical, economic, or social considerations. How different countries use science, incorporate scientific uncertainties and nonscientific considerations into their regulatory frameworks for biotechnology, and resolve the conflicts that stem from different approaches, may be critical to not just trade and international relations regarding trade issues, but also to the path of scientific innovations.

# 16

# The Economics of Agricultural Biotechnology: Differences and Similarities in the US and the EU

**Tassos Haniotis[1]**

*European Commission, Member of Cabinet of Commissioner Fischler*

## Introduction

In recent years, agricultural biotechnology has become one of the most difficult and contentious issues in transatlantic relations. While in the United States (US) the rapid expansion of the use of genetically modified organisms (GMOs) led to a significant share of crop area being allocated to genetically altered crops, in the European Union (EU) developments followed a different path. Increased concerns over a whole range of public health and safety issues relating to food resulted in a slower rate of GMO approvals, and an even lower rate of acceptance of the use of agricultural applications of biotechnology.

To date, the transatlantic debate on this issue is often perceived in terms of a general and vague *yes* or *no* position. Yet issues are more complex on both sides of the Atlantic, and understanding current differences appears to be a prerequisite for bridging the existing gap between opinions. This chapter

---

[1] The author was Agricultural Counselor, European Commission Delegation to the US when this article was written. Views expressed here are personal, and do not reflect an official position of the European Commission. This article reflects the situation as of July 2000. It expands the author's presentation at the GMO Conference organized by the EU Center of the University of Illinois in Chicago, IL, October 22–23, 1999.

Genetically Modified Organisms in Agriculture
ISBN 0-12-515422-4

attempts to shed some light on the underlying causes of EU and US differences in this area. Considering agricultural biotechnology as a further step in the never-ending process of innovations aimed at improving food supply, it will try to explain why the opportunities and risks of agricultural biotechnology weigh differently today in the EU and the US.

The main points are as follows. Supply factors, linked to the development of output-enhancing and/or cost-reducing GMOs, dominate agricultural applications in the US, and provide a strong push for their rapid development and commercial use, with emphasis on crops expected to exhibit strong export demand growth. Food demand factors, linked to food safety concerns, dominate the agricultural biotechnology debate in the EU, while the orientation of agricultural policy reform provides incentives for lower output, and emphasis on quality aspects, both of products and of production methods.

Although these differences may result in different regulatory approaches, this should not be considered a reflection of differences in the underlying objectives of both sides, which are fundamentally the same. Rather, different approaches are a reaction to different market forces and indicate how wider issues of preference affect attitudes towards risk, and could result in different levels of risk aversion and consumer protection.

The differences in GMO acceptance are also a reflection of the fact that advances in science outpace regulatory changes in this field, and exert unique pressures for harmonized adjustments worldwide.

## Agricultural Biotechnology – the US Case

The presentation of an exclusive list of economic factors determining developments on the US side is beyond the scope of this chapter and the ranking of factors presented below is random and not indicative of relative importance. However, the leading US position in agricultural applications of biotechnology (demonstrated in terms of both product approval and commercialization) can be attributed to the following factors.

Concerns similar to those expressed today in Europe about the potential risks of GMOs belong to an earlier phase in the development of the industry in the US, which is currently characterized by a widespread acceptance of agricultural biotechnology among producers, consumers, and policy makers. Aided by the above consensus, the US regulatory framework was adapted to meet and facilitate the emerging needs of the agricultural biotechnology industry. (The fact that awareness about GMO applications in agriculture has been growing in the US in recent months does not alter significantly the above assertion, although it may influence developments in the future.)

The Federal Agricultural Improvement and Reform (FAIR) Act of 1996 fundamentally changed US farm policies for major crops by introducing production flexibility that encouraged area shifts towards crops where yield-improving GMOs were entering commercial use (mainly corn and soybeans). Strong export orientation, already a major feature of US crop production, received an additional boost from the fact that exports were until recently considered the 'ultimate safety net' for US farmers, thus providing further incentives for crop improvements to increase competitiveness.

In addition, strong growth in world food demand was expected to benefit either the same crops where biotechnology has made major advances (again, corn and soybeans) or products where these crops are used as main inputs (pork and poultry). These are certainly not the only economic factors that contributed to the growth of agricultural biotechnology in the US. In fact, for some of the above factors, such as the prospects for growth in US exports, reservations and concerns about their impact were already being expressed in the aftermath of the financial crisis in Southeast Asia.

Other factors tend to present a more mixed picture, and will probably increase in significance in the years to come. These factors include issues related to intellectual property rights, to the relationship between public and private research, to the rate of new technology adoption, to technology transfer, to concentration in the biotech industry, or to potential risks as commercial applications spread.

However, the point that is stressed here is that during this brief period of the past few years, when EU–US agricultural trade relations were – and continue to be – repeatedly tested in this field, the US found itself in a unique situation. A series of factors combined in providing a strong supply push that encouraged the development of agricultural biotechnology in the US.

## Agricultural Biotechnology – the EU Case

Developments in the European Union followed a different path. Both in the area of farm policy and in the area of consumer acceptance, things tended to move in the opposite direction. These developments can be summarized as follows.

EU farm policies went through a major overhaul in 1992, and then again in 1999. On both occasions, the objective of the reform of the Common Agricultural Policy (CAP) was to move EU agriculture towards more market orientation. Given the previously high level of domestic support, this could only be done by providing incentives for a decrease in output (and, consequently, for less dependence on surplus disposal through export subsidies). Therefore, on both production and exports, EU emphasis is not on increased quantity, but on enhanced quality.

But the most important development affecting attitudes towards the acceptance of GMOs in the EU was not coming from the supply, but from the demand side. It stemmed from the BSE crisis ('mad cow' disease), and affected consumer attitudes in the EU in a whole series of areas related to food safety. Not only was this crisis considered by EU consumers to be a failure of the regulatory system. More importantly, this crisis was also considered a failure of science, since the risks of transmission across species were not identified earlier.

As a result, concerns that are natural among consumers in the early stages of new technology applications, in terms of both food safety and environmental impacts, were exacerbated in the EU in the aftermath of the BSE crisis. To respond to these concerns, which extend to areas as diverse as the use of antibiotics in animal feed and the use of hormones as growth promoters in animals, a new approach to food safety regulations was introduced in the EU. This resulted in a complete overhaul of the food safety regulatory system both in the area of scientific evaluation and in that of consumer protection.

Contrary to what is customarily believed, a detailed analysis of developments in EU food safety regulations indicates a clear move towards a more transparent and independent science-based decision-making process. At the same time, however, there has been a clear trend (justifiable under the circumstances) to increase the overall level of consumer protection, defined as a reply to consumer preference for a high level of certainty about food safety factors.

## US Supply and EU Demand – Searching for a Difficult Equilibrium

In a sense, EU–US differences in agricultural biotechnology are another reflection of the increasing interdependence of economies on a world scale. In this particular field, this interdependence has taken the rather peculiar form of a supply–demand interaction crossing borders (literally, oceans). While the strongest supply factor is undoubtedly in the US, the strongest demand factor lies with the EU. But the balance of market forces in this area still seems elusive.

Normally, the supply curve of a particular GMO crop is expected to shift to the right because of an increase in yield and/or a decrease in costs. In the absence of changes in demand, consumers can then reap the benefit of a lower price, even if no other demonstrable benefits materialize. Yet the demand for products derived from GMO crops shifted to the left in the EU as a result of many factors, including non-GMO factors, that raised concerns about the usefulness and safety of GMO products.

This resistance of EU consumers to accept GMO applications in agriculture was probably best reflected by moves of the retail industry to procure its products from non-GMO sources. This was often depicted on the US side of the Atlantic as an attempt by the EU to impede trade by erecting artificial barriers in the form of various regulatory measures that are not justified by scientific evidence. It is true that quantities of certain commodities with GMO applications that are exported by the US are significant, and delays in the EU approval system have had trade impacts (e.g., on US corn exports to the EU). Concerns about trade implications are thus natural. But it is also true that trade in commodities where approval was granted on both sides of the Atlantic (e.g., soybeans) is continuing.

Given that the move of EU retailers away from GMOs was market-driven, responding to specific consumer demands, it is essential for the comprehension of GMO attitudes in the EU to look to the overall approach on biotechnology, as reflected in Table 16.1. This table summarizes attitudes of EU consumers towards biotech applications as these were measured in the Fall of 1999. It includes four categories of questions that are comparable with similar questions asked in a previous survey in the Fall of 1996.

As is evident from the table, EU consumer attitudes towards biotechnology are not uniform. Acceptance of applications of biotechnology increases as we move away from food towards medical applications. But figures in this table reflect a significant decline of positive responses with respect to responses in the 1996 survey. It is against this background that differences in regulatory approaches between the EU and the US need to be considered. Therefore, some thoughts, from a European perspective, on EU regulatory initiatives seem pertinent.

**Table 16.1**: EU attitudes towards GMO applications in 1996 and 1999 (percentage of positive response)

| GMO application | Useful | | Risky | | Ethical | | Should be encouraged | |
|---|---|---|---|---|---|---|---|---|
| | *1996* | *1999* | *1996* | *1999* | *1996* | *1999* | *1996* | *1999* |
| Use of biotech in the production of food | 54 | 43 | 61 | 59 | 50 | 37 | 44 | 31 |
| Plant-to-plant gene transfer to increase pest resistance | 69 | 55 | 48 | 49 | 62 | 47 | 58 | 42 |
| Human genes into bacteria for vaccines | 80 | 68 | 47 | 44 | 70 | 57 | 71 | 57 |
| Genetic testing to detect hereditary diseases | 83 | 72 | 40 | 38 | 74 | 63 | 75 | 63 |

*Source*: 'The Europeans and Biotechnology', *Eurobarometer* 52.1. Report by INRA (Europe)–ECOSA on behalf of the Directorate General for Research of the European Commission, March 15, 2000.

*Note*: Percentages reflect combined results of 'mostly or totally agree' responses.

## EU Regulatory Issues

There is a strong belief in the EU that a technology should not be used just because it is available. Under this critical approach to the advance of science, consumers will judge each case on its merits (as they have done so in the case of medical applications of biotechnology in the EU, both widely used and widely accepted). But in order to make an informed judgement they must be aware of the possible risks and benefits of such a technology.

The policy question in the EU, therefore, is not whether or not to use biotechnology in agriculture. It is rather how to use this technology in a safe and fruitful way that will be accepted by the consumer. Guaranteeing the safety and quality of food is, therefore, fundamental and a top priority for industry, producers and traders, and for policy makers.

This is also true for the Commission proposals on labeling, which is considered by EU consumers as a fundamental right to informed choice. The best argument in support of this claim comes from the fact that those that have a direct interest in promoting the very products of agricultural biotechnology, food retailers and processors and the biotech industry in the EU, are also supporting labeling for GMOs.

Thus meeting consumer concerns with the introduction of mandatory labeling in an environment of widespread reluctance to accept a new technology is not an attempt to impede trade. On the contrary, it is the only way to enhance trade in the long term, since it is a precondition for an increase in overall demand for a certain group of commodities incorporating this new technology. Producers of novel foods and other genetically modified products will have to adapt to consumer concerns if they want their products sold.

The question that then arises is whether this need for adjustment to consumer requirements constitutes a threat to the biotechnology industry. This is a pertinent question, since it is not only the consumer who presses for reliable legislation. The biotechnology industry also needs a clear regulatory framework that is an essential condition for further investment. In that respect, both the EU authorization procedure and the Commission labeling proposals have received strong criticism, and have been at the heart of the EU–US differences in this area.

With respect to GMO approval procedures, the revision of the EU's approval process under Directive 90/220 is in its final stages of conciliation between the Council/Commission proposal and the European Parliament's amendments. Given the narrowing of existing differences, the final outcome (whatever this eventually will be), will reflect the strong EU orientation towards a system of approval that increases efficiency, effectiveness, and transparency.

Does this imply then that such a system introduces an approval process that is essentially political, and not based on science? This argument, again stressed on the US side, seems to oversimplify the role of science in the decision-making process on issues related to food safety. Throughout the world scientists do not take decisions; this is simply not their role. They do, however, make a very significant contribution to the policy-making process, with the final word left to those whose role is exactly that – to 'make' policy.

In this process, *weighing risk factors on the basis of scientific evaluation is not done in a vacuum, but in the concrete environment of societies with specific preferences, and differences among societies are sometimes also reflected in different choices on the basis of the same scientific evidence*. This is not a betrayal of science. It is, on the contrary, an indication of respect for what science really is, the provision of knowledge with a 'confidence interval', that is with a degree of uncertainty. Treating science as a provider of the absolute truth (in essence, as a religion) is stretching its limits to an extent that undermines its essential role in policy making.

If this point of view is taken into consideration, then regulatory steps taken in the EU will start being viewed as what they really are. Not as an attempt to impede the progress of agricultural biotechnology, but as a necessary stage to allow the full utilization of its potential in a difficult environment.

## Concluding Comments

The thrust of this attempt to explain differences in EU and US approaches to biotechnology has been that the supply of agricultural biotechnology in the US has been assisted by a combination of factors that all contributed towards the fast development of agricultural applications of GMOs. The change in farm policies encouraged production flexibility and export, while widespread approval among producers, consumers, and policy makers alike allowed the development of a regulatory framework that relaxed previous constraints.

In the EU, on the other hand, farm policy changes pointed towards the need for output controls, and did not provide incentives for commercial applications of GMOs in agriculture. In addition, EU policies were dominated by a complete rethinking of food safety policy that resulted in regulatory changes whose basic aim was to build up consumer confidence in food safety.

The above difference in priorities was compounded by the fact that US supply and EU demand for GMOs interacted in the context of regulatory processes that, although aiming at the same objectives, applied different approaches to meet these objectives.

Yet as we move ahead, I believe that focus will shift away from the current, often simplistic emphasis on presenting differences as an indication of potential conflict, towards a situation where we could look into these differences as sources of useful insight into how agricultural biotechnology can best advance by increasing its potential for benefits while at the same time safeguarding against potential risks. On both sides of the Atlantic, it can be expected that the debate will gradually center on the full scale of actual benefits and risks.

As the commercialization of GMO applications advance in the US, their supply impact will be judged more accurately on the basis of real performance. And as we move towards adapting our regulatory framework into developments in the EU, benefits and costs of agricultural biotechnology will also have to be judged on a more concrete basis. Respecting and understanding present differences appears to be the only way to arrive at an equilibrium that allows the full potential of agricultural biotechnology to be utilized, and the (often neglected) common objectives to come to the forefront.

# 17

# Modern Agricultural Biotechnology and Developing Country Food Security

**Per Pinstrup-Andersen**
*IFPRI, Washington DC, USA*

**Marc J. Cohen**
*IFPRI, Washington DC, USA[1]*

Fierce debate swirls around the application of modern biotechnology to food and agriculture in industrialized countries. Far less attention has focused on the potential contributions of biotechnology to poverty alleviation, enhanced food security and nutrition, or sound natural resources management in developing countries. This chapter offers input into such discussion, which should be led by the people of the developing countries.

## State of World Food Insecurity

One of every five people (1.2 billion in all) lives in absolute poverty, on the equivalent of $1 a day or less (World Bank, 1999). Poor people in developing

---

[1] Per Pinstrup-Andersen is Director General and Marc Cohen is Special Assistant to the Director General at the International Food Policy Research Institute (IFPRI), Washington DC. IFPRI is one of the 16 Future Harvest international agricultural research centers supported by the Consultative Group on International Agricultural Research.

---

Genetically Modified Organisms in Agriculture
ISBN 0-12-515422-4

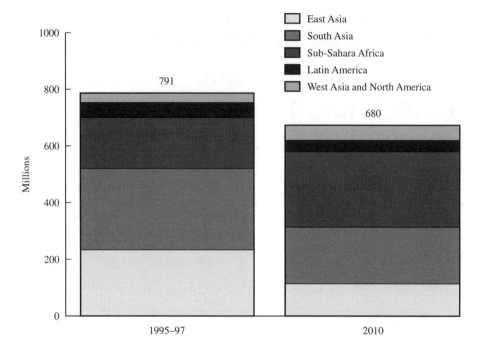

**Figure 17.1**: Number of food-insecure people, 1995–1997 and 2010

*Source*: FAO (1996a, 1999).

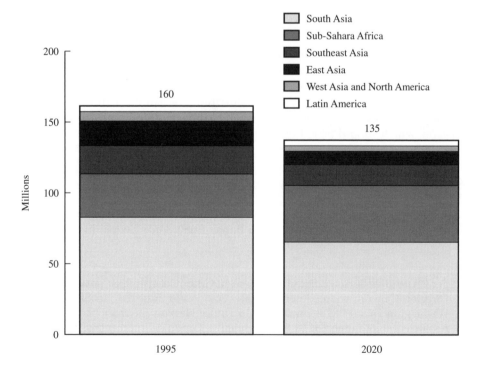

**Figure 17.2**: Number of malnourished children 1995 and 2020

*Source*: Pinstrup-Andersen *et al.* (1999).

countries frequently cannot afford all the food they need, although they spend up to 50–70% of their income trying (Deaton, 1997), and generally lack access to land to produce their food. Nearly 800 million people in the developing world are food insecure, with average calorie intakes below minimum requirements (FAO, 1999). Real people with names and faces stand behind hunger statistics: people such as Kone Figue, who weeds and harvests her small Côte d'Ivoire rice farm by hand, and seldom produces enough to feed her family of eight for a whole year (Schioler, 1998). Without significant changes in national and international policies, by 2010, 680 million people will remain chronically undernourished, and the 1996 World Food Summit goal, cutting the food insecure population by half by 2015, will remain unrealized (Figure 17.1) (FAO, 1999).

Of particular concern are the 150 million malnourished preschool children in developing countries (Figure 17.2) (ACC/SCN and IFPRI, 2000). Malnutrition is a factor in 5 million child deaths annually, and those who survive face impaired physical and mental development (WHO, 1999). Food insecurity robs humanity of countless scientists, artists, community and national leaders, and productive workers.

Micronutrient malnutrition is widespread in developing countries. Iron deficiency anemia affects about a billion people, including 56% of pregnant women, 42% of preschoolers, and 53% of school-aged children (Figure 17.3). It increases maternal and newborn mortality, and impairs child health and development. Anemia reduces national income by 1% annually in many developing countries, but few public health programs address this scourge (ACC/SCN, 1997; ACC/SCN and IFPRI, 2000; WHO, 1999).

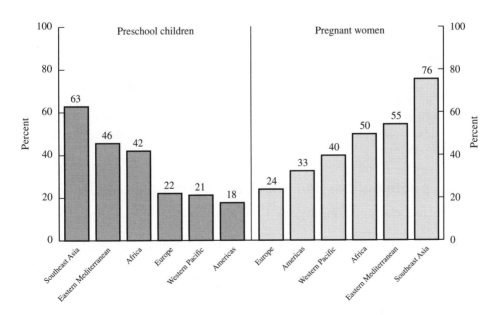

**Figure 17.3**: Prevalence of anemia in preschool children and pregnant women by region, 1999.

*Source*: ACC/SCN and IFPRI (2000).

Insufficient vitamin A intake among developing country children causes blindness and contributes to infections and deaths. Vitamin A deficient pregnant women face increased risk of mortality and HIV transmission to children. Responses include food supplementation and fortification, promotion of increased green vegetable consumption, and breeding of provitamin A rich rice and cassava (WHO, 1999; ACC/SCN and IFPRI, 2000). These approaches are complementary, not either–or choices.

## Agricultural Development Crucial for Food Security

Seventy percent of poor and food insecure people reside in rural areas, depending directly or indirectly on agriculture for their livelihoods (McCalla and Ayres, 1997). Small farmers in developing countries face many problems. Pre- and post-harvest crop losses from pests and weather result in low and fluctuating yields, incomes, and food availability, and reduce potential output value by up to 50% (Oerke *et al.*, 1994). Low soil fertility and lack of access to plant nutrients, along with acid, salinated, and waterlogged soils also contribute to low yields, production risks, and natural resource degradation. Sixteen percent of global agricultural land is seriously degraded, including 75% in Central America and 20% in Africa (IFPRI, 2000). Poor farmers often must clear forests or farm marginal land (Scherr, 1999). Inadequate infrastructure, land tenure biased against poor people, poorly functioning markets, and lack of access to credit and technical assistance are further impediments.

Low agricultural productivity results in high unit costs of food, poverty, food insecurity, poor nutrition, low farmer and farm worker incomes, little demand for goods and services produced by poor nonagricultural rural households, and urban unemployment and underemployment. Agricultural growth catalyzes broad based development. Agriculture's linkages to the nonfarm economy in poor countries generate employment, income, and growth economy-wide. In sub-Saharan Africa, each dollar of new farm income increases total household income by up to $2.57 (Delgado *et al.*, 1998).

Agricultural growth and development also help meet growing food needs driven by population growth and urbanization. World population will rise 25% to 7.5 billion in 2020 with 97% of growth in developing countries. By 2020, 52% of the developing world's people will live in cities (United Nations Population Division, 1998).

### Appropriate policies and public investment are critical

Poverty reducing agricultural growth requires accelerated public investment in:

- environmentally friendly, yield increasing crop varieties and improved livestock;
- access to resources – land, water, credit, and inputs – for small farmers;
- extension services and technical assistance;
- improved infrastructure and effective markets;
- programs for women farmers, who grow much of the locally produced food in developing countries; and

- basic education and health care, clean water, safe sanitation, and good nutrition for all.

These investments must be supported by good governance – rule of law, transparency, sound public administration, and respect for human rights – as well as trade, macroeconomic, and sectoral policies that do not discriminate against agriculture. Policies must also provide incentives for sustainable natural resource management, such as secure property rights for small farmers.

Development efforts must engage low-income people as active participants, not passive recipients. To assure responsive policies, poor people need accountable organizations that articulate their interests (Cohen, 1994).

Current public investment patterns in developing countries must change. On average, they devote 7.5% of government expenditures to agriculture, and the figure is even lower in sub-Saharan Africa, where agriculture's contribution to gross domestic output is 30–80% (FAO, 1996b, 1999, 2000). Donors cut aid to agriculture and rural development by nearly 50% in real terms over 1986–1997. Donors' two decade emphasis on reducing government's economic role has contributed to public agricultural disinvestment (FAO, 1996b, 1998, 1999; Michel, 1999).

## Agricultural research and food security

Public investment in agricultural research that can improve small farmers' productivity in developing countries is crucial. It must combine all appropriate scientific tools with better utilization of indigenous knowledge. For example, in India, efforts to intensify hill tribe agriculture and boost incomes combine traditional soil and water conservation techniques with cultivation of new, high value cash crops (Hazell *et al.*, 2000).

The private sector is unlikely to undertake much research needed by small farmers in developing countries because expected gains obtainable by the private sector will not cover costs. However, gains to society and to poor people are high. Social rates of return to agricultural research investment exceed 20% per year, compared to long-run real interest rates of 3–5% for government borrowing (Alston *et al.*, 2000; Rosegrant *et al.*, 1995). Yet low income developing countries invest less than 0.5% of the value of farm production in agricultural research, compared to 2% in higher income countries (Pardey and Alston, 1996).

There are several ways to expand small farmer oriented research. Governments can increase public sector research and convert some social benefits to private gains, e.g., by purchasing exclusive rights to new technology and providing it to small farmers (Anonymous, 1999d). Other arrangements include public–private joint ventures, research foundations, competitive funds, and production levies.

Research and technological change will only support sustainable poverty alleviation if policies are appropriate. For example, in Tamil Nadu State in India, adoption of high yielding Green Revolution grain varieties meant not only increased yields and cheaper, more abundant food, but income gains for small and large farmers alike, as well as nonfarm poor rural households. Higher incomes improved nutrition. State antipoverty policy included social welfare programs; investment in human capital, agriculture, and rural development; and measures to assure equitable access to resources.

Where increased inequality followed adoption of modern varieties, this resulted not from factors inherent in the technology, but from policies that did not focus on equity and human capital development. And even in these areas, rural landless laborers usually found job opportunities due to increased agricultural productivity, particularly if physical infrastructure and markets were well developed (Hazell and Ramasamy, 1991).

Successful adoption of Green Revolution varieties depended on access to water, fertilizer, and pesticides. Inequality between well-endowed and resource-poor areas increased. Excessive or improper use of inputs sometimes harmed the environment. Productivity increases offset this somewhat by preserving forests and marginal lands (Lipton with Longhurst, 1989).

## Agricultural Biotechnology and Food Security

Very little transgenic seed material has been grown in the developing world. Argentina, Brazil, China, Egypt, India, and South Africa account for most current research. *Ex post* assessment of risks and benefits, and of biotechnology's food security relevance, is virtually impossible. While a great deal is known about conventional plant breeding's risks and benefits, some modern agricultural biotechnology differs significantly from conventional breeding, i.e., where it involves transfers of one or more genes between species. While all breeding arguably involves 'genetic modification,' conventional breeding crosses varieties *within* species.

### Privatization of research

The public sector has traditionally carried out conventional crop and animal research, especially in developing countries. In contrast, private firms undertake most agricultural biotechnology research, and seek intellectual property rights protection over results. Farmers cannot legally plant or sell for planting the crop produced from patented seeds without permission. The so-called 'terminator' gene could be used as an intellectual property protection technology. Seeds containing it produce plants with sterile seeds. It is inappropriate for small farmers in developing countries because existing infrastructure and production processes may be unable to segregate fertile and infertile seeds. Small farmers could face severe consequences from inadvertently planting infertile seeds. The Consultative Group on International Agricultural Research (CGIAR), which sponsors the 16 Future Harvest international agricultural research centers, has rejected use of such technology.

The private sector is patenting biotechnology research processes as well as products, whereas conventional breeding technology lies in the public domain. Limiting access to basic knowledge and techniques constrains public agencies' research relevant to poor farmers and consumers. At the same time, public agencies, such as the Future Harvest centers, face pressure to protect their own intellectual property, notably their extensive germplasm collections, at least for defensive purposes.

So far, Future Harvest centers have acquired access to techniques, equipment, materials, and information subject to intellectual property rights

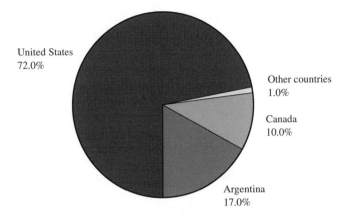

**Figure 17.4**: Area with GM crops by country, 1999.

*Source*: James (1999) with permission.

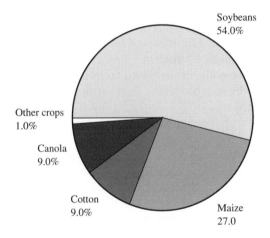

**Figure 17.5**: Area with GM crops by crop, 1999.

*Source*: James (1999) with permission.

protection through licenses, formal agreements, and informal arrangements. Some proprietary technology owners permit research but not distribution of resulting products. As small, noncommercial institutions, centers cannot negotiate based on expected profits.

Patents are country specific, and most related to agricultural research are held in industrialized countries. If patents are not held in a developing country, public agencies there may use processes and traits in research, adaptation, and release to farmers. The country cannot, however, export commodities produced from patented seed to countries where patents are held. Since patenting is costly, corporations will probably seek them only where they expect sizeable demand for products or access to techniques.

Some firms have transferred proprietary technologies to developing countries without charge. Monsanto provided virus resistance technology to Kenyan and Mexican government agricultural research institutes and placed its rice genome map in the public domain. But such arrangements are exceptional (Monsanto, 2000; Qaim, 1999b; Serageldin, 1999).

So far, modern agricultural biotechnology has focused overwhelmingly on production on developed country farms and for developed country markets. In 1999, 82% of genetically modified (GM) crop area was in developed countries, with 72% in the United States alone (Figure 17.4). Herbicide-tolerant soybeans and maize and cotton containing insect resistance and/or herbicide tolerance genes comprised over 90% of plantings (Figure 17.5) (James, 1999).

Little private sector biotechnology research has focused on developing country food crops other than rice, maize, and cassava. Little adaptation of research to developing country crops and conditions has occurred through the public and philanthropic channels prominent in conventional breeding. In 1998, the Rockefeller Foundation granted $7.4 million for biotechnology research relevant to developing nations, mainly through international agricultural research centers and developing country public research institutes. This sum pales by comparison with Monsanto's 1998 $1.3 billion research and development budget, much of which funded agricultural biotechnology research (Monsanto, 1999; Rockefeller Foundation, 1999b).

Direct transfers of currently available biotechnology research results to the developing world may be inappropriate. Poor farmers may not be able to afford herbicides. Appropriate research would employ biotechnology and conventional breeding to develop weed resistance instead of herbicide resistance. The West African Rice Development Association (WARDA), a Future Harvest center in Côte d'Ivoire, used tissue culture to cross African and Asian rice, producing a leafy plant that denies weeds sunlight. It is high yielding and reduces women's weeding time, allowing more attention to caring practices essential for child nutrition (WARDA, 1999).

Insect-resistant crops could benefit poor farmers. Presently, however, crops containing *Bacillus thuringiensis* (*Bt*) bacterium genes, which produce a natural insecticide, are developed for North America. They require knowledge intensive cultivation, such as planting non-*Bt* refuges to manage resistance in pests.

## Weighing benefits and risks

Consumers outnumber farmers by a factor of more than 20 in the European Union, and Europeans spend less than 20% of income on food. US farms account for less than 2% of households, and the average US consumer spends under 12% of income on food (US Bureau of Labor Statistics, 1999; US Census Bureau, 1999; US National Agricultural Statistics Service, 1999). Developed country consumers can afford to pay more for food and subsidize domestic agriculture. In contrast, poor developing country consumers depend on agriculture for a living and spend most of their income on food.

Modern biotechnology is not a silver bullet for achieving food security, but, used in conjunction with traditional knowledge and conventional agricultural research, it may offer a powerful poverty reduction tool. It may contribute to increased productivity, lower unit costs and prices for food, preservation of bio-diversity, and improved nutrition. This depends on the relevance of the research to poor farmers and consumers, the economic and social policy environment, and intellectual property rights arrangements.

Biotechnology research aimed at reducing input requirements and increasing input use efficiency could lead to crops that utilize water more efficiently, fix nitrogen from air, extract phosphate from soil more effectively, and resist pests without synthetic pesticides. Reduced dependence on purchased inputs such as pesticides would make technology more accessible to poor farmers, with environmental and health payoffs.

Biotechnology may also help achieve productivity gains needed to feed a growing global population and preserve wild or fragile land from cultivation. Biotechnology may heighten crops' tolerance to adverse weather and soil conditions, enhance crops' durability during harvesting or shipping, and aid development of drought-tolerant maize, insect-resistant cassava, micronutrient-dense staples, high protein foods, and edible vaccines, to the benefit of small farmers and poor consumers (Anonymous, 2000; Pinstrup-Andersen and Cohen, 2000a).

## Future Harvest Biotechnology Activities[2]

Most Future Harvest centers engage in biotechnology research. The largest program, at the International Livestock Research Institute in Kenya, includes work on characterization, conservation, and use of genetic resources; development of disease-resistant livestock; and vaccine development. It accounts for 23% of the center's budget. Other centers use biotechnology in breeding programs for poor people's staple food crops. The share of center budgets devoted to such research ranges from 1.5% at WARDA to 12% at the International Rice Research Institute (IRRI) in the Philippines. The centers' combined investment in biotechnology runs at $25 million annually, or 7.7% of overall resources.

The centers' approach is pragmatic, using molecular biology based techniques where they offer time, cost, or reliability advantages over conventional methods for germplasm improvement. For example, IRRI has used biotechnology to develop salt-tolerant and bacterial blight-resistant rice. Centers have generally focused on applied research, but the International Maize and Wheat Improvement Center in Mexico conducts basic research on apomixis, i.e., developing hybrids capable of asexual reproduction. This could allow farmers to reuse high yielding seed without yield reduction, reducing dependence on purchased seed.

In addition to genetic engineering, Future Harvest centers carry out tissue culture, molecular marker assisted studies, and gene discovery. Centers also train developing country scientists in research techniques, designing and implementing biosafety regulations, laboratory safety, and field testing transgenic crops.

---

[2] See Pinstrup-Andersen and Cohen (2000b) and Morris and Hoisington (1999) for additional details.

## Potential Risks

Genetic engineering is quite new compared to classical plant breeding. Potential risks need careful assessment. If developing countries are to utilize genetic engineering effectively, they will need to have in place an appropriate regulatory framework to assure food safety, biosafety, and intellectual property rights enforcement. This will require financial and technical assistance.

### Food safety and biosafety

Food safety and biosafety regulations should reflect international agreements and a society's acceptable risk levels, including the risks associated with not using biotechnology. GM foods are not intrinsically good or bad for human health. Health effects depend on content. Iron rich GM foods are likely to benefit iron deficient consumers. But transferring genes between species may also transfer allergens, so GM foods must be tested before commercialization. Such testing prevented commercialization of soybeans with a Brazil nut gene. GM foods with possible allergy risks should be labeled. The European Union requires removal of antibiotic resistance marker genes used in research before commercialization of GM food due to potential although unproven health risks.

Labeling may also be needed to identify content for cultural and religious reasons or because consumers want to know. While the public sector must design and enforce public health and labeling standards, informational labeling related to research and production processes might be left to the private sector.

Effective biosafety regulations must be in place before genetically engineered seeds are introduced (Juma and Gupta, 1999). Policy makers need to assess ecological risks, including the spread of traits such as herbicide resistance from genetically modified plants to unmodified plants (including weeds), resistance in target pests, harm to nontarget species, and the potential threat to biodiversity of monoculture of bioengineered crops. These risks are particularly significant in centers of origin and diversity of major food crops.

### Socioeconomic risks

Introduction of agricultural biotechnology could lead to increased inequality, as larger farmers capture benefits through early adoption (Leisinger, 1999). Appropriate policies can reduce this risk. For example, an *ex ante* assessment of Monsanto's transfer of transgenic virus resistance technology to Mexico found that all classes of potato growers would likely benefit. Government programs to distribute transgenic potatoes to small and medium sized farmers would significantly increase their share of benefits. Higher output and lower prices would benefit consumers (Qaim, 1999).

A handful of transnational 'life sciences' companies dominate agricultural biotechnology research and development. Rapid consolidation resulted in 25 major acquisitions and alliances worth $15 billion over 1996–1998. This may engender reduced competition, oligopoly profits, exploitation of small farmers and consumers, and extraction of government favors. Effective antitrust regulations are needed, particularly in small developing countries where one or only a few seed companies operate. Global competition standards are also required; regulatory capacity has not kept pace with economic globalization.

Developing countries must protect intellectual property rights in order to benefit from biotechnology. Legislation must harmonize protection of farmers' rights to save, exchange, and improve seed and plant breeders' rights to benefit from innovations.

## Trade related risks

If the 'precautionary principle' is accepted as the basis for sanitary and phytosanitary standards and technical barriers to trade, then the EU could discriminate against GM imports without scientific evidence of harm. Low income developing countries wishing to pursue agricultural export led growth would have to choose between adopting agricultural biotechnology or producing GM-free food for export to the EU. They could differentiate and label GM foods and non-GM foods, and if they can manage such a system, they would be able to use biotechnology for domestic consumption while exporting GM-free foods. A large share of developing country food imports originates in the United States, so importing countries must decide whether to demand differentiation and labeling of imported food as well. Development of GM substitutes for developing country export crops, such as low fat canola (rapeseed) oil as a tropical oil substitute, could harm developing country farmers.

## Conclusion

Modern agricultural biotechnology, including genetic engineering, can contribute to food security in developing countries as part of a comprehensive sustainable poverty alleviation strategy focused on broad based agricultural growth. Biotechnology is not, as some critics have charged, 'a solution looking for a problem.' The problems are genuine and momentous.

Increased public sector research is essential for assuring that molecular biology based science serves the needs of poor people, as is increased public–private cooperation. Poor people must be included in debate and decision making about technological change. It is also essential to strengthen regulatory capacity within developing countries.

The biggest risk is that technological development will bypass poor people in a form of 'scientific *apartheid*,' focusing exclusively on industrialized countries and large-scale farming (Serageldin, 1999). If opposition in developed countries leads to moratoria or outright bans on agricultural biotechnology research, developing countries will be unlikely to receive scientific or financial support for their research. This would preclude most research except in a few large developing countries, such as China, India, South Africa, and Brazil, and negatively affect opportunities to reduce poverty, food insecurity, child malnutrition, and natural resource degradation.

# 18

# GMOs: A Miracle?[1]

**Vandana Shiva**
*Research Foundation for Science, Technology and Ecology, New Delhi, India*

Genetically engineered crops and foods were to have been the miracle solution to world hunger and the ecological crisis created by industrial agriculture. However, GMOs in agriculture create new hunger and new ecological risks.

## The Myth of Saving the Planet

As the Green Revolution miracle fades out as an ecological disaster, the biotechnology revolution is being heralded as an ecological miracle for agriculture. It is being offered as a chemical-free, hazard-free solution to the ecological problems created by chemically intensive farming. The past forty years of chemicalization of agriculture has led to severe environmental threats to plant, animal and human life. In the popular mind 'chemical' has come to be associated with 'ecologically hazardous'. The ecologically safe alternatives have been commonly labeled as 'biological.' Biotechnology has benefited from its falling under the 'biological' category which has connotations of being ecologically safe. The biotech industry has described its agricultural innovations as 'Ecology Plus.'

It is, however, more fruitful to contrast the ecological with the engineering paradigm, and to locate biotechnology in the latter. The engineering paradigm offers technological fixes to complex problems, and by ignoring their complexity, generates new ecological problems which are later defined away as 'unanticipated side effects' and 'negative externalities.' Within the engineering ethos it is impossible to anticipate and predict the ecological breakdown that an engineering intervention can cause. Engineering solutions are blind to their own

---

[1] Originally published in the online journal 'Slow: the international herald of tastes', Issue no. 17, April–June 2000. Available on the web at *http://www.slowfood.com/publications/slow17/slow17.html*. Reprinted with permission.

Genetically Modified Organisms in Agriculture
ISBN 0-12-515422-4

impacts. Biotech, as biological engineering, cannot provide the framework for assessment of its ecological impact on agriculture.

Genetic engineering is creating the risks of a new form of pollution or 'contamination' called genetic pollution or biopollution. In certain cases, biopollution can have major health and environmental impacts and create bio-hazards. Introduction of new species into ecosystems had led to the phenomenon of bioinvasion, which is one form of biohazard. Introduction of exotic genes into crops can have unpredictable ecological impacts. Some organisms can be pushed to extinction by crops that release toxins. Other organisms can become invasive species, dominating ecosystems and displacing native biodiversity. Genetic engineering introduces new ecological risks in the form of biopollution or biohazards. Unlike toxic hazards, biohazards multiply and have no recall.

There are multiple ways in which GMOs will increase the ecological vulnerability of agriculture. Firstly, GM crops will increase rather than decrease chemical use. Chemical corporations pushing genetic engineering have used various strategies to make the public believe that biotech in agriculture implies the end of chemical hazards and that biotechnology 'protects the planet.'[2] However, there are four reasons why biotechnology will lead to an increase in chemical use in agriculture.

1.  The predominant agricultural application of genetic engineering is herbicide resistance. This will increase rather than reduce herbicide use.
2.  The use of chemicals will spread to new regions of the world which were free of intensive chemical use in agriculture.
3.  The applications such as *Bt* resistance which are supposed to get rid of pesticides can actually lead to increased pesticide use through build-up of *Bt*-resistance and the destruction of ecological alternatives for pest control.
4.  The engineering of a toxin into the plant itself can have the potential impact of increasing toxins in the plant and the ecosystem.

The Roundup Ready Soybean (RRS) is the most widespread genetically engineered crop introduced so far. The strategy for RRS is basically to sell more Roundup. Monsanto's Roundup accounts for 95% of all glyphosate sales. Glyphosate is the world's best selling herbicide. The worldwide sales of glyphosate products are currently worth approximately $1200 million annually and represented about 60% of global non-selective herbicide sales in 1994.

Roundup sales are nearly $1 billion. The patent for Roundup expired in 2000. Roundup Ready crops are a strategy for Monsanto to protect its sales of the chemical Roundup through its life sciences division, which is separated from the chemical division as a PR exercise, but which is still involved in pushing chemicals.

The second strategy used by Monsanto is to spread Roundup in countries where it was not used before. Monsanto has greatly increased its manufacturing

---

[2] Speech delivered by Hendrik Verfaillie, President of Monsanto, at the Forum on Nature and Human Society, National Academy of Sciences, Washington DC, October 30, 1997.

capacity of glyphosate, investing US$200 million in production and formulating technology in Australia, Brazil, Belgium, India, and China. The use of Roundup in biodiversity-rich regions and where biodiversity is the resource of the poor will destroy species as well as the livelihoods of the poorest.

The advertisements of the genetic engineering corporations, which call themselves the life sciences corporations, are all directed at making the public believe that genetically engineered crops are ecologically safe and reduce the use of chemicals. In an advertising campaign in Europe in 1998, Monsanto claimed that Monsanto's genetically engineered crops reduce pesticide use and provide a safe and sustainable method of weed control. As one of the advertisements stated, 'More biotechnology plants means less industrial ones.' However, while Monsanto was selling its Roundup Ready crops, it was also setting up new Roundup factories in India, China, and Brazil. More biotechnology plants mean more industrial ones, since herbicide-resistant crops are designed to be tolerant to the proprietary herbicides of the company, which makes money selling both seeds and chemicals. Monsanto makes farmers buying its Roundup Resistant crops sign a contract that they will not buy chemicals from any other company and will not save seed. Monsanto has thus retained its monopoly through genetic engineering at a time when its patent for Roundup was expiring.

Herbicide-resistant crops are designed to be tolerant to the herbicide of the company, and their spread will mean an increase in herbicide use, especially in those regions of the world where farms are small, labor is abundant, polycultures control weeds and women use the weeds for food and fodder. Weeds are part of biodiversity in small farms, and a useful resource. The spread of herbicide-tolerant seeds would destroy biodiversity, destroy sources of food and fodder, destroy women's livelihoods and introduce toxic chemicals.

Farming systems in the Third World depend on 100–200 plant species. The introduction of herbicide-resistant varieties will push the rich diversity of small farms to extinction. Monsanto's Roundup advertisements in remote villages of India declare: 'Are your hands tied up by weeds? Roundup will set you free.' The sole purpose of Roundup Ready crops is to increase the use of Roundup herbicide, so Monsanto's claim that their use will reduce herbicide use is sheer deception. Roundup is Monsanto's largest selling product, accounting for 17% of the total annual sales of $9 billion.

GMOs can pollute ecosystems at the biological and genetic level. Genetically engineered transgenic crops can contribute to genetic pollution or biological pollution in many ways.

1. In herbicide-resistant varieties, transgenes can spread to wild and weedy relatives, creating superweeds.
2. Contamination or pollution of biodiversity can destroy the unique characteristics of diverse species.
3. Transgenic crops engineered to produce pesticides can lead to evolution of resistance in major insect pests.
4. Toxins from the genetically engineered crop can kill beneficial species.

These processes will not control weeds and pests, they will create superpests and superweeds.

In addition to creating superweeds and superpests, GMOs can also spread disease.

Breeding plants resistant to viral infections by inserting virus genes in the plant genome can create new superviruses which have new hosts and new properties (Greene and Allison, 1994).

Genetic pollution or biopollution can also occur through horizontal gene transfer. Horizontal or lateral gene transfer is defined as the non-sexual transfer of genetic information between organisms.

Horizontal gene transfer is therefore different from the ordinary form of gene transfer which takes place vertically from parent to offspring. One such horizontal gene transfer is the case of a genetic parasite belonging to yeast, a favorite subject for genetic engineers, which has suddenly jumped into many unrelated species of higher plants.

Our knowledge at the genetic level is too immature to assess the probability or consequences of such horizontal gene transfer, and genetic pollution resulting from it. Little has been done to understand the ecology of genes, though much effort has gone into the engineering of genes without any knowledge of the impact of genetically engineered organisms on other organisms and the environment. The lack of knowledge concerning ecological impact has been taken as proof of safety, when it is in fact evidence of ignorance of biohazards.

## The Myth of Feeding the Hungry

GMOs will increase hunger because they will introduce a chemical and capital-intensive monoculture agriculture, displacing small farms and small farmers who use biodiversity to feed themselves and their families.

Small biodiverse farms are more productive than industrial monocultures.

Yields usually refer to production per unit area of a single crop. Planting only one crop in the entire field as a monoculture will of course increase its yield. Planting multiple crops in a mixture will produce low yields of individual crops, but a high total output of food. In the biodiversity perspective, biodiversity based productivity is higher than monoculture productivity (Shiva, 1996; Rosset and Altieri, 1999). The Mayan peasants in the Chiapas are characterized as unproductive because they produce only 2 tonnes of corn per acre. However, the overall food output is 20 tonnes per acre.

In the terraced fields of the high Himalaya, women peasants grow jhangora (barnyard millet), marsha (amaranth), tue (pigeon pea), Urad (black gram), gahat (horse gram), Soybean (*Glycine max*), bhat (*Glycine soja*), Rayans (rice bean), swanta (cow pea), koda (finger millet) in mixtures and rotations. The total output is nearly 6000 kg per acre which is nearly six times more than industrially farmed rice monocultures.

The work of the Research Foundation for Science, Technology and Ecology has shown that farm incomes can increase threefold by giving up chemicals and using internal inputs produced by on-farm biodiversity including straw, animal manure and other by-products.[3]

---

[3] D. Deb, Case study – Bengal and V. Bhatt, Case Study – Himalaya, RFSTE, 1999.

Research done by the Food and Agriculture Organization (FAO) has shown that small biodiverse farms can produce thousands of times more food than large, industrial monocultures. Indigenous farmers of the Andes grow more than 3000 varieties of potatoes and in Papua New Guinea, as many as 5000 varieties of sweet potatoes are under cultivation, with more than 20 varieties grown in a single garden (Heywood, 1995). In Java, small farmers cultivate 607 species in their home gardens, with an overall species diversity comparable to a deciduous tropical forest. In sub-Saharan Africa, women cultivate as many as 120 different plants in the spaces left alongside the cash crops. A single home garden in Thailand has more than 230 species, and African home gardens have more than 60 species of trees. Rural families in the Congo eat leaves from more than 50 different species of trees. A study in eastern Nigeria found that home gardens occupying only 2% of a household's farmland accounted for half of the farm's total output. Similarly, home gardens in Indonesia are estimated to provide more than 20% of household income and 40% of domestic food supplies.[4]

The assumption of the inverse relationship between biodiversity and productivity has guided all technological change in agriculture, destroying biodiversity. Yet it does not hold when one takes diversity of crops and their diverse outputs into account. The increased yields from genetically engineered crops is the most important argument used by the genetic engineering industry. However, genetic engineering has actually led to a decline in yields. Bill Christianson, a soybean farmer in the US who participated in the first conference on 'Biodevastation' held at St Louis, Missouri, the Headquarters of Monsanto, says that in Missouri, genetically engineered soybean had a five bushels per acre decrease in yield.[5] According to Ed Oplinger, Professor of Agronomy at the University of Wisconsin, who has been carrying out yield trials on soybean for 25 years, genetically engineered soybeans had 4% lower yields than conventional varieties on the basis of data he collected in 12 states which grow 80% of US soybeans (Holzman, 1999). In a study by Marc Lappe and Britt Bailey, in 30 out of 38 varieties, the conventional soybeans outperformed the transgenic ones, with an overall drop in yield of 3.34 bushels per acre, or a 10% reduction compared to conventional varieties (Lappe and Bailey, 1999, pp. 82–83). Dr Charles Benbrook has reported a 6.7% decline in yields in soybeans engineered to be resistant to Roundup on the basis of 8200 university-based soybean varietal trials in 1998. In his view, 'If not reversed by future breeding enhancements, this downward shift in soybean yield potential could emerge as the most significant decline in a major crop ever associated with a single genetic modification (Benbrooke, 1999). The yield of *Bt* cotton was found to be dramatically reduced in the first *Bt* trials undertaken in India.

## Speed and Biohazards

Speed has been at the root of ecological crisis in agriculture. Speed is the justification for shifting from organic manures to synthetic fertilizers. Speed is the

---

[4] FAO, Women Feed the World, World Food Day, October 16, 1998.
[5] Statement of Bill Christianson at First Grassroots Gathering on Biodevastation, St Louis, July 18, 1998.

justification for shifting from harvesting the biodiversity of 'weeds' to destroying them with herbicides like Roundup. Speed is the justification for using genetic engineering on crop breeding. Speed is the justification for rushing GMOs to the marketplace without tests and evaluations. GMOs were introduced in agriculture without assessment of productivity or risks. They were rushed into the marketplace on the declaration of safety, not tests for safety.

The safety debate has been repeatedly suppressed by bad science parading as 'sound science.' One of the unscientific strategies used to extinguish the safety discussion is to define a novel organism or novel food created through genetic engineering as 'substantially equivalent' to conventional organisms and foods. However, a genetically engineered crop or food is different because it has genes from unrelated organisms – it cannot, therefore, be treated as equivalent to a non-genetically engineered crop or food. In fact, the biotechnology industry itself gives up the claim of 'substantial equivalence' when it claims patents on GMOs on grounds of novelty. When industry wants to avoid risk assessment and issues of liability, the argument used is that the genetically engineered organism is 'substantially equivalent' to the non-engineered parent. However, when industry wants property rights, the same GMO becomes 'novel' or substantially inequivalent to the parent organism. When the claims for safety and for intellectual property rights in the genetic engineering industry are put together, what emerges is an unscientific, incoherent undemocratic structure for total control through which absolute rights are claimed and all responsibility is denied and disclaimed. This ontological schizophrenia is based on and leads to incoherence, which is a characteristic of bad science. Good science is based on coherence. The inconsistency and incoherence between the discourse on property rights and the discourse on issues of safety contributes to an undemocratic structure in which there are no mechanisms to protect citizens from corporate irresponsibility.

The global citizens' call for a five-year freeze on release of GMOs into our agriculture and food systems is a necessary pause that should induce responsibility, sanity and safety in the insane rush to sell GMOs at any cost, including irreversible risks to biodiversity and human health. Slowing down the GMO madness is an essential part of the Slow Food Movement because the GM culture will destroy the ecological and economic foundations on which Slow Food cultures have been based. In this pause we can build and strengthen food cultures based on small farms, biodiversity, and ecological time in place of the culture of giant corporations, monocultures, and Wall Street time.

# 19

# GMOs in Agriculture: An Environmentalist Perspective

**Richard Caplan**
*US Public Interest Research Group, Washington DC, USA*

G enetically engineered crops have been introduced despite inadequate demonstrations of their safety or need. Government and many scientists have downplayed or completely ignored the negative social, ethical, and economic impacts this new technology may bring. Issues related to food safety, the rights of indigenous peoples, and numerous other areas related to genetic engineering are serious enough to warrant an immediate moratorium on genetically engineered foods and crops. However, these issues are out of the scope of this chapter, which will focus on environmental issues. The conclusions about the merits of this technology based on its ecological risks, however, remain the same.

According to the US Environmental Protection Agency, 4.5 billion pounds (2.04 billion kilos) of pesticides are applied to US fields every year (USEPA, 1999). The worldwide damage caused by pesticides has been estimated at $100 billion per year (Myers, 1998). This harmful agricultural paradigm has been brought on by largely unchecked industry-driven changes in farming practices, and tolerated through denials of risk by government and scientific authorities.

Now, the same institutions that have created this unsustainable morass are peddling genetically engineered crops as the solution to the problems they themselves created. Even Monsanto's outgoing Chairman has admitted his company is guilty of 'condescension or indeed arrogance' because 'too often we have

---

Genetically Modified Organisms in Agriculture
ISBN 0-12-515422-4

forgotten to listen' to the public (Thayer, 1999). This arrogance includes holding on to the outdated notion of 'substantial equivalence,' which has been adroitly criticized (Millstone *et al.*, 1999). In fact the foolhardiness of such a misguided philosophy was demonstrated recently when new gene sequencing on Roundup Ready soybeans revealed that the varieties had more foreign DNA than Monsanto, or the government, thought (Meikle, 2000). In addition, the idea that genetically engineered plants are merely a more precise 'extension' of traditional plant breeding is a claim that falls flat in the face of critical examination (Hansen, 2000).

## Unpredictable Effects of Gene Mixing

The mixing of genes from very different sources may introduce new elements of unpredictability. Conventional breeding, including hybridization and wide crosses, only permits the movement of an extremely tiny fraction of all the genetic material that is available in nature. Because it only allows mixing, and recombination, of genetic material between species that share a recent evolutionary history of interaction, one would expect that the products of conventional breeding would be more stable and predictable. With a complex ecosystem, such as a genome, the introduction of a new gene through genetic engineering can have a range of effects, from virtually nothing to a catastrophic effect. These changes cannot be reliably predicted knowing the biology of the introduced species or gene.

A clear example of this unpredictability was seen in an experiment performed with *Arabidopsis thaliana*, a plant in the mustard family that is frequently used for biological experiments (Bergelson *et al.*, 1998). *Arabidopsis thaliana* is normally a self-pollinating species with very low rates of cross-pollination. An experiment was designed to test the effect, if any, of genetic engineering on gene flow. The results were quite surprising. The per-plant outcrossing rate was 0.30% for mutant fathers (i.e., containing the herbicide tolerance gene from mutation breeding) and 5.98% for transgenic fathers (i.e., containing the herbicide tolerance gene from genetic engineering). Further genetic investigation found that the outcrossing rates in the two genetically engineered lines were very different: 1.2% versus 10.8%. Thus the two lines of genetically engineered *A. thaliana* demonstrated 4-fold and 36-fold higher rates of outcrossing compared to traditional breeding. Since the herbicide tolerance gene was the same, the differences between genetic engineering and mutation breeding appear to be associated with the overall process of genetic engineering. The differences between the two genetically engineered lines appears to be due to the location where the insertions happened, as the entire genetic construct was the same.

This experiment clearly demonstrates that a dramatic variation can exist between conventional breeding (including mutation) and genetic engineering as a result of the haphazardness of the location of the inserted gene. As the authors note elsewhere, this 'has great implications for biotechnology and the controversy surrounding the risk of releasing transgenic crops into the environment' (Wichmann *et al.*, 1998).

Unexpected effects resulting from the genetic engineering of foods have already been observed, but have been largely ignored by the biotech industry and even the US government. Only through legal pressure (Alliance for Bio-Integrity *et al. v.* Shalala *et al.*) did the US Food and Drug Administration recently release documents demonstrating that their own scientists raised a number of concerns about genetically engineered foods. Among the concerns expressed by scientists at the agency (and ultimately ignored in FDA's 1992 Statement of Policy (US FDA, 1992)) were the creation of new toxins or the elevation of existing levels of toxins in unexpected ways, possibly by switching on or off another nontarget gene. We know very little about this phenomenon, also known as gene silencing; in fact it was recently discovered in transgenic plants and initially thought to occur only in the case of transgenes. A review of this topic in 1994 made the point clear: 'While there are some examples of plants which show stable expression of a transgene these may prove to be the exceptions to the rule. In an informal survey of over 30 companies involved in the commercialization of transgenic crop plants, which we carried out for the purpose of this review, almost all of the respondents indicated that they had observed some level of transgene inactivation. Many respondents indicated that most cases of transgene inactivation never reach the literature' (Finnegan and McElroy, 1994).

We have seen the unpredictability of results of genetic engineering due to silencing and other impacts with severe crop failures of genetically engineered cotton (Fox, 1997) and genetically engineered soybeans (Coghlan, 1999). In the former case, unexpected deformed bolls were formed and boll drop occurred. This problem occurred in part because 'the companies hurried Roundup Ready cotton to market without allowing them [state and Federal cotton experts] to test it' (Myerson, 1997). As a result of the losses suffered, compensation was paid to farmers in (at least) Mississippi, Arkansas, Tennessee, Missouri, and Texas (Lambrecht, 1998 and Hansen, 2000). In the soybean case, it was discovered that Monsanto's genetically engineered soybeans grown in hot climates are more likely to grow shorter and have stems split open. Genetically engineered soybeans grew an average of 15 centimeters in hot climates compared to a conventional height average of 20 centimeters, and 100% of the genetically engineered plants had split stems compared to 50–70% for conventional varieties. There are many other examples, including genetically engineered petunias that unexpectedly failed to express the desired color (Meyer *et al.*, 1992).

## Damage to Soil Ecosystems

The results of applying this unpredictable new science to agriculture present serious ecological risks. One profound but largely unexplored area is the damage genetically engineered crops may cause to soil ecosystems. Work published by Saxena *et al.* (1999) demonstrated that *Bt* toxin is released into the rhizosphere soil in root exudates from *Bt* corn. They concluded that 'there may be a risk that non-target insects and organisms in higher trophic levels could be affected by the toxin.' In response to Saxena *et al.*'s research, the Biotechnology Industry Organization astoundingly claimed that, 'It's hard to find anything here that's surprising' (National Public Radio, 1999). If the news that a toxin retains its insecticidal properties for at least 234 days is not surprising, it is doubly troubling. Saxena's work is reinforced by Donegan and Seidler (1999) who state that 'pesticidal proteins produced in transgenic plants can persist in soil and that binding of the proteins to soil particles can protect them from biotic degradation. We also found that plant genomic DNA in transgenic plants can persist in a field environment for several months.' In contrast to the *laissez faire* attitude of the Biotechnology Industry Organization, the authors point out that 'it is crucial that risk assessment studies on the environmental use of transgenic plants consider the impacts on microbial communities. Research in this area has been quite limited, however, as demonstrated by the few available references.'

## The Creation of 'Superweeds'

Another frequently mentioned ecological concern posed by the introduction of genetically engineered crops is that genes designed to give crops a competitive advantage may be passed to related wild plants with which they interbreed, spawning new 'superweeds.' In fact the current reliance on just a few broad-spectrum herbicides makes it likely that resistance will develop even faster. Already canola weeds resistant to three herbicides have been found in a field in northern Alberta, Canada (MacArthur, 2000). And Snow *et al.* (1999) recently reported that the physiological costs of this new trait are 'negligible,' suggesting that it may persist and spawn more troublesome weeds. In fact the few studies of the relative fitness of hybrids between genetically engineered crops and wild relatives show that they are not necessarily less fit than their wild parent (Ellstrand, 2000). This problem is particularly troubling in light of the pell-mell rush into international commercialization of these crops (Rissler and Mellon, 1996). Without regulatory oversight, genetically engineered plants will continue to hybridize with wild relatives, and potentially create serious problems as invasive species. The costs imposed on the United States by non-native species are already estimated at $138 billion annually (Pimentel *et al.*, 1999a).

## Damage to Nontarget Species

Research regarding the potential impacts on monarch butterfly larvae (Losey *et al.*, 1999) is fairly well known as a result of media attention. Yet arguably more striking than the news from John Losey and colleagues that higher mortality for monarch larvae was encountered with *Bt* corn plants than their conventional counterparts was how, when the article was published, proponents of the technology had such superficial data with which to counter the findings. Monsanto had to admit that it had 'not yet conducted its own research on *Bt*'s impact on monarch butterflies' (CNN.com, 1999). When a conference was hastily convened in November 1999 – organized and paid for in part by biotechnology companies (Yoon, 1999a) – its attempt to conclude that risk to the monarch was minimal was justifiably pilloried as a 'manipulation' (Ferber, 1999).

Despite the high profile Losey's work has received, unfortunately far fewer people have heard about other research demonstrating adverse effects on nontarget species from genetically engineered crops. Giroux *et al.* (1994) reported that the ladybug predator of the Colorado potato beetle consumed fewer potato beetle eggs when potato *Bt* levels were high. And in work reported from the

Swiss Federal Research Station for Agroecology and Agriculture, Hilbeck *et al.* (1998a, 1998b, 1999) have reported that lacewing larvae reared on prey that were fed *Bt*-producing corn took longer to develop and had a strikingly elevated mortality rate. Other studies have produced similar results, including a four-year study in Wisconsin (National Research Council, 2000) and another in Ohio (Hoy *et al.*, 1998). The National Research Council (2000) recently stated: 'It is important to ask whether such indirect effects will have a harmful effect on the agroecosystem.' Unfortunately, the question is being asked and studied far too late. As Hilbeck has stated, 'We risk disrupting the regulatory mechanisms that naturally keep pests in check' (Tangley, 2000).

## Insecticide Resistance

Plants engineered to kill insects are likely to hasten the creation of pesticide-resistant species, already a major problem (Halweil, 1999). *Bt* crops, now widespread, threaten to accelerate the emergence of *Bt*-resistant pests. This will result in a cost that organic farmers will disproportionately absorb because chemical alternatives are forbidden under organic rules (Schulz, 2000). Although EPA's resistance management strategies have been predicated on high-dose crops plus a small refuge, applications have been approved for crops that produce only moderate doses (Union of Concerned Scientists, 1999). A study published last year confirms that *Bt* crops expressing only moderate doses of *Bt* toxin can elicit resistance to *Bt* (Huang *et al.*, 1999). And Liu *et al.* (1999) suggest that the assumption of random mating that undergirds the refuge strategy may not hold. With moderate dose crops approved, and the news about insects potentially developing resistance to *Bt*, simply relying on a high-dose-plus-refuge strategy may be inadequate. In January, 2000, EPA announced new rules for resistance management with *Bt* corn. According to the rules no more than 80% of a field can be planted in *Bt* corn varieties, and in cotton-growing areas no more than 50% can be planted in *Bt* corn varieties. While this announcement is a further admission of inadequacies in the initial oversight of the technology, farmers dependent on the use of *Bt* can now only wait and see if irreparable damage has not already been done.

## Threats Posed by Plants Engineered to be Virus-resistant

One last major ecological concern to be raised is the threat posed by plants engineered to be virus-resistant. Three main concerns are that new viral strains may arise, viral host ranges may be broadened, or that existing viral diseases may be more severe. Schoelz and Wintermantel (1993) and Greene and Allison (1994) have both reported instances of viral recombination. And concerns have been raised about one particular promoter – the cauliflower mosaic virus – used in nearly every genetically engineered plant either in commercialization or field trials (Ho *et al.*, 2000). USDA's decision to deregulate virus-resistant crops is currently being challenged by public interest groups (Natural Resources Defense Council *et al. v.* Dan Glickman).

## Inadequate Data

Inadequate data about the ecological risks of genetically engineered plants have been collected. When US government regulations were first proposed for field experiments, they were sharply criticized as inadequate by Congress's own investigative arm (General Accounting Office, 1988). Despite this, USDA has decidedly weakened regulations over time (USDA, 1993, 1997). Their ability to do so was without scientific merit, as two important analyses from 1995 make clear. In reviewing the 85 most recent reports of field trials then available, Mellon and Rissler (1995) note that none mentioned experiments to assess weediness, none of the 19 reports on virus-resistant crops mentioned experiments measuring the production of new virus strains, and none of the reports on *Bt* crops mentioned experiments on the likelihood of adverse impacts on nontarget insects. Findings in a similar vein by Purrington and Bergelson (1995) concluded that, 'It seems clear . . . that both industry and government must improve their collection and interpretation of data.' Thus regulations were weakened before having adequate empirical evidence. According to ecologists, 'It still is the case. A lot of the key experiments don't seem to be being done' (Yoon, 1999b).

## The Myths of Genetically Engineered Crops

Despite these and other ecological risks, proponents of genetically engineered crops continue to advance two major mythologies about the technology. The first is that this technology is an integral part in a strategy to help feed the world. That a genetic quick fix from a laboratory could solve the world hunger problem was debunked many years ago. For example, Dawkins (1987) has written that, 'Increases in food production may temporarily alleviate the problem, but it is mathematically certain that they cannot be a long-term solution; indeed, like the medical advances that have precipitated the crisis, they may well make the problem worse, by speeding up the rate of the population expansion.' Rosset (1999) has pointed out that we have enough food already available to provide 4.3 lb (1.95 kg) to every person every day, but that the real problems are poverty and inequality. Tewolde Berhan Gebre Egziabher of the Institute of Sustainable Development in Addis Ababa, Ethiopia, has stated the problem perhaps better than anyone. 'There are still hungry people in Ethiopia, but they are hungry because they have no money, no longer because there is no food to buy . . . we strongly resent the abuse of our poverty to sway the interests of the European public.'

Another regular excuse offered for the need for genetically engineered crops is that they will eliminate the need for dangerous farm chemicals. Companies involved in genetic engineering are also heavily invested in agrochemicals (RAFI, 2000). As noted above, there is a great irony that the same companies that have profited so long from selling these chemicals are now claiming that their latest technology will solve the problems caused by their older technologies. And in contrast to their claims, a series of studies has indicated that the adoption of genetically engineered crops is frequently not reducing the use of harmful chemicals, and in some cases has increased it by making spraying easier (Benbrook, 1999; Carpenter and Gianessi, 2000; USDA Economic Research

Service, 1999). One exception has been the reduction in insecticide applications on *Bt* cotton, but even this success will undoubtedly be a short-term phenomenon that is not likely to last. Recent evidence indicates that biotech companies are even admitting that genetically engineered plants will need more pesticides, including carbamates, in their patent applications (Coghlan and Fox, 1999).

The lack of benefit for consumers of genetically engineered crops has been admitted by biotech proponents like AgrEvo, which has conceded that, in terms of genetically engineered organisms with any benefit to consumers, 'none is commercially developed at this stage' (Thayer, 1999). In contrast, alternatives to genetically engineered crops are available and effective (Altieri *et al.*, 1998). According to Miguel Altieri, a leading figure in the movement towards a sincere sustainable agriculture, 223 000 farmers in southern Brazil are using cover crops and green manures and have doubled yields of maize and wheat to four to five tonnes per hectare. Forty-five thousand farmers in Guatemala and Honduras have tripled maize yields in their systems in the hillsides and survived Hurricane Mitch, as opposed to Green Revolution and commercial systems that suffered 98% damage. More than 300 000 farmers in dry lands of India are using a range of water and salt management technologies and have tripled sorghum and millet yields to some 2 to 2.5 tonnes per hectare. And there are thousands of examples like this.

## Conclusion

An agriculture dependent on short-term fixes, toxic chemicals, and inputs from chemical companies sets us on an unsustainable course. The regulatory framework governing genetically engineered crops is weak, and the technology is inherently unpredictable and risky. Perhaps most significant, alternatives to genetic engineering that produce sustainable solutions have been demonstrated to work. To continue in the direction we are headed with this technology can only be done in ignorance of an objective benefit–cost analysis. As recently stated in the *New York Times*, one would 'have to be crazy to buy this stuff' (Pollan, 1999).

# 20

# Genetically Modified Organisms Can Help Save the Planet

**Dennis T. Avery**

*Hudson Institute, Indianapolis, Indiana, USA[1]*

Europe is opposed to genetically modified crops and foods. That's remarkable. Never before in history has Europe rejected an increase in its food security.

- When primitive Europeans discovered spears, they eagerly used them to kill larger animals to feed their families.
- When men learned to plant seeds, European travelers like Marco Polo brought 'alien species' from Asia in the urgent search for higher-yielding, more nutritious crops.
- When Columbus brought Europe potatoes and corn (maize) from America, crops which yielded more food per acre than the wheat Europe had previously imported from the Middle East, Europeans eagerly planted them. They even began eating the American tomato – a member of the sometimes-deadly nightshade family.

[1] Dennis T. Avery directs the Center for Global Food Issues, a project of the Hudson Institute of Indianapolis, IN. He has served as a policy analyst for the US Department of Agriculture, for President Johnson's National Advisory Commission on Food and Fiber, and the US Commodity Futures Trading Commission. From 1980 to 1988, he was the senior agricultural analyst for the US Department of State, and won the National Intelligence Medal of Achievement in 1983. He is the author of *Saving the Planet with Pesticides and Plastic*.

Genetically Modified Organisms in Agriculture
ISBN 0-12-515422-4

- About 1860 Gregor Mendel, an Austrian monk, revealed the basic principles of cross-breeding, which allowed us to create hybrid plants that, in his
  words, 'nature never intended.' Hybrid seeds swept across Europe and the
  world, carrying the yields of corn, sorghum, sunflower, rice, and many
  fruits and vegetables to record levels.

After World War II, the never-ending search for more productive crop genes led
Europeans to create mutant plants. They bombarded seeds with radioactivity. Or
soaked them in chemicals harsh enough to destroy parts of their genetic material. One of the most famous 'mutant crops' is a wheat–rye hybrid called triticale.
Northern Europe is now growing millions of tonnes of triticale per year. The
British used radioactive bombardment to create one of their most famous brew-
ing barleys in the 1950s; it was even used in organic beer! The United Nations
Food and Agriculture Organization has a list of more than 1300 forced-mutant
crop plants that have been created and used around the world.

Then public health advances such as clean water and vaccinations began radically reducing death rates, in the Third World as well as in the affluent
countries. History's biggest population surge gathered steam. Many of the
world's most respected thinkers predicted a massive famine that would sweep
over densely populated Asia first, and then engulf Latin America and Africa.

But the famine never happened. The Green Revolution happened instead.

The Ford and Rockefeller Foundations started an international wheat
improvement center in Mexico and a rice-breeding station in the Philippines.[2]
These early small efforts turned into the Green Revolution, which essentially
tripled the yields on the best farmland all over the Third World – and Europe.
(America had begun its plant-breeding revolution 20 years before.)

Short-strawed wheat and rice varieties were bred from natural plant mutations. These semi-dwarf plants could accept irrigation and chemical fertilizer
without growing so tall that they 'lodged,' that is, literally fell over from the
weight of their own grain heads and dropped their grain to rot on the ground.
Planted in such densely populated places as China, India, and Mexico, the new
seeds and farming inputs doubled food production. Primitive traditional farmers in 40 countries were suddenly able to feed their families adequately, and
even supply more food to nearby cities.

The humanitarian achievement was appropriately celebrated by awarding the
1970 Nobel Peace Prize to Dr Norman Borlaug, one of the Rockefeller's original
plant breeders.

Unfortunately, however, this unprecedented plant breeding success changed
Europe's perception about food and farming as its traditionally high farm price
supports stimulated more food than Europe could eat. Mountains and lakes of
surplus grain, sugar, wine, meat, and milk piled up, costing its governments billions of dollars per year in extra subsidies and creating the illusion that the
world needed less, not more, agriculture.

---

[2] These original plant-breeding stations turned into today's International Maize and
Wheat Improvement Center in Mexico, and the International Rice Research Institute
in the Philippines, and then grew into the Consultative Group on International
Agricultural Research, with about 20 centers researching agriculture, forestry, and
marine resource development throughout the Third World.

# The Threat of Overpopulation

In the midst of the Green Revolution triumph, we frightened ourselves. The scary story went like this: There was no longer any restraint on human population growth. If we could produce all the food we needed, what was to prevent human numbers from spiraling upward until we literally covered the earth? And thus the environmental movement was born, opposing both population growth and affluence.

Deep down, the eco-activists blamed high-yield farming for what they perceived as the end of the world as they idealized it should be. It was bad enough that the world's population had doubled to nearly 2 billion. Now this burgeoning multitude was demanding cars and highways, overcrowding parks and the wilderness. Fiberglass brought the cost of owning boats within reach of the average family, and the noise of outboard motors sullied the once-quiet lakes. Camper trailers proliferated. Second homes began to line the coasts of lakes, rivers, and bays.

It is no surprise that the environmental activists are opposed to biotechnology in food. Eco-activists have opposed virtually every element of the high-yield agriculture that prevented massive world famine. Paul Ehrlich began writing his famous book, *The Population Bomb*. Garret Hardin began to think about 'lifeboat ethics,' in which the most able humans should save themselves at the expense of the rest.

The environmental movement says it wants only 2 billion people on earth instead of the current 6 billion – or perhaps only 300 million (Pimentel *et al.*, 1999b). Eco-activists also decry the affluence that lets people feed their children resource-costly hamburgers and buy power-thirsty cars and air conditioners, though virtually every measure of environmental health seems to improve with per capita incomes. Dr Indur Goklany, a senior policy analyst at the Department of the Interior, documents this in two papers, 'Richer is Cleaner: Long-Term Trends in Global Air Quality' and 'Richer is More Resilient: Dealing With Climate Change and Other More Urgent Environmental Problems' (Goklany, 1995, 2000). There is no question that much of the Third World is still in the most polluting phase of economic growth, burning lots of coal, smelting lots of iron, and cutting trees that are not replanted.

The campaign against modern farming began with the baseless charge that farmers' pesticides would give us cancer. We are still looking in vain for the cancers. Looking back, we now know that the cancer risks of First World non-smokers began to decline sharply just about the time that we began using pesticides widely.

Now, the eco-opposition has expanded the campaign against high-yield farming to virtually all aspects of modern agriculture. Included in their charges:

- *monocultures* – vulnerable to pest epidemics;
- *high-yield seeds* – displace old landrace varieties and reduce biodiversity;
- *animal antibiotics* – will weaken our ability to treat human disease;
- *confinement feeding* – dumps heavy loads of animal waste into rivers.

It's a remarkable campaign, and has garnered remarkable success seeing that none of the charges pass the test of truth. (1) Monocultures produce far more food per acre, and we've kept ahead of the pest epidemics with a flow of new varieties and new pest control systems. (2) High-yield seeds also save room for nature with higher yields, and we can't environmentally afford to turn half the world's cropland into a gene museum. Instead, we're keeping the landrace germplasm in seed banks. (3) We know why we're getting resistance to antibiotics; people have so much faith in them that half the prescriptions are written for ailments that antibiotics won't cure. Equally bad, we quit taking the medicine when we feel better, leaving the toughest bacteria alive and able to reproduce. No evidence has linked antibiotics in feed to bacterial resistance in humans (Avery, 2000). Confinement feeding is saving perhaps 3 million square miles of land for wildlife, and the wastes are carefully saved for use as organic manure on feed crops; it's the little outdoor hog farms that let their wastes wash into the nearest stream.

Now, the eco-activists are even campaigning against nitrogen fertilizer. No plants can grow without nitrogen, and virtually all of the world's fish and animal life are equally nitrogen-dependent. Most of today's fertilizer is simply taken from the air, which is 78% N. Modern medicine says nitrogen causes no cancers, and does not even cause the rare 'blue baby syndrome' (Avery, 1999; National Research Council, 1995). But the press, even the science press, is full of vague warnings about the alleged dangers of nitrogen. A White House Task Force is seriously considering radical constraints on farm nitrogen in the US Midwest 'to protect the Gulf of Mexico,' though the study group admits it can find no negative ecological or economic impacts on the Gulf (White House Committee on Environment and Natural Resources, 1999).

## The Environmental Triumphs of High-yield Farming

Oddly, amid all this talk of saving the environment, there is almost no mention of humanity's biggest environmental triumphs.

First, the high yields from modern farms have allowed the world to produce its crops on the same six million square miles of land that we formerly used to feed 2 billion people. Without the yield increases and the confinement feeding, another 20 million square miles of land would have been taken from nature (Avery, 1997). That's more than the total remaining forests on the planet (16 million square miles).

Modern agriculture even saved the correct land for nature. All over the world, the best farmland has usually had relatively little biodiversity (Huston, 1994; Connell and Orias, 1964). (America may have only about 200 000 above ground species, most of them in warm parts of Florida, Texas, and California; but middle America has the biggest chunks of prime farmland anywhere in the world.) Most of the world's wild species are in the tropical forests, swamps, coastal marshes, and other places where we don't and shouldn't farm. Biologists say we may have 30 million wild species, with 25 million of them in the tropical forests (Hartshorn, 1992).

Michael Huston, an ecologist with the Oak Ridge National Laboratory, says we need to farm the best cropland for all it's worth, and leave the rest for nature. That's what modern farming allows us to do.

Greenpeace never says a word of thanks.

Second, modern farming is the most sustainable in history. Soil erosion has long been the biggest threat to human society. Soil degradation had become a significant problem for the Mediterranean Basin by Roman times, and probably caused the downfall of the Aztec and Mayan cities. America's Dust Bowl occurred less than a century after we plowed the Great Plains. But farming three times as much land would more than triple our soil erosion (because we'd be farming lots of steeper and more fragile land than we do now).

However, modern farming's contribution to sustainability goes far beyond higher yields. Conservation tillage now lets farmers virtually throw away their plows and their old bare-earth farming systems. Disking the heavy residues from high-yield crops into only the top layer of soil creates billions of tiny dams that prevent wind and water from carrying away the soil.

Dr Stanley Trimble of UCLA recently performed 'soil archeology' on the erosion-prone Coon Creek watershed in Wisconsin, and found it is suffering only 6% as much soil erosion as in the Dust Bowl days of the 1930s (Trimble, 1999). Coon Creek is now creating topsoil faster than it is losing it. (Some erosion is inevitable as soil is never in a static condition – it is always being created and being moved.) Conservation tillage is already widely used in the United States, Canada, Europe, South America, and Australia. It also works in Asia and Africa.

The environmental movement still complains that conservation tillage depends on safety-tested herbicides for weed control. Do activists dislike technological abundance even when it's sustainable?

## Why Do We Need More Food?

Europe and the eco-activists say the world already has enough food. That's a foolish falsehood. *Europe* has enough food. The world, however, still has 800 million hungry people and billions eating low-quality diets. By 2050, a more populous, affluent world will need nearly three times as much farm output than it harvests today.

The good news is that the population ogre has been slain. Births per woman in the Third World have come down from about 6.4 in 1965 to less then 3.0 today – and stability is 2.1 (UN Development Program, 1999; World Bank, 1988). Current projections indicate a peak human population of about 8.5 billion, about 2035.

However, we still have to meet the challenge of affluence. There is no vegetarian trend in the world, and livestock calories are two or three times as expensive as cereal calories. In fact, there has never been a voluntarily vegetarian society. Instead, there seems to be an urgent human hunger for the high-quality protein in meat and milk, and a strong physiological need for such micronutrients as calcium, iron, and zinc.

Newly affluent Asia is just beginning to reach out for high-quality diets. China has doubled its meat consumption (to nearly 70 million tonnes per year) in the 1990s. Surveys of India's Hindus, an historically vegetarian society, show that most of them will eat meat when they can afford to do so (IFPRI, 1999; *Poultry Annual Report, India*, 1995). In Bangladesh, more than half the women and nearly half the preschool children are anemic, in part because meat consumption is low and bioavailable iron is scarce (IFPRI, 1999, p. 6).

Africa must intensify its farming to support twice as many people in 2050 and still have room for its unique wildlife. Dr Calestus Juma, a Kenyan now at Harvard's International Development Center, says genetically modified crops are uniquely suited to Africa's farming problems; the technology package is delivered in a seed that can reproduce itself, rather than as separate inputs which Africa's weak foreign exchange position and poor infrastructure might have difficulty importing and transporting annually.

There will even be a pet challenge. America has 113 million companion cats and dogs; a rich, one-child China in 2050 may have 500 million. And those pet owners will demand high-quality diets for their beloved animals.

We dare not pit the needs of the wildlife against the needs of people's children and their pets. The wildlife will lose every time and in every place.

## The Myth of Organic Farming

The 'organic food' solution is a dangerous myth. Organic farming takes too much land. Its yields of field crops are only about half as high as modern farmers' yields, mainly because organic farmers refuse to use synthetic nitrogen fertilizer. Instead, they use animal manure, and/or keep large tracts of land in green manure crops like clover and rye. Mainstream farmers take their N from the air, which is 78% nitrogen. (Plants cannot use nitrogen until it reverts to its chemical form.) Organic farmers also accept higher losses to pests, which further limits yields.

The animal manure used on organic food crops is also an important health problem. It contains dangerous bacteria, such as *Salmonella* and *E. coli*, which sicken millions of people every year and kill thousands. The new *E. coli* O157 is a particularly deadly mutant; cattle manure is its main reservoir and composting does not reliably kill it.

## The Environmental Potential of Genetic Engineering

High-powered seeds (many with bred-in pest resistance), irrigation, chemical fertilizer, and modern pesticides all helped to advance the Green Revolution. These technologies are already widely used on the world's better farmlands (except in Africa). Extending them fully on the rest of the world's arable land might double world food output but not much more. Recently, the upward trend in world cereal yields has slumped from 2.2% per year (1967–1982) to 1.5% annually (1982–1994) (Mann, 1999).

If we cannot improve the nutritional value of foods in poor countries, millions of children will continue to suffer and die from such diet-related diseases as kwashiorkor and rickets.

If yield gains fail to match farm product demand in a more populous and increasingly affluent world, the result would almost certainly be massive wild-lands destruction.

If the world falls short of food for people, what will Greenpeace tell the pet owners?

Charles Mann, in a thoughtful 1999 *Science* article, quoted a Chinese crop scientist in the thick of the fight to boost farm yields. 'We may be able to create the new [rice] plant type without biotech,' said Shaobing Peng, 'but that is where new opportunities will have to come from in the future.' Nor are Peng and his colleagues talking about the little single-gene transformations that are currently in the fields; they say that to break the current yield barriers, the plants will have to be 'thoroughly re-engineered' (Mann, 1999).

Talking about the potential of agricultural biotechnology today is like trying to project the future of the airplane from the Wright brothers' first biplane. However, we have enough concrete achievements to know that genetic engineering is potentially the most important tool that agriculturists have ever touched:

- About 80% of the cheese produced in the US and Great Britain, and 40% of the world's cheese, is now made with genetically engineered chymosin (the coagulating agent for milk).[3] This is allowing millions of calves to grow to adulthood which would otherwise be slaughtered at about three months of age so we could 'harvest' the coagulating agent from their stomach linings. If America had to raise another 5 million suckling calves per year for chymosin, it would take the equivalent of 8000 square miles of Lake States forestland to provide the additional pasture for the calves and their mothers.
- 'Golden rice,' funded by the Rockefeller Foundation and the European Union may well eliminate the severe shortage of vitamin A that currently causes millions of poor children in rice cultures to go blind – or even die – every year. The genetically engineered rice will also overcome the chronic iron deficiency of rice diets that puts hundreds of millions of women (and their babies) at risk of birth complications (Gura, 1999).

Two Mexican researchers have produced the first crop plants adapted to the highly acid soils characteristic of the tropics (Barinaga, 1997). The researchers developed plants that secrete natural citric acid from their roots – and fend off the aluminum in the soils that otherwise stunts their growth. Acid soils currently cut yields by up to 80% on nearly half the arable land in the tropics. What could be a more powerful tool in protecting tropical forests from the pressures of low-yield farming than such higher yields?

Biotechnology has now produced virus-resistant crops (rice, sweet potatoes, squash, and bananas). Conventional plant breeding had failed against viruses. The virus-resistant sweet potatoes will apparently yield half again as much food per acre in Africa, where they are a major source of calories and food security.

Gene mapping allows us to identify useful genes from wild relatives of our crop plants. Experiments with these wild relatives produced a 50% increase in tomato yields, and probably a one-third increase in rice yields (Tanksley and McCouch, 1997).

A new rice variety from Washington State University contains a gene from the corn plant to give it a higher rate of photosynthesis – and thus 35% more yield (Walter, 2000).

---

[3] David Berrington, of Chris Hansen, Inc., Milwaukee, a major maker of fermentation agents.

Of course, such a powerful technology must be carefully and effectively regulated. In theory, it might be possible to cross the worst features of crabgrass and poison ivy to create the world's most noxious weed. Such an unlikely mix must be avoided. But we cannot afford to ban biotechnology in food.

## Biotechnology in Forestry

Nor can we ban it in forestry.

A major study by the World Bank and the World Wildlife Fund ('Global Vision 2050 for Forestry') concludes that the world forest product demand will rise by 66% by 2050, and that a much larger percentage of that harvest needs to come from high-yield tree plantations (Bazett, 2000).

The average yield of an American wild forest is only about 1.4 cubic meters of wood per hectare per year. The yields of fast-growing trees under intensive management can be 10 or 20 times that high.

Roger Sedjo, forestry expert for Resources for the Future in Washington DC, suggests that putting 3.5–7% of the wild forest area into high-yield tree plantations might well eliminate the need to even log, let alone clear-cut, the other 95% of this critical wildlife habitat (Sedjo, 1995).

Sedjo says biotechnology is already helping forestry. Through cloning and tissue culture, foresters are able to make copies of their best trees by the thousands. Cloned and tissue-culture Georgia yellow pine is now being planted in coastal Brazil (a high-sunlight environment) and yielding 50 cubic meters of pulpwood per hectare per year.

Given the low value per acre and the often-remote locations of industrial forests, biotechnology could have a major impact by putting more of the 'tree management' (pest control, fertilization, etc.) into the tree seed itself.

## The Drawbacks of Biotechnology in Agriculture

For the future, it is not hard to imagine genetic engineering providing us with better-tasting off-season fruits and vegetables, more nutritious foods (such as a soybean with a complete set of amino acids, like meat), a full set of virus-resistant crops, more tender beef (a 'tenderness gene' has already been located) (Bedsworth, 2000), fast-growing pest-resistant trees, and much more.

The environmental movement could be faced with its worst nightmare, the ultimate success of technological abundance over managed scarcity with 9 billion people living happily, their abundant lifestyles produced not from stolen natural resources but simply from new knowledge. The eco-activists had to counter-attack and they did.

They attacked first in Europe, where the modest farm surpluses created by guaranteed high farm prices made additional food production a tax burden rather than a benefit. It helped that 'mad cow disease' and the contamination of the French blood bank with the AIDs virus had lately raised Euro-consumers' fear of inept government regulation to new heights.

The eco-activists first complained about biotech escapes into nature. This is certainly one of the most important real concerns about agricultural biotechnology, and one of the major reasons for its close regulation. However, it should not be difficult to prevent escapes – by making all of the transgenic plants sterile. This idea (cast as the 'terminator' gene) raised great opposition among activists, when it was suggested by the US Department of Agriculture. Nevertheless, sterile transgenics may make a great deal of sense for the environment.

So far as 'superweeds' are concerned, farmers think they're already fighting them (crabgrass, pigweed, thistle, etc.). The main constraint on wild weed populations is the competition from other plants, along with each weed's own natural pests and diseases. Being resistant to a synthetic chemical that is never sprayed in the wild hardly turns an existing weed species into a superweed able to invade new habitats.

Moreover, it is clear that the activists are really more concerned about blocking biotechnology in food production than about escapes into nature. They have long advocated controlling farm pests with 'biological weapons' – parasitic insects, fungi, diseases – rather than safety-tested and precisely controlled pesticides that cannot proliferate on their own. The world has never documented a species lost to farm inputs, including pesticides and GMOs, but has lost many species to 'alien species' which have been introduced either by accident or by man's misguided intentions. Biocontrols can take on a life of their own. The European flower-head weevil, released in America to prey on European thistle species, is threatening to cause the extinction of two native thistle species and the pictured-wing fly that feeds on one of them (Strong, 1997).

Activists say they are concerned about small farmers who would not be able to afford the terminator-type seeds, but high-yielding seeds normally pay for themselves several times over in one growing season. It took only ten years for hybrid corn seed to saturate the cornfields of small-farm America in the 1930s. It didn't take much longer for the high-yield wheat and rice seeds of the Green Revolution to spread across the Third World when those countries were much poorer, and their farmers less literate, than today. In the real world, if a biotech seed is not profitable for farmers, there is no way a corporation can force them to plant and nurture it.

Concerns about food safety are legitimate but radically overblown. If ever a new food technology was potentially dangerous, it was the radioactive and chemical mutation of seeds. There, we were truly dealing with the unknown – cracking DNA chains and reassembling them without even knowing what they were. However, the regulatory system worked in exemplary fashion; not a single evil genetic monster was unleashed upon people or nature in 50 years of forced plant mutations.

By comparison, modern genetic engineering is vastly more precise. In fact, it looks to be even safer than cross breeding, which typically produces a dozen different results, eleven of which the researchers didn't want and often didn't expect.

Biotech testing is needed, of course, and biotech testing is done. In fact, the real problem for GMO opponents is that there's so little difference in the foods because the changes are so precise and limited. Still, the US Food and Drug Administration looks hard for any differences in the foods from modified organisms. Any differences found are tested more thoroughly than any other foods in all history.

That testing works was exemplified in the famous case of the research designed to place a gene from the Brazil nut into soybeans to provide a more complete roster of amino acids. Testing before approval showed that the small number of people who are allergic to Brazil nuts would be also allergic to the modified soybean. The gene was dropped from the research.

That isn't enough to satisfy the activists, of course. They are demanding sub-molecular testing of each of thousands of natural compounds in each GMO. They're trying to block biotech foods with the same regulatory overkill they used on the nuclear power industry (which, except in Communist regimes which had little regard for their people's welfare, has an outstanding safety record).

Did testing 'fail' in the case of the monarch butterfly? No. The natural toxin from the soil bacteria *Bacillus thuringiensis* is toxic to a wide variety of cater-pillars, which is why it has been used as a pesticide by organic farmers for many years. That is why it was bred into corn, to kill corn borers. (It is not toxic to the butterflies themselves, but to their offspring, the caterpillars.) Laboratory tests at Cornell University deprived monarch caterpillars of any food except milkweed leaves heavily coated with *Bt* pollen; some of the cater-pillars died. The real question, however, was whether the use of *Bt* crops in fields would kill significant numbers of monarch offspring in the real world. The developers of the technology knew that corn pollen is heavy and seldom gets far from the field; that milkweed leaves are shiny, and the pollen tends to fall off; and that few of our monarchs feed or lay their eggs in cornfields because the farmers have already eliminated most of the milkweeds from their cornfields.

One monarch expert says the best thing America could do for the monarch is to mow its roadsides only one or two times per summer instead of four ('Chasing Butterflies', 2000). Nevertheless, eco-activists are still dressing up their kids like butterflies to provide photo-ops at demonstrations.

Martina McGloughlin, now director of the biotechnology program for the University of California/Davis, grew up hand-weeding potatoes in Ireland. She knows first-hand both the harsh realities of a subsistence farm and the fab-ulous power of modern high-yield farming. In a broad-based defense of biotechnology's potential and safety for the *Washington Post*, she recently noted, 'Without any doubt the biggest problem with our robust, logical, science-based regulatory system is that not enough people understand how it works' (McGoughlin, 2000).

Should biotech foods be labeled? If consumers want that, so be it. But biotech-nology is a tool, not a product. It's a bit like having a tag on a new car that says, 'This car was designed with the aid of computers.' The label doesn't tell you why biotech was used, or what difference it makes in the food. (In the herbicide-resistant soybeans and *Bt* corn, in fact, no difference has been found in the edible part of the plant.) The activists want labeling because they hope to demonize biotech foods. But a biotech label would tell the consumer less that's useful than an organic label which read, 'Warning: This organic food may have been grown with animal manure containing dangerous pathogenic bacteria.'

Activists worry that biotechnology will produce new allergens. If it does, they will be quickly labeled (if the foods have unique value) or withdrawn. In the meantime, researchers are working to engineer the natural allergens out of such foods as wheat, milk, and peanuts, to which millions of people are allergic now. Thus biotech is almost certain to reduce our allergy problems, not aggravate them.

The world can't afford Europe's precautionary principle until all of the humans have adequate food, all of our kids have complete nutrition, our pets are well cared for, and all of our wild creatures are assured of keeping their homes. To demand that we adopt the precautionary principle now looks like an incredibly arrogant strategy, put forward by an affluent elite with little real concern for people, pets or wildlife.

21

# Genetic Engineering: A Technology Ahead of the Science and Public Policy?[1]

**Dennis J. Kucinich**
*Representative, Ohio 10th District*

## From the Congressional Record of November 1

Mr. Speaker, genetically engineered (GE) food is and should be controversial. However, one voice has tended to dominate official discourse on the subject— that of the agri-business industry. These corporations and their paid public relations spokespersons have claimed: that GE food is identical to foods bred by selective (traditional) breeding; GE food is safe; GE food is associated with good environmental practices; and GE food will cure world hunger. Federal regulators have largely left these claims unchallenged, permitting the industry to introduce GE food rapidly and widely without producing scientific evidence to back their claims.

The public is skeptical. There is a growing popular movement that is critical of GE food promises and suspicious of its industry proponents. In other countries, consumers have flatly rejected GE food, and opposition to GE food is growing in this country. I believe that GE food is an example of a radically new technology, the massive commercialization of which has out-paced science and public policy.

In this article, I wish to examine the industry's claims and scrutinize federal actions. I will then present alternatives.

## Is GE food just like traditional food?

There are significant and obvious differences between the genesis of traditional food and the manufacturing of GE food. Scientists note that conventional breeders rely on processes that occur in nature (such as sexual and asexual

---

[1] An Extract Reprinted from the US Congressional Record of November 1 (pages E2054–E2055) and November 2 (pages E2072–E2073) 2000.

reproduction) to develop new plants. By contrast, genetic engineers use "gene guns" and bacteria among other methods to forcibly insert or "smuggle" foreign genetic material into a plant or animal. Genetic engineers also use genetic elements such as viruses which "turn on" the foreign genes in the new host organism as well as genes for antibiotic resistance that mark which cells have accepted the foreign genetic material.

Conventional breeders are bound by species boundaries that allow them to transfer genetic material only between related or closely related species. By contrast, the very purpose of genetic engineering is to allow scientists to transfer genes from completely unrelated life forms, creating such concoctions as corn that exudes toxins found in soil bacteria or tobacco that glows due to the insertion into its genome of a firefly gene.

Scientists warn that genetic engineers cannot always accurately predict the outcome of their experiments. Many scientists argue that the genetic engineering process is inherently unpredictable and that genetic engineers are operating with incomplete knowledge about how genes interact with each other and with their external environment. While genetic engineers can with some precision locate and isolate a trait or gene to be inserted, they cannot control with any precision where that gene will be inserted into the host plant or how it will interact with other genes in the host plant. The new gene may disrupt the function or regulation of a plant's existing genes.

Field trials and lab research have documented the unpredictable nature of GE plants. In a 1990 study, scientists attempted to suppress the multiple colors of petunia flowers by turning off pigment genes in the plant. Researchers predicted that all the engineered flowers would be the same color. The flowers, however varied in terms of the amount of color in their flowers and in the pattern of color in individual flowers. Some flowers also changed color as the season changed.

The unpredictability of GE crops was further highlighted in 1997, when farmers growing GE cotton reported that the plants had stunted growth, deformed root systems and produced malformed cotton bolls.

## Is GE food safe?

Despite endless reassurances by biotechnology companies and the Food and Drug Administration (FDA) that GE food is safe to eat, several concerns have arisen. Genetic engineering has the potential to introduce new allergens and toxins into food, increase levels of natural toxins, reduce the nutritional quality of food and increase the rate of antibiotic resistance in bacteria. Yet, our experience with GE crops is limited. They have only been growing on a wide scale for five years and, consequently, have only been part of the American diet for the same amount of time. The long-term consequences of a diet of GE food are therefore unknown. To date, not a single peer-reviewed study has been conducted on the long-term consequences for humans of eating a diet of GE food. Moreover, without segregation and labeling protections in place to inform consumers about what they are eating, it will be difficult to pinpoint and monitor whether the presence of GE material in food products is impacting human health.

The lack of long-term safety studies has correctly led the Environmental Protection Agency (EPA) to not approve Starlink corn for human consumption because of concerns with potential allergens. Unfortunately, this corn was found in Taco Bell taco shells found on our grocery stores. Kraft, the maker of these taco shells, recalled 2.5 million boxes of these contaminated shells.

## Environmental impacts associated with GE food

Despite claims that GE crops will help the environment, to date, the main focus of biotechnology has been to generate herbicide resistant crops and pest and disease resistant crops—crops that encourage more intensive use of pesticides. The failure of GE to move agriculture in a more sustainable direction is a serious threat to the environment.

Equally serious is the threat of genetic pollution which is potentially irreversible. Studies are revealing that predictions of gene flow, harm to beneficial insects, insect resistance, and the possibility that GE crops could become weeds are already coming true. Early experiments showed that pollen from GE herbicide resistant canola could spread to their wild relatives—radish plants—in nearby fields, highlighting the possibility of new "superweeds." More recently, a Canadian farmer, who had planted three different GE herbicide-tolerant crops, reported that a canola plant in his field was resistant to the three different herbicides. Cross pollination by GE crops has contaminated organic crops, in one instance forcing an organic tortilla manufacturer to recall 80,000 bags of tortilla chips. The threat of cross pollination has also prevented organic farmers from planting certain crops in some parts of the country.

Numerous studies have shown the potential fallout of transgenic "insect-resistant" crops on the environment. Both lab and field studies have confirmed that pollen from B.t. corn is lethal to monarch butterfly larvae. Swiss entomologists have found that lacewings and lady bugs are negatively impacted when they feed on organisms that have ingested the GE corn. Research undertaken at the New York University shows that contrary to expectation, B.t. toxins bind to soil particles and can persist in the soil for up to 250 days. These toxins have been shown to harm soil microorganisms that break down organic matter.

Given that half of our cotton crop and nearly one-third of our corn crop are GE "insect resistant" varieties, it is alarming that such studies were not conducted earlier, underscoring the fact that the experiment with GE crops is taking place in farmers' fields and on consumer plates rather than in controlled, laboratory settings.

Insect resistance to the B.t. toxin poses a serious threat for organic farmers who use the toxin in a natural spray as part of an integrated pest management scheme. A study published in Science found that a common pest of cotton was able to build up resistance to insect resistant varieties very quickly. If the toxin is rendered useless, organic farmers will be deprived of an essential tool.

Not content with simply engineering food crops, biotechnology companies are introducing new test tube "products." GE engineered salmon that are close to commercialization may be able to "outcompete" wild salmon in reproduction and further deplete this endangered species. Genetically engineered trees are also in the product line and may introduce ecological threats to our national forests.

## Can biotech feed the world?

There is no question that the nations of the world must take action to stop global hunger. It is a travesty that 800 million people go hungry each day. Biotech proponents argue that genetic engineering is the solution to the problem because it will increase crop yields to feed a growing population. A techno-fix, however, ignores the root causes of hunger.

Hunger persists today despite the fact that increases in food production during the past 35 years have outstripped the world's population growth by 16 percent. Indeed, the United Nations Food and Agriculture Organization recently stated that growth in agriculture will continue to outstrip world population growth. The Institute for Food Policy notes that there is no relationship between the prevalence of hunger in a given country and its population. The real causes of hunger are poverty, inequality and lack of access. Too many people are too poor to buy the food that is available (but poorly distributed) or lack the land and resources to grow it themselves.

The much heralded "Green Revolution" was an example of the failure of new technology applied to farming to reduce hunger. Using the technology, developing countries significantly increased crop yields, but they nevertheless failed to eliminate hunger, because they failed to address the root social and economic causes of hunger. Furthermore, the Green Revolution exacerbated poverty and social inequality. It favored larger, wealthier farmers who could afford the new high yielding crop varieties and the chemical fertilizers, pesticides, and irrigation systems that accompanied them. Left behind were poorer farmers unable to afford such inputs. In the meantime, the heavy use of chemical fertilizers and pesticides generated resistant pests and degraded the fertility of the soil, undermining the very basis for future production.

The growing use of patents to "protect" biotechnology innovations also threatens subsistence farmers in the developing world and could exacerbate hunger. Patents have been taken out on plants, animals, bacteria as well as genes, cells and body parts. Sanctioned and imposed by the global trading system, this "commodification of life" has allowed multinational companies to patent staple crops in developing countries such as yellow beans in Mexico, South Asian basmati rice as well as medicinal herbs, livestock and marine species. Such a predatory system threatens to enable companies to maximize their control over farming processes and the world's food resources.

Landmark studies are showing that traditional farming methods, including multi-cropping and small scale techniques are proving to be just as effective in producing high yields as conventional farming. Most recently, in one of the largest agricultural experiments ever, thousands of rice farmers in China were able to double the yields of their crops simply by planting a mixture of two different rices—a practice that did not require using chemical treatments or investing any new capital. Clearly, these types of farming methods are suited to local needs and ecosystems. They will protect the environment and increase an affordable food supply. Biotechnology, however, will likely repeat the failure of the Green Revolution's fertilizers and pesticides. Biotech will not solve the problem of world hunger but may exacerbate it.

# From the Congressional Record of November 2

Mr. Speaker, Federal regulatory review of biotechnology products is patchy and inadequate. Spread out over three regulatory agencies—the Food and Drug Administration (FDA), the U.S. Department of Agriculture (USDA) and the Environmental Protection Agency (EPA)—the system is characterized by huge regulatory holes that fail to safeguard human health and environmental protection. Furthermore, independent scientific advice available to the agencies is severely limited.

Despite the fact that GE food may contain new toxins or allergens, the FDA determined in 1992 that GE plants should be treated no differently from traditionally bred plants. Consequently the FDA condones an inadequate premarket safety testing review and does not require any labeling of GE food products. The FDA has essentially abdicated these responsibilities to the very companies seeking to market and profit from the new GE products. FDA's recent proposed rule for regulating biotechnology will hardly change the present system. Although the proposal requires that companies notify the Agency before marketing new GE products, it still fails to require a comprehensive pre-market safety testing review or mandatory labeling.

The FDA's 1992 decision to treat GE food as "substantially equivalent" to conventional food (thereby exempting most GE food on the market from independent premarket safety testing or labeling) is a violation of the public's trust and an evasion of the Agency's duties to ensure a safe food supply. The concept of "substantial equivalence" has been challenged in numerous scientific journals. FDA's failure to label GE foods led a 1996 editorial in the New England Journal of Medicine to conclude that "FDA policy would appear to favor industry over consumer protection."

EPA's regulation of environmental hazards is equally inadequate. Under the nation's pesticide laws, EPA regulates biological pesticides produced by plants. It does not, however, regulate the plants themselves, leaving that duty to the USDA. Consequently, EPA regulates the B.t. toxin, but not the corn, cotton or potato plants exuding the toxin. EPA has allowed B.t. crops to come to the market without conducting a comprehensive environmental review. Much further research is needed on the impacts of "pest protected" crops as outlined by a National Academy of Sciences report. For plants engineered for other traits, such as herbicide tolerance or disease tolerance, EPA does no environmental review at all.

The USDA's Animal Plant and Health Protection Service (APHIS) is charged with evaluating potential environmental impacts of field tests of GE crops. However, having virtually abandoned its original permit system which registered an environmental impact assessment before a field test, the Agency can no longer claim to be doing its job. APHIS has adopted a much less rigorous "notification" system which permits researchers to conduct field trials without conducting an environmental risk assessment and without submitting specific environmental impact data.

The National Academy of Sciences (NAS), the premier scientific body in our nation, has recently published a scientific assessment of GE foods. Unfortunately, many of the scientists on the NAS review committee had financial links to the biotech industry. The failure of the NAS to find an unbiased panel is problematic because their mission to supply decision makers and the public with unbiased scientific assessments cannot be achieved. This reduces the lack of independent science for our regulatory agencies to rely upon.

## Popular demand for an evolution in policy regarding GE food

A strong testament to consumers' desire for labeling and greater safety testing of GE food is the flurry of legislative activity and ballot initiatives that have taken place at the state and local levels. Over the past year, the city councils of Boston, Cleveland and Minneapolis have passed resolutions calling for a moratorium on GE food, and Austin has called for the labeling of all GE food. Boulder, CO has banned GE organisms from 15,000 acres of city-owned farmland. Bills requiring labeling of GE food were introduced in the state legislatures of New York, Minnesota, California and Michigan. The state legislature in Vermont considered legislation that would require farmers to notify the town hall if they were planting genetically engineered seeds. In California, a task force is exploring whether schools should be serving GE food and in 1999 a petition signed by over 500,000 people demanding labeling was submitted to Congress, President Clinton and several federal agencies including the FDA.

In survey after survey, American consumers have indicated that they believe all GE food should be labeled as such. Consumers have a right to know what is in the food they eat and to make decisions based on that knowledge. While some observe strict dietary restrictions for religious, ethical or health reasons, others simply choose not to be the first time users of these largely untested foods.

The failure to label GE crops and food is short-sighted and could close off key markets for U.S. farm exports. Labeling protections have been established in Europe, Japan, South Korea, Australia and New Zealand. The Cartagena Biosafety Protocol drafted early this year allows nations to refuse imports of GE organisms.

## Other impacts of GE foods deserving attention

The gene revolution is being led by the agribusiness industry. These are a handful of multinational companies which own much of the world's supplies of seeds, pesticides, fertilizers, food and animal veterinary products. The result of numerous acquisitions and mergers, the agri-business conglomeration has spent millions of dollars on research and development of GE products. Given such heavy investment, it should come as no surprise that its primary goal is to recover its expenses and turn a profit.

It is to profit-seeking companies, therefore, that we are ceding the right to re-engineer the earth—our plants, our food, our fish, our animals, our trees, even our lawns. Genetic engineering in agriculture should be considered a commercial venture that includes the privatization of agriculture knowledge through the patenting system and the increasing concentration of key agricultural resources in a handful of multinational agricultural companies.

Marketed by agrichemical companies, genetic engineering in agriculture promises to perpetuate the present industrialized system of agriculture—a system characterized by large farms, single cropping, heavy machinery and dependence on chemical pesticides and fertilizers. Such a system has consolidated acres into fewer and larger farms, marginalizing small farmers and reducing the number of people living on farms and in rural communities.

With a goal of marketing GE seeds worldwide, genetic engineering will continue the trend of industrialized farming to reduce crop diversity, making our food supply increasingly vulnerable to pests and disease. The Southern Corn Leaf Blight which in 1970 destroyed 60 percent of the U.S. corn crop in one summer, clearly demonstrates that a genetically uniform crop base is a disaster waiting to happen. The linkages of genetically engineered seeds and pesticides, such as Monsanto's GE Roundup Ready Seeds will ensure continued use of agricultural chemicals.

Genetic engineering is likely to further diminish the role of the farmer. GE seeds are designed to be grown in a large scale agricultural system in which farmers become laborers or "renters" of seed technology. Desperate to increase their yields to make up for low prices, many U.S. farmers have adopted the "high-yielding" GE seeds. In doing so, they have been forced to sign contracts legally binding them to use proprietary chemicals on their transgenic crops and in some cases to permit random inspections of their fields by biotechnology company representatives who check that farmers are not saving and reusing the licensed seed. Despite the premium farmers pay for high tech seeds, they receive no warranty for the performance of these seeds as the contracts protect biotechnology seed companies in the event of seed failures.

## A protective regulatory structure

Despite the uncertainties associated with genetic engineering, nevertheless, GE crops covered 71 million acres of U.S. farmland last year, and GE ingredients are present throughout the food supply. Ranging from ice-cream and infant formula to tortilla chips and veggie burgers, foods produced using genetic engineering line our supermarket shelves. These foods are unlabeled and have not been appropriately assessed for safety. Consumers, therefore, are unwitting subjects in a massive experiment with their food.

Our regulatory system has clearly failed to ensure the protection of human health, the environment and farmers. In response I have authored legislation in the 106th Congress that would fill the regulatory vacuum.

To ensure food safety, I have introduced a bill that requires that GE food go through the FDA's current food additive process, acknowledging that a food is fundamentally altered when a new gene is inserted into it. The review process would look at concerns unique to GE products including allergenicity, unintended effects, toxicity, functional characteristics and nutrient levels.

To date, the public has been largely left out of the biotechnology regulatory process, and that needs to change. Consequently, I propose that the FDA conduct a public comment period of at least 30 days once a completed safety application is available to the public. All studies performed by the applicant must be made available including all data unfavorable to the petition. The FDA should also maintain a publicly available registry of the GE foods for which food additives are pending or have been approved.

When the FDA was called upon to confirm the Taco Bell taco shell contamination for a possible regulatory enforcement action, it was unable to do so because it lacked the necessary testing protocols. The FDA should correct this failure by immediately creating testing protocols for all GE foods and test for potential contamination in these foods. Until then, the FDA cannot determine the ingredients in our food supply, it is unlikely that the FDA can ensure the American public that other foods are not contaminated.

I have also introduced a bill requiring mandatory labeling of GE foods or foods containing GE ingredients so that American consumers can make informed choices about what they are eating. Packaged foods carry nutritional labels, drugs and medications come with descriptions of their contents. There is no reason that GE food should not also be labeled granting consumers their fundamental right to know what is in their food.

Clearly, environmental regulations for the release of the GE organisms need to be strengthened. Similarly, the USDA allows field trials of all GE plants that prevent adequate assessments of the environment risks posed by these plants. Though genetically engineered fish are predicted to be commercialized by 2001, it is still unclear which agency will regulate them. The US Fish and Wild Life Service as well as the National Marine and Fish Service must play a role in developing regulations for GE fish.

Finally, Congress should hold hearings on the failure of the regulatory agencies in protecting the American public.

## Conclusion

The controversy surrounding genetically engineered food should not be a surprise to anyone. The mechanical manipulation of genes in the food one eats instinctively raises questions of health and safety. We instinctively trust farmers to grow and raise our food, but we must question the motivation of large corporations who want to create impure food for pure profit. When we feed our family, we don't take chances. If we are not sure how old the leftovers in the back of the fridge are, we throw them out. And as long as we are not convinced that this new technology is flawless, people should be hesitant to serve genetically engineered food to their children. New technologies always have unforseen effects. The American consumer does not want to be a part of an experiment at their dinner table.

# 22

# Food Industry Perspective on Safety and Labeling of Biotechnology

**Gene Grabowski**

*Vice President, Communications, Grocery Manufacturers of America, Washington DC, USA*

## Safety of Modern Biotechnology

The decision as to whether or not modern biotechnology will be accepted and used in foods, medicines, clothing, and household goods in the United States and around the world ultimately rests with consumers. In this regard, the food and consumer products industry serves consumer demand and remains neutral on the use of biotechnology to enhance whole and processed foods, animal feeds and apparel derived from natural plants like cotton.

But from the perspective of industry researchers, government regulators and independent scientists who have scrupulously studied the safety and applications of modern biotechnology, the overwhelming verdict is that the technology is safe and offers a multitude of benefits to consumers everywhere. By using biotechnology, producers, in fact, believe they can provide a more abundant, better quality and more nutritious food supply to the world's consumers.

Even so, some opponents of modern biotechnology have charged that biotech crops and food ingredients have not been adequately studied for safety. That accusation is totally false. In fact, no other foods in history have been tested and observed as diligently as the foods developed from modern biotechnology. For more than two decades before they were approved for the consumer market, biotech crops and their derivatives were studied in the laboratory and field to ensure their safety for human and animal consumption and to determine their impact on the environment.

After lengthy discussion and some vocal dissent, industry researchers and government regulators concluded in 1992 that biotech crops and food ingredients do not differ in any substantial way from ordinary food products in their composition or impact on the environment and thus should be regulated in the same manner. For that reason, the US Food and Drug Administration, the US Department of Agriculture and the US Environmental Protection Agency regulate biotech crops and foods by the same standards used for other crops and foods.

To date, no evidence exists to suggest that a single person or animal has ever been harmed by biotech foods or that it has a negative impact on the environment. Nevertheless, the food and consumer products industry supports continued vigorous study and oversight of modern biotechnology to ensure its continued safe use.

The food industry believes that if consumers do eventually decide to turn away from biotech foods, they should do so only after careful consideration of the facts about the safety of technology and of the consequences of abandoning a science that offers hope to so many in the developing world.

Some relevant facts and figures:

World population has topped the 6 billion mark, an increase of 1 billion people in the last 12 years alone. The United Nations estimates that figure could reach up to 10.7 billion by the year 2050 – with 95% of that growth in some of the world's poorest regions. The International Service for the Acquisition of Agri-biotech Applications (ISAAA) estimates that some 800 million people are already victims of malnutrition daily. Because the world's arable land is finite, the only known way to feed such a growing population is through the application of modern biotechnology.

No technologies will be able to approach biotechnology's potential to help counter world hunger in the next century. According to the 1997 World Bank and Consultative Group on International Agricultural Research (CGIAR), biotechnology is expected to help increase food productivity by up to 25% in the developing world.

Farmers and other experts have conducted more than 20 years of outdoor testing for each crop variety enhanced through biotechnology. Fifty varieties of biotechnology-enhanced crops have been approved in the United States, as of the year 2000.

At the time of this writing, researchers working at state colleges and universities around the United States are close to developing fruits and vegetables that contain more beta-carotene and vitamins C and E, which may help to reduce incidences of cancer and heart disease.

Scientific facts and growing nutritional needs in the developing world seem to point firmly in the direction of allowing biotech foods and other consumer products to develop. But our industry believes that in the end, it is inevitable that consumers – not scientists or government regulators or activist groups – will decide the fate of biotech foods in a free market.

# Health and Nutritional Benefits of Modern Biotechnology

It is the position of the food and consumer products industry that the ability to identify and encourage beneficial traits in food crops represents nothing new. Farmers have crossbred and hybridized plants for centuries. What modern biotechnology brings to that process, however, is the ability to target very specific characteristics for enhancement, providing farmers and researchers with the latest tools in the search for better, more healthful foods. In fact, the industry believes biotechnology makes the process safer and more precise.

The most common crops used today to develop ingredients for processed foods are grains, fruits, and vegetables containing pesticide-resistant and herbicide-tolerant characteristics that result in the reduced need for chemical applications and which ease erosion and wear on cropland. These more resilient plants can tolerate farmers' application of very specific herbicides for weed control, thus reducing the overall need for chemical applications and stress to the world's natural resources.

While the food and consumer products industry is now doing little measurable research itself into biotechnology, researchers working at colleges and universities, as well as in laboratories at life sciences companies, are working to develop new crops and ingredients that will yield improved health and nutritional benefits, including:

- Grains, fruits, and vegetables that contain more nutrients, such as proteins, vitamins, and minerals, and have reduced fatty acid profiles.
- A new rice under development to counter vitamin A deficiency, the leading cause of blindness in children in the developing world.
- Modified potatoes with more solid content, permitting less oil to be absorbed during cooking and making for a healthier french fry or potato chip.
- Improved nutrients in strawberries, and vitamin-enhanced sweet potatoes and canola (rapeseed) oil.
- Allergen-free rice and peanuts.
- Fruits and vegetables that contain more beta-carotene and vitamins C and E.
- A banana that can be used to deliver vital oral vaccines for diseases such as hepatitis B.
- Tomatoes with even more naturally occurring antioxidants.

Once consumers are able to enjoy the more concrete and direct benefits of biotechnology now under study, most experts in the food and consumer products industry believe there will be little or no debate about the risks and benefits of the helpful technology. Of course, much may depend on the swift and successful application of these beneficial traits if criticisms about the 'unknown effects' of biotechnology are to be countered.

## Regulatory and Labeling Issues

Perhaps the most controversial argument in the debate over modern biotechnology is whether or not biotech ingredients should be included on food labels. Indeed, for food and consumer product manufacturers, this is the key issue.

Food labeling, in general, serves a vital purpose. Labels list essential information such as ingredients and quantities, describe features, give instructions, explain benefits, and deliver advisories and warnings. Federal law mandates that any information considered critical to health and safety appear on the label.

Labels must also be effective signals. None of the information on the label can accomplish anything if consumers do not notice the product and consider its suitability for their purposes. In stores where hundreds and thousands of competing products are sold, labels are the means by which producers win the consumer's attention.

For these reasons, the label is the single most important resource for consumers buying packaged foods. Unnecessary information on a label can drown out critical messages, or worse, confuse consumers.

Designing an effective label is complicated because the amount of material that can be incorporated on a label is limited by its size and by the ability of consumers to absorb information. The label cannot tell every consumer everything he or she might want to know about every product, because different consumers care about different things.

A fundamental challenge facing the food manufacturer is to design labels that communicate effectively the information that matters most to the consumers of its products. Labels that carry too many messages may fail to deliver the most important ones, and there is no question as to the importance of the information about contents, ingredients, nutrients, and quantities that labels must convey now.

The question at hand is whether information about modern biotechnology should be included on every label of a food produced with it.

From the industry perspective, there is no scientific reason to include biotech ingredients on food labels unless the ingredients constitute a significant change to the food. Labeling without scientific reasons undermines sound public policy. This is because such labeling can easily be confusing and alarming to consumers. Biotech information may be of interest, but by FDA standards it is not essential and thus could detract attention from other more important information on the label.

While protection of the consumer requires essential information to be displayed on labels of foods, the effectiveness of the label requires that other claims remain voluntary, subject only to the requirement that they be accurate and substantiated.

A food labeling policy that strikes this balance not only protects consumers but also preserves their ability to choose, because it permits manufacturers to communicate effectively to them. Indeed, consumers and manufacturers are legally entitled under free speech guarantees such as the First Amendment of the US Constitution to a policy that does not require labels to deliver unnecessary information.

Consumer product regulators have the final word on which information is mandatory for food labels. Their objective is to maintain a legal system that protects consumers from harm and requires manufacturers to describe accurately the products they sell. In order to achieve that purpose, regulators must apply both the physical sciences, like biology and chemistry, and the social sciences, like economics and consumer behavior, when devising the rules that govern product labels.

Nearly a century of experience at regulatory agencies in the United States and around the world has yielded the basic elements of successful consumer protection regulation. These elements include prohibiting product claims that deceive consumers, expecting manufacturers to have the evidence appropriate to back up the types of claims that they make, and requiring disclosure of information that is necessary to inform consumers about such basics as the quantity, ingredients and safety of the contents of a package.

The safety and nutritional value of foods enhanced through biotechnology are carefully regulated prior to commercialization. There are several agencies involved in the inspection and regulation of new plants and food products derived from biotechnology, ensuring their safety before they are incorporated into foods for the commercial market.

Biotech foods and food products must conform with state and federal marketing standards, including state seed certification laws, the Federal Food, Drug, and Cosmetic Act (FFDCA), the Federal Insecticide, Fungicide and Rodenticide Act (FIFRA), the Toxic Substances Control Act (TSCA) and the Federal Plant Pest Act.

Under the Food, Drug, and Cosmetic Act, the FDA usually focuses on the food, rather than the processes used in its production, as the basis for regulation of food products, including those derived from biotechnology. The FDA offers a decision-tree approach for companies developing plant-based biotech foods. These decision-trees represent a series of testing procedures that enable food processors to anticipate safety concerns and to consult with the FDA as necessary for regulatory review of new plant varieties and product testing under development.

As noted in detail elsewhere in this book (see Chapter 9), the FDA assessment process focuses on the following areas:

- The safety and nutritional value of newly introduced proteins.
- The identity, composition and nutritional value of modified carbohydrates, fats or oils.
- The concentration and bioavailability of important nutrients for which a food crop is consumed.
- The potential for food allergens to be transferred from one food source to another.
- Toxins characteristic of host and donor plants.

To ensure safety, a variety of toxicological and other product safety data may need to be supplied to the FDA for food products or ingredients produced through the use of new plant varieties and biotechnology. USDA currently has primary responsibility for assessing the ecological effects of new plants developed through biotechnology. The Animal and Plant Health Inspection Service (APHIS) within the USDA is the primary agency regulating the safety testing of biotechnology-enhanced plants. APHIS approval generally must be obtained before proceeding to field-test or market a biotechnology-derived plant.

In order to test a biotechnology-derived plant in the field, applicants petition APHIS for an environmental release. The applicant must include information on the plant, the origin of any new genes or gene products in it and the purpose and method of conducting the test. APHIS evaluates the application and any potential environmental impact of the proposed test field. Once testing is approved, APHIS and state agriculture officials may inspect the test field throughout the testing process to ensure that tests are conducted safely.

Before biotech-enhanced crops can be grown commercially, even more information must be submitted to APHIS, including scientific details about the plant, results of field tests and any indirect effects on other plants. This petition is published in the Federal Register, giving the public time to comment. APHIS will approve a petition for commercialization only when it has determined that the plant poses no significant risk to the environment and is as safe as traditional varieties.

The EPA has jurisdiction over crops that are insect and disease resistant under the Federal Insecticide, Fungicide and Rodenticide Act (FIFRA). EPA regulates environmental exposure to pesticide substances produced in crops to ensure that there are no adverse effects on the environment, including nontargeted insects, birds, fish, deer, and other mammals.

The abundant oversight and regulatory safeguards for biotechnology already in place in the United States ensure that consumers are well protected. Nevertheless, the food and consumer products industry is working with the government to strengthen existing regulations and develop explicit guidelines for the application of voluntary labeling to bolster consumer confidence in the regulatory structure. This approach will make certain that consumers maintain their right to choice among biotech and non-biotech products, while researchers can safely explore the potential benefits of modern biotechnology.

## The Food Industry's Position

The Board of Directors of the Grocery Manufacturers of America (GMA), the world's largest organization representing food manufacturers, strongly supports the current US government labeling policy for foods derived through biotechnology. This means GMA supports the labeling of biotech foods where there is a significant compositional change, where the food is nutritionally different from its traditional counterpart, or where a potential allergen has been introduced.

Some critics of US labeling policy have urged the government to mandate the disclosure of genetic techniques used in the development of a product, even if the food that results is substantially equivalent to its traditional counterpart and even though it presents no demonstrated health or safety risk to the population. These critics seem to be dissatisfied with policy based on whether a substantial difference exists between the modern food and its traditional counterpart. Their argument is that some consumers prefer foods produced without modern science, so its use should be disclosed on the label.

Labeling policy based on this rationale is not only unnecessary, but actually detrimental to consumers. It is unnecessary, because competitive, free markets offer consumers the products they want and the information they need to find them. Consumers who desire to purchase traditional or specialty foods create opportunities for companies eager and able to serve them. For example, consumers today may choose from a bounty of organic, natural, or kosher offerings that are identified and labeled voluntarily under federal guidelines. Moreover, organic food marketers and other specialty food stores are highly adept at labeling (voluntarily) products that contain no biotech ingredients, thus attracting and informing consumers who desire non-biotech foods.

More importantly, mandating special labeling of modern biotechnological methods used to develop foods would require expensive testing and handling of all foods to determine which ones needed a special disclosure. Farmers and food processors would require separate grain silos, trucks, manufacturing sites and expensive testing systems to ensure the separation of biotech and non-biotech crops and ingredients.

Consumers would ultimately pay the costs for such meaningless separation. Statistical surveys conducted by the food industry and the government also show consumers would likely assume that the government had imposed the regime because it was necessary to alert the public to potential safety risks that technology had introduced to the labeled foods.

Because this message would not be intended, and would be contrary to the conclusion that the government had reached about the safety of those foods, the policy could have the unintended consequence of alarming consumers without reason rather than helping them make informed choices.

Overall, the United States' current approach to consumer protection is perfectly compatible with modern science and technology. In a market system driven by consumer choice, government regulation should protect consumers from real risks while helping them make informed decisions. In this light, it is important to remember that FDA can stop a food product – biotech or non-biotech – from being sold at any time if it determines that a product or ingredient is unsafe.

In fact, it is the rational, science-based labeling system employed by US food regulatory agencies that is the source of consumer support and trust for our food agencies. Trust in our system would erode dangerously if US regulatory agencies mandated slapping meaningless labels on food and consumer products because some activists, hiding behind the 'right to know' mantle, called for them.

And where would such mandatory labeling end? It would be nice to know, for example, if the tomatoes in a can were hand-picked or machine-picked, but because such information is irrelevant to the safety or content of the food, the government has no business mandating a label indicating the difference. Of course, it is the genius of the current US system that a food manufacturer may publish that kind of information on a label if it so chooses – as long as the information is truthful and non-misleading.

A policy that raises unwarranted suspicion of research and development, as mandatory labeling of modern biotechnology could do, would ultimately deny the public the benefits of innovation. The value of biotechnology, like the legacy of Gregor Mendel and the horticulturist Luther Burbank, should be judged by the fruits of the methods, not by an unsupported suspicion of the methods themselves. It would be a great loss to humankind to allow activists to drive a technology with proven benefits and even greater potential into extinction.

By mandating only essential information, allowing voluntary claims about modern biotechnology, and demanding accuracy in all labeling, the US government's policy governing food labels safeguards the rights and safety of consumers.

# 23

# GMO Regulations: Food Safety or Trade Barrier?

**Malcolm Kane**

*Head of Food Safety, Sainsbury's Supermarkets Ltd (1980–1999)*

I cannot think of a more important subject matter facing the US soya (soybean) and maize (corn) export industries than the current controversy about GM foods and the European attitudes and responses to them. If I can serve to throw a little light upon this fraught subject here in the US, I will be professionally very satisfied.

Food is a subject of enduring passion, and has been for many generations. Not only that, but the international reputation for culinary excellence has been a matter of pride between nations for as long as humans have traveled.

Now this reinforces my personal belief that in the whole business of international relations in all sectors of the food trade there resides a degree of natural suspicion, which is entirely understandable when one reflects upon the various examples of venerable disputes that have exercised our respective diplomats . . . dare I mention bananas?

My task is to dispel your suspicions that the issue of genetically modified foods falls into this category of diplomatic barter, and convince you that there is a real and serious consumer issue in Europe that may well spill over here. When first approached by the conference organizers, I was given six guideline questions to address which at first sight led me to think I would have the quickest and easiest paper to deliver of all time.

The questions asked were:

1.  What changes have there been in European consumers' attitudes towards specific products?
    Answer?
    From an ill-informed indifference to an outright hostility.

*Genetically Modified Organisms in Agriculture*
ISBN 0-12-515422-4

2.   What products have European consumers rejected?
     Answer?
     Everything . . . doubled and redoubled.
3.   What has been the effectiveness of consumer advocacy groups?
     Answer?
     100%. They have won the argument . . . period.
4.   What are the different strategies for meeting consumer demands? (GM
     labeling is effectively a health warning)
     Answer?
     One strategy . . . total rejection of GM foods.
5.   What are the costs of consumer assurances?
     Answer?
     Businesses have been severely damaged.
6.   What specific examples illustrate consumers' objections to GMOs and
     retailers' responses to these objections?
     Answer?
     Soya, maize, and tomato paste, on the latter of which I will now give more
     specific details.

## Background

Three years ago, Sainsbury's and Safeway (UK) supermarkets were the first to
launch a canned tomato paste, boldly and properly labeled made from genetically modified tomatoes. It was sold at a 25% price premium, and achieved a
150% market share, over ordinary (Italian) tomato paste. Although this may
seem a good ratio of price premium to market share, those of you familiar with
the marketing realities will know this result was modest.

But it was on the positive side, and the prevailing sentiment at the time was
to persevere with the promotion of GM foods as the vision of the future. In my
view there is a tendency among a certain section of food scientists to become
over-enthusiastic with what I call the 'appliance of science' to food. This is incau-
tious as it risks upsetting some very deep, fundamental human attitudes and
responses to the food we eat.

It is not for nothing our shared language is peppered with clichés like bread
and butter, mother's milk, motherhood and apple pie, milk of human kindness,
have your cake and eat it, etc.

I will return to this theme, but to continue the history of GM tomato paste in
Sainsbury's and Safeway, the product was launched well (i.e., it was clearly
labeled and offered with a non-GM alternative choice). In short, no supermarket
chains besides Sainsbury's and Safeway had taken such a bold, positive, supportive position on GM foods.

All that changed in February 1999 when the combined efforts of consumer
advocacy groups and the press created a consumer awareness of the GM foods
issue that resulted in what today is being described, in the UK, as the greatest
ever victory for consumerism. This is a victory that may well set a precedent for
consumer advocates as they learn to develop this newfound power to influence
consumer purchasing behavior.

From sales of 150% over conventional Italian tomato paste, the sales slumped relative to the conventional product. In other words, sales fell off a cliff. And they never recovered.

Additionally, customers stopped coming through the doors. They positively shifted their trade to smaller chains that were taking advantage of the furore by advertising that non-GM foods only were sold in their stores. They may or may not have had justification for making these claims, but they were not challenged by the advocacy groups about that. They were in fact supported by the advocacy groups.

## A Non-GM Strategy

At this time I was then instructed to find a way of delivering a credible commercial non-GM strategy to our customers. Our previous promotion of GM tomato paste was unfortunately actually working to our disadvantage. The focus was upon GM soya in the first instance, with GM maize not far behind, and animal feed next in priority.

Conventional wisdom, supported at the time by advice from the industry, was that separation of GM soya from non-GM soya was impossible.

We had a very simplistic approach. The GM issue was more a problem of food safety perception than of food safety reality. But the marketing dictum is that perception is reality.

To understand the forces of perception and reality at work in this complex area, we prepared a mind-map. This maps out not our personal views, but the totality of views, opinions, facts and fallacies that are firmly embedded in consumers' minds and influencing their behaviors.

We therefore applied the basic principles of food safety control that we had been applying for 20 years to real food safety issues, to this new 'perceived' food safety issue. And it worked.

The statistics bear the evidence. From the European perspective, it was evident that two-thirds of world soya production came from Brazil and the US. Of this about 50% of the US production, and 95% of the Brazilian production was non-GM, a total of at least 50 million tonnes. In all, more than double the total European soya imports was available in non-GM soya from either the US or Brazil.

The issue was not *if* but *how* we could deliver non-GM soya. In other words it was a management problem of how to ensure security of supply by establishing systems of traceability which paralleled those we were already using to secure Italian olive oil as distinct from Greek olive oil. Or Florida orange juice as distinct from Brazilian or Spanish orange juice.

What was needed was a management system of control. And that is precisely what we developed. We arranged for a British company, Law Laboratories, who collaborated with an American company, Genetic ID, to produce a management control standard which delivered three elements:

- It defined what we meant by GM and non-GM food.
- It described a HACCP based audit and traceability scheme.
- It defined GM analysis to achievable standards of accuracy, reproducibility and repeatability.

In addition, we organized a consortium of European supermarkets who all had similar commercial imperatives, to deliberately get the weight of numbers publicized and a momentum of thinking within the industry, that it was possible to deliver a credible non-GM sourcing policy.

Together we agreed to reformulate soya products out of our own brand lines in a deliberate move to signal the seriousness of our intentions. The simplest product was soya oil. This could easily be substituted with rapeseed (canola) oil or sunflower oil. Similarly, we found that soya protein products could in many instances be substituted with non-soya proteins. This may upset the soya protein scientists, but the fact of the matter is that in many markets the benefits of soya protein over alternative proteins had been oversold. Actually this is not unique to soya – many food ingredients are substitutable.

As a result, a concerted effort by seven of the leading European supermarkets to formulate soya derivatives out of their own brand products had a dramatic commercial effect.

This commercial support was reinforced by the recruitment of the major European international food manufacturing groups in support. The reason for their support is significant. Firstly they had been experiencing severe cross-branding effects with GM labeling. The legal requirement to label foods containing GM ingredients impacted adversely not only upon the individual product line, but also upon all product lines under the same brand umbrella regardless of whether they all had GM ingredients or not.

Secondly, they were unaware of the 'Due Diligence' requirement to implement chemical residue surveillance of herbicide residues in GM herbicide-resistant crops.


## Herbicide Resistance

Why would anyone put weedkiller onto their food crops? Conventional food crops would self-evidently be destroyed, and in fact historically, weedkillers have not been approved for use on food crops. The answer is of course that weeds compete with crops for soil moisture and nutrients, reducing crop yields. So the farmer's need for weed control is understandable.

GM herbicide-resistant crops allow the easy and effective application of weedkiller without fear of crop damage and economic loss.

But this means that weedkillers are being used in a new way, i.e., they are being sprayed onto food crops for which new approval has been required, and has been obtained, a fact which has been poorly appreciated by professional managers within the food industry till now.

As with all pesticides, weedkillers (or herbicides) will leave residues within the food crops they are applied to. And as with all pesticides, it will now be necessary for the food industry to conduct regular, 'Due Diligence' herbicide residue analysis. The cost of this has never been introduced into the GM debate, and the UK has been importing herbicide-resistant soya for three years, with little herbicide residue analysis being done in any UK laboratories anywhere.

More importantly, if we care to consider the severe consumer confidence issues surrounding any question of chemical residues in foods, then the only conclusion can be that food professionals have been ill-informed at best, to have ignored or neglected this as a potentially serious 'consumer perception' issue. Glyphosate, the active agent involved with the main herbicide-resistant soya product, is at the lower end of the toxicity spectrum of herbicide chemicals, but that is not likely to change the perception of this issue by the average European consumer, i.e., the principle of applying weedkiller to foods.

The biotechnology companies cannot be blamed for focusing on their product's advantages but they are not likely to focus on potential disadvantages; no commercial organization will. The law of caveat emptor applies. It is up to food industry professionals to protect their business interests by thinking through such issues and highlighting the potential disadvantages regarding adverse consumer safety perceptions. Had the food industry scientists done so, instead of being blinded by the elegance of the GM technology, we all might have been spared the stresses of the last couple of years.

The dilemma in which European food businesses (i.e., food retailers and manufacturers alike) found themselves is whether they should preemptively react to these ultra-sensitive issues, or do nothing. The overwhelming prevailing response of the food industry was to act openly in accordance with their understanding of consumer sensitivities. The food industry also calculated that the only way to rescue the GM issue from the grip of the consumer advocacy groups was by being seen to take the lead on behalf of the consumer. A similar approach was taken with GM insect-resistant crops.

## Insect Resistance

Pesticide residue surveillance data confirms that in the UK, and very probably in the US, we usually experience an incidence of pesticide residues in food crops of less than 1%. The level of residue contamination is typically less than 0.5 ppm within that 1% incidence, and we accept such statistics as good evidence of well-managed agricultural operations. Farmers generally manage their crops well.

In introducing the *Bt* gene into food crops, scientists have created crops with *Bt* toxin throughout the crop tissue fluids. All crops will be affected, i.e., the incidence of toxin presence within the crop will be 100%, by definition. This is 100 times the incidence of conventional pesticide presence.

With well-established integrated crop management systems, and effective adherence to pre-harvest intervals (PHI), i.e., where crops are not harvested for set periods of several days after the last pesticide application, there is an acceptable level of control, which results in acceptable residue levels as above. However, there can be no pre-harvest interval with GM insect-resistant crops. Genes cannot be programmed to switch off a set number of days before an unpredictable harvest date. GM insect-resistant crops will be harvested and consumed with the biologically active concentration of toxin present within their tissues. This can apparently be 10 ppm or more, i.e., many times the residue level of conventional pesticide. Combining these two points means that consumers of GM pesticide-resistant crops will consume many times as much (natural) toxin as they will of conventional pesticides from conventional crops. The claim that the *Bt* toxin is natural is contentious as frankly there is nothing natural about the process of taking a bacterial gene and transferring it into a plant, to express a toxin in concentrations for human consumption which are several orders of magnitude greater than could possibly be the case in nature. Additionally there is no established maximum residue level (MRL) for the *Bt* toxin, because it is classified as a natural toxin.

As an aside, should professionals not have asked why this new use of *Bt* toxin in GM insect-resistant crops should not have triggered the introduction of MRLs for *Bt* toxin?

## Summary

There is comparatively little concern within the European food industry about the fundamental safety of GM foods in principle (though there are some notable and very respectable exceptions). We are aware of the widespread use of GM technology in the production of enzymes and other processing aids, which has so far attracted little concern.

The potential benefits of GM technology in human and animal health care are immense and there is a real danger that the crisis of consumer confidence over GM food and food ingredients will act to the detriment of advances in healthcare and medicine. This would be a tragedy. The medical benefits of GM technology must be fully supported, rather than undermined by hasty commercial GM food developments.

Most of us, including myself, would defend the use, for example, of GM-derived enzymes as delivering real customer benefit. GM-derived cheese coagulant for instance, delivers both animal welfare benefits in the eyes of the customer, and hygiene benefits in the eyes of the food safety professional. In short it is important in this debate to distinguish between the different applications of GM-derived food materials.

The key issue is that any new food product/ingredient/processing aid can only hope to be commercially successful in today's market if it is seen to deliver a tangible customer benefit. GM insect resistance and GM herbicide resistance have failed to be seen to deliver this. There may be tangible customer benefits associated with them, but if so they have not been seen to be so. Which means they have not been communicated.

This failure to employ effective customer communication, coupled with an insufficiently critical appraisal of customer benefits by the food industry before accepting GM foods, is the final conclusion that must be drawn over the events of the last couple of years.

The lessons must be learned.

1. We must critically appraise all novel food, GM or otherwise, for a real customer benefit before accepting them.
2. We must develop robust defenses for the use of GM-derived food materials such as enzymes, where we are confident of the customer benefit.
3. We must communicate the food industry view of both those GM foods we approve of and those we do not. Only in this way will we wrest control of the agenda from out of the hands of the press and consumer advocacy groups, into the hands of the food professionals, where it belongs.

24

# Genetically Engineered Food: Make Sure It's Safe and Label It

**Michael Hansen[1]**

*Consumers Union, Yonkers, New York, USA*

## Summary

Genetic engineering of food raises a host of questions about effects on the environment, economic impacts, and ethics. However, perhaps the most fundamental question about such food is whether it is safe and wholesome to eat.

In this chapter, we argue that the US Food and Drug Administration (FDA) should establish a mandatory process to assure the safety of genetically

---

[1] Michael K. Hansen PhD, a Research Associate with the Consumer Policy Institute, a division of Consumers Union (publisher of Consumer Reports magazine), currently works on biotechnology issues. He is the author of Biotechnology and Milk: Benefit or Threat?, published in 1990. He has been largely responsible for developing CU positions on safety, testing and labeling of genetically engineered food. Dr Hansen has testified at hearings in Washington DC, many states, and Canada, and has prepared comments on various proposed US governmental rules and regulations on biotechnology issues. He has been quoted widely in national media on safety of genetically engineered foods. He served on a Biotechnology Advisory Committee set up to advise the CGIAR (Consultative Group on International Agricultural Research) on biotechnology issues for the public sector International Agricultural Research Centers (IARCs). Dr Hansen served as an international expert in 1998 to the FAO/WHO Joint Expert Committee on Food Additives. Dr Hansen currently serves on the USDA Advisory Committee on Agricultural Biotechnology. During 1999 and through 2000 he has traveled extensively in South Asia, Europe, Eastern Europe and the US as an invited expert on potential human health and environmental concerns related to genetically engineered crops.

---

Genetically Modified Organisms in Agriculture
ISBN 0-12-515422-4

engineered food. That process must ensure the safety of engineered foods developed in other counties as well as in the United States. The review process should be designed to achieve a standard of 'reasonable certainty of no harm.'

FDA should also require mandatory labeling of genetically engineered food. In polls, some 70% to 90% of people say they want labeling, and mandatory labeling would be consistent with existing FDA policy, which requires labeling of irradiated and frozen food, as well as food ingredients. In addition, new scientific data about unexpected effects in genetically engineered plants argues for labeling as a precautionary measure. In the event that some unexpected difficulty should develop, labeling would facilitate identification of the problem. Labeling is also essential to the health and well-being of individuals with food allergies and sensitivities.

## Scientific/Safety Issues

The 'splicing' of genes into food plants from completely unrelated plants and animals – indeed even the introduction of virus and bacteria genes – raises new and unique food safety issues. Natural toxic substances and proteins that trigger food allergies may unexpectedly appear in an engineered plant. Levels of key nutrients may change. Some genes may also produce substances in food plants that can lead to antibiotic resistance. The effects of gene splicing are in fact in many ways unpredictable.

The FDA acknowledges these problems, but since the mid-1990s it has only encouraged companies to come forward for voluntary safety consultations before bringing products to market.

Such policies are not sufficient to protect the public. It is essential that the current FDA voluntary consultation process be mandatory so that consumers in the US and our trading partners abroad can be assured that all US genetically engineered foods have gone through a safety review. All major transnational corporations marketing genetically engineered (GE) crops in the US assert that they have brought all products forward to FDA for review. It should also be tougher, so that the standard of safety achieved is comparable to that for food additives. However if the review is voluntary consumers have no guarantee that all developers of GE crops will always come forward.

Of particular concern at this time is the question of assuring the safety of genetically engineered food imported from developing countries. Increasingly, developing countries are developing GE crops in their own laboratories. Many of these crops are likely to be grown for the export market – indeed this may already be the case. Among the Asian countries that have genetic engineering

---

In addition to the work outlined above, Dr Hansen has also written reports for the Consumer Policy Institute and Consumers Union on household pest control, alternatives to agricultural pesticides in developing countries, and the pesticide and agriculture policies of the World Bank and the UN Food and Agriculture Organization.

Dr Hansen received his undergraduate degree with Highest Distinction from Northwestern University and his doctorate in Ecology and Evolutionary Biology from the University of Michigan. He did post-graduate study at the University of Kentucky on the impacts of biotechnology on agricultural research.

research programs are China, India, the Philippines, and Thailand. Among the crops they are working on, according to presentations at the October 1999 Conference of the Consultative Group on International Agricultural Research in Washington DC are rice, wheat, corn, papaya, bananas, mangos, coconut, potatoes, tomatoes, peppers, cucumbers, and tobacco. China reported that 53 transgenic varieties are in commercial production or have been field tested (Zhang, 1999). Other countries with research programs include Iran, Brazil, Mexico, Egypt, South Africa, and Malaysia. Systems for assuring the safety of these crops vary widely.

The FDA needs to develop a comprehensive program for assuring the safety of genetically engineered imported food. This should include requirements that all GE crops developed in other countries and sold in the US go through an FDA safety review. To achieve this, FDA will have to begin testing imported food for presence of GE foods. It will also be essential to make mandatory its current system of FDA review, for all products sold in the US market, regardless of country of origin.

The current FDA review process also needs to be strengthened and made more transparent. The 'decision-tree' approach outlined in the FDA 1992 policy appears to give wide latitude to companies to self-evaluate the safety of their GE products. One approach advocated by Environmental Defense and others is to require GE foods to go through a food additive petition review – the review required for such things as artificial sweeteners and preservatives. From the public's point of view, GE foods should meet the standard for food additives of 'reasonable certainty of no harm.' Like additives, genetically engineered varieties are not essential to the food supply, and there is no reason to introduce any GE food that in any way makes food more hazardous either to the general population, or to particular subpopulations.

While FDA's review system needs to be strengthened, it is possible that a food additive review may not be the optimal system for evaluating safety, and that a review process specifically designed for genetically engineered food could yield both greater assurance of safety for consumers and more efficient use of FDA resources. FDA should develop specific proposals in this area and propose them for public comment.

FDA should have clear protocols for evaluating the known risks of GE foods, which include introduction of toxins, allergens, and nutritional changes; for catching any unexpected effects that have health implications; and for addressing any public health risks that GE foods may pose, such as exacerbation of antibiotic resistance. In addition, it should establish clear benchmarks for how it defines 'safety' in each case, which begin from an assumption of 'reasonable certainty of no harm' from the product. FDA should give notice and request comment of its policies in this area. We recommend that FDA consider, as benchmarks for safety, that a food not be considered 'safe' if the genetic engineering process has introduced a known common allergen, and if the final product includes an antibiotic marker gene.

## Toxins and unexpected effects

Information has appeared in the scientific literature related to the safety of foods derived from GE plants which collectively suggests that the FDA's present regulatory approach is insufficient to ensure that GE foods will not pose health risks to those who consume them. This information relates to unexpected and unpredicted effects of gene insertions, and instability of the genetic characteristics that are introduced. This information leads to the view that FDA must scrutinize genetically engineered foods more closely than it has so far, and in particular should require long-term (one to two year) animal feeding studies of the whole engineered food. Requiring a more detailed molecular characterization for each genetic transformation event will also help FDA evaluate the potential for risk and may provide a means for FDA to decide how much additional testing is needed.

The studies which lead to greater concern about unexpected effects can be put into two categories: unpredictability of the location and expression of transgenic DNA inserts; and differences resulting from post-translational processing (e.g., proteins from the same gene are not identical in differing organisms).

## Unpredictability of the location and expression of transgenic DNA underlines need for long-term toxicity tests of engineered food

The FDA maintains that GE is more precise than traditional breeding because *just* the desired gene(s) can be transferred without extra unwanted genetic material and that this increased precision 'increase[s] the potential for safe, better characterized, and more predictable foods' (US FDA, 1992). We disagree. Although rDNA techniques may be more precise than traditional plant breeding in terms of the identity of genetic material transferred, they are less precise in terms of where the material is transferred. Conventional plant breeding shuffles around aberrant versions (alleles) of the same genes, which basically are fixed in the chromosomal locations as a result of evolution. With GE (or rDNA techniques), one inserts genes on essentially a random basis, using a gene 'gun' or other techniques (e.g., use of Ti-plasmid, chemoporation, electroporation, etc.) into a plant's pre-existing chromosomes. Frequently the genetic material comes from living things with which the host organism(s) would never cross in nature.

The process of insertion of genetic material via GE is unpredictable with regard to a number of parameters, including: the number of inserts of transgenic DNA, their location (chromosome, chloroplast, mitochondria), their precise position (i.e., where and on which chromosome), their structure, and their functional and structural stability. While all of these parameters can have consequences, perhaps the most important is the random or semi-random nature of the physical location of the genetic insert. The inability to control where the insertion happens is of key importance. This means that each transformation event is unique and cannot be replicated because the precise location of the insertion of genetic material always will be different.

The variable insertion site can have a number of unpredictable, and potentially negative, consequences (Doerfler *et al.*, 1997). The insertion site can affect expression of the inserted transgene itself as well as the expression of host genes (i.e., genes in the recipient organisms). The former is known as the 'position effect.' A classic example involved attempting to suppress the color of tobacco and petunia flowers via the transfer of a synthetically created gene designed to turn off (via anti-sense technology) a host pigment gene (van der Krol *et al.*,

1988). The expected outcome was that all the transformed plants would have the same color flowers. However, the transformed plants varied in terms of the amount of color (or pigmentation) in their flowers as well as the pattern of color in the individual flowers. Not only that, but as the season changed (i.e., in different environments), some of the flowers also changed their color or color pattern. The factors contributing to the position effect are not fully understood.

The expression of host genes can be influenced by the location of the genetic insertion as well. If the material inserts itself into 'the middle' of an important gene, that gene would functionally be turned off. In one experiment, insertion of viral genetic material into a mouse chromosome led to disruption of a gene which resulted in the death of the mouse embryos (Schnieke *et al.*, 1983). If the 'turned off' gene happened to code for a regulatory protein that prevented the expression of some toxin, the net result of the insertion would be to increase the level of that toxin.

The genetic background of the host plant can also affect the level of expression of the transferred gene, which explains the common observation that varieties of the same plant species varied widely in the ease with which they can be genetically engineered (Doerfler *et al.*, 1997; Traavik, 1998). In some varieties, the trait can be expressed at high enough levels to have the desired impact. In others, the expression level is too low to have the desired impact. In general though, scientists do not really understand why some plant varieties yield more successful results in GE than other varieties.

To get around the common problem of an insufficient level of expression of a desired gene product, powerful regulatory elements – particularly promoters/enhancers – are inserted along with the desired transgene and used to maximize gene expression. The promoter has numerous elements that enable it to respond to signals from other genes and from the environment which tell it when and where to switch on, by how much and for how long. When inserted into another organism as part of a 'genetic construct,' it may also change the gene expression patterns in the recipient chromosome(s) over long distances up- and downstream from the insertion site. If the promoter (plus associated transgenes) is inserted at very different places on a given chromosome or on different chromosomes, the effects may be very different; it will depend on the nature of the genes that are near the insertion site. This uncertainty of insertion site, along with the promoter means that for all transgenic plants, there will be a fundamental unpredictability with regard to: expression level of the inserted foreign gene(s); expression of a vast number of the recipient organism's own genes; influence of geographical, climate, chemical (i.e., xenobiotics) and ecological changes in the environment; and transfer of foreign genetic sequences within the chromosomes of the host organism, and vertical and/or horizontal gene transfer to other organisms. Such unpredictability explains the common observations that different insertion events in the same variety can vary greatly in terms of the level of expression of the desired transgene and that the majority of transformation events do not yield useful results (i.e., the transgenic plant is defective in one way or another).

The unpredictable influence of the environment may explain what went wrong in Missouri and Texas with thousands of acres of Monsanto's glyphosate-tolerant cotton and *Bt* cotton, respectively. In Missouri, in the first year of approval, almost 20 000 acres of this cotton malfunctioned. In some cases the plants dropped their cotton bolls, in others the tolerance genes were not properly expressed, so that the GE plants were killed by the herbicide (Fox, 1997). Monsanto maintained that the malfunctioning was due to 'extreme climatic conditions.' A number of farmers sued and Monsanto ended up paying millions of dollars in out-of-court settlements. In Texas, a number of farmers had problems with the *Bt* cotton in the first year of planting. In up to 50% of the acreage, the *Bt* cotton failed to provide complete control (a so-called 'high dose') of the cotton bollworm (*Helicoverpa zea*). In addition, numerous farmers had problems with germination, uneven growth, lower yield and other problems. The problems were widespread enough that the farmers filed a class action against Monsanto. In 1999, Monsanto settled the case out of court, again by paying the farmers a significant sum (Schanks (plaintiff's attorney), personal communication). If there could be this unexpected effect on the growing characteristics of the cotton, it is theoretically possible that there could be changes in the plant itself which affect the nutritional or safety characteristics of the plant (used as cattle feed) or the seed (the oil from which is used in a number of food products). *This raises the question of whether FDA should establish procedures for assuring safety in the long term.*

The unpredictability associated with the process of genetic engineering itself could lead to unexpected effects such as the production of a toxin that does not normally occur in a plant or the increase in a level of a naturally occurring toxin. An example of the former occurred in an experiment with tobacco plants engineered to produce gamma-linolenic acid. Although the plants did produce this compound, another metabolic pathway ended up producing higher quantities of a toxic compound, octadecatetraenic acid, *which does not exist in non-engineered plants* (Reddy and Thomas, 1996).

An example of the latter occurred in an experiment involving yeast where genes from the yeast were duplicated and then reintroduced via genetic engineering (Inose and Murata, 1995). The scientists found that a three-fold increase in an enzyme in the glycolytic pathway, phoshofructokinase, resulted in a 40-fold to 200-fold increase of methylglyoxal (MG), a toxic substance which is known to be mutagenic (i.e., tests positive in an Ames test). This unexpected effect occurred even though the inserted genetic material came from the yeast itself. As the scientists themselves concluded, 'Although, except for the case of microbes, we have no information as to the toxic effect of MG in foods on human beings, the results presented here indicate that, in genetically engineered yeast cells, the metabolism is significantly disturbed by the introduced genes or their gene products and the disturbance brings about the accumulation of the unwanted toxic compound MG in cells. Such accumulation of highly reactive MG may cause a damage in DNA, thus suggesting that the scientific concept of 'substantially equivalent' for the safety assessment of genetically engineered food is not always applied to genetically engineered microbes, at least in the case of recombinant yeast cells. . . . Thus, the results presented may raise some questions regarding the safety and acceptability of genetically engineered food, and give some credence to the many consumers who are not yet prepared to accept food produced using gene engineering techniques' (Inose and Murata, 1995).

A highly controversial study is that of Ewen and Pusztai published in *Lancet* in late 1999 (Ewen and Pusztai, 1999). That study used potatoes that were genetically engineered to contain a chemical from the snowdrop plant (a lectin, *Galanthus nivalis* agglutinin (GNA)) to increase resistance to insects and nematodes. Feeding experiments with rats demonstrated a number of potentially negative effects. The study found variable effects on the gastrointestinal tract, including proliferation of the gastric mucosa. Interestingly, the potent proliferative effect on the jejunum was seen only in the rats fed GE potatoes that contained the GNA gene but not in rats fed non-transgenic potatoes to which GNA had been added. Indeed, a previous feeding study utilizing GNA with a 1000-fold higher concentration than the level expressed in the GE potatoes had found no proliferative effect (Pusztai *et al.*, 1990). The authors proposed 'that the unexpected proliferative effect was caused by either the expression of other genes of the construct or by some form of positioning effect in the potato genome caused by GNA gene insertion' (Ewen and Pusztai, 1999, p. 1354). Such a fine-grained feeding study, which involved utilizing young rats that were still growing and involved weighing various organs and looking very carefully for effects on various organ systems and the immune system, is far more detailed than the general feeding studies done utilizing GE plants. While many criticisms have been leveled at this study, we believe it raises important questions that merit further research.

Because of the unexpected effects that are theoretically possible and which have been seen in various experiments, we feel FDA should require long-term animal feeding studies using the whole food product. Such testing should be done on growing animals, so that effects on various organ systems can be readily observed. In addition, fairly extensive data should be taken on the weights of various organs and on histopathology and immunology. In addition, there should be follow-up feeding studies if any data from the laboratory or field demonstrate that the genetic insert is unstable. FDA should propose its procedures for public comment so that it can get further input from the scientific community and others.

The most commonly used promoter in plant genetic engineering is one from the cauliflower mosaic virus (CaMV); all GE crops on the market contain it. A promoter has numerous elements that enable it to respond to signals from other genes and from the environment which tell it when and where to switch on, by how much and for how long. A CaMV promoter is used for a number of reasons: because it is a very powerful promoter, because it is active in all plants – monocots, dicots, algae – and in *E. coli* and because it is not greatly influenced by environmental conditions or tissue types. CaMV has two promoters, 19S and 35S, but the 35S is the one most frequently used because it is the most powerful. The powerful nature of the CaMV 35S promoter means that it is not readily controlled by the host genes that surround it and often yields a high expression level of the transgene next to it. This is not unexpected as CaMV is a virus that is designed to hijack a plant cell's genetic machinery and make many copies of itself. This also means that it is designed to overcome a plant cell's defensive devices to prevent foreign DNA from being expressed. In the case of transgenic crops, however, the CaMV promoter is used to put the transgenes outside the normal regulatory circuits of the host organism and have them expressed at very high levels. Being placed outside of normal regulatory circuits may be one of the reasons why GE foods are known to be so unstable (Finnegan and McElroy, 1994). The questions raised by the extensive use of the CaMV 35S promoter in engineered crops should be investigated with further research (Ho *et al.*, 1999).

## Post-translational processing

Another area of study that raises serious questions about the safety of transgenic traits is the phenomenon of post-translational processing, which consists of the modification of a protein after it has been translated from the genetic message. And such post-translational processing can have a significant impact on the structure and function of a gene. Furthermore, post-translational processing can differ between organisms, so that the same gene expressed in different genetic backgrounds may have the same amino acid sequence but may differ in structure and function. Examples of such processing include glycosylation, methylation and acetylation.

Glycosylation consists of the addition of sugar groups (usually oligosaccharides) and can dramatically affect the three-dimensional structure and thus, function of a protein. Indeed, glycosylation is thought to be connected to allergenic and immunogenic responses (Benjuoad *et al.*, 1992). The different proteins produced from the same gene are called glycoforms. Research with recombinant human tissue plasminogen activator (rt-PA) revealed that different glycoforms were created depending on whether the rt-PA gene was expressed in human, Chinese hamster ovary, or mouse cells (Parekh *et al.*, 1989a, 1989b). Different glycoforms were even produced when different human cell lines were used (Parekh *et al.*, 1989a). The activity (or behavior) of these glycoforms differed. Further work demonstrated that when the rt-PA gene was inserted into tobacco, although it was expressed and the protein had the normal amino acid sequence, it had no physiological activity whatsoever. Parekh *et al.* (1989b) argue that recombinant glycoproteins produced in plants could be allergenic as it is known that many allergens are glycoproteins.

But perhaps the most dramatic example of how glycosylation can affect the structure and function of proteins and have negative results occurs with the prion protein, which is thought to be the causative agent for transmissible spongiform encephalopathies (Scott *et al.*, 1999). Prion proteins are normally found attached to the surface of cells in the nerve and immune system. Research has demonstrated that the prion proteins in people suffering nvCJD – a particularly severe form of Creutzfeldt–Jakob disease (CJD) that has been recently strongly linked to bovine spongiform encephalopathy (BSE) – have a glycosylation pattern that differs significantly from that of prion proteins from people suffering other forms of CJD and is identical to the glycosylation patterns of prion proteins from cows with BSE (Hill *et al.*, 1997; Scott *et al.*, 1999). This occurs despite the fact that the amino acid sequence from normal prion proteins and those suffering nvCJD is identical. In this case, the altered glycosylation pattern has had a catastrophic effect on the behavior of the prion protein.

Given that glycosylation patterns can dramatically change the structure and function of proteins and may affect antigenicity and allergenicity, FDA should require information on the glycosylation patterns of all transgenes expressed in GE foods.

Acetylation of proteins consists of the addition of acetyl groups to certain amino acids, thereby modifying their behavior. Although incompletely understood, acetylation of the amino acid lysine has been most studied in certain groups of proteins that bind with DNA – histones and high-mobility group proteins – and such acetylation appears to be involved with the regulation of interaction of these proteins with negatively charged DNA molecules (Csordas, 1990). However, it has been discovered that some of the lysine residues in recombinant bovine growth hormone (rbGH) are acetylated, to form epsilon-N- acetyllysine when it is produced in E. coli. Harbour et al. (1992) found this to occur at lysine residues 157, 167, 171, and 180 or rbGH, while Violand et al. (1994) found it at residues 144, 157, and 167. The creation of this mutant amino acid may be overlooked because '(T)he identification of this amino acid cannot be determined by simple amino acid analysis because the acetyl group is labile to the acidic or basic conditions normally used for hydrolysis' (Violand et al., 1994; 1089). The effect this has on the safety, structure, and function of rbGH is not known as it has not been actively studied.

The differences in glycosylation and acetylation that can happen when transgenes are expressed in plants or bacteria can possibly affect toxicity and therefore lend further support to the need for toxicity testing using the whole engineered food. At present, to test for acute toxicity of a given transgene, the companies invariably do not use the protein that is produced in the plant itself. Rather, in order to obtain large enough quantities of the protein for testing, the companies put the transgene into a bacteria (invariably E. coli), isolate the expression product (i.e., the protein) and use that for the acute toxicity testing. However, the protein produced in the bacteria may be glycosylated differently than the same protein produced in the plant. Even if there are no differences in glycosylation, acetylation of lysine residue(s) could cause differences. The presence of such mutant lysine residues could easily be missed as routine amino acid analysis will remove the acetyl group; to find if there are mutant lysine residues, one must conduct a special test to detect them. Thus, whenever possible, FDA should require the companies to use material derived from the transgenic plants themselves in toxicity studies rather than bacterially derived proteins.

Methylation is the process of putting methyl groups on a molecule. Methylation of DNA, which occurs with the nucleotide bases cytosine and adenosine, is important as this appears to prevent that piece of DNA from being expressed (or 'turned on'). Methylation is one of the mechanisms behind the phenomenon of 'gene silencing,' whereby a cell 'turns off' a gene. Transgenic work has found that if you try to insert multiple copies of a gene into a plant, the plant will frequently turn off all, or all but one, of the copies of the transgene (Finnegan and McElroy, 1994). Indeed, some scientists now think that gene silencing is an important defense mechanism that plants use to prevent foreign DNA from being expressed (other mechanisms exist to try to degrade the foreign DNA before it can enter the nucleus of the cell) (Ho, 1998; Traavik, 1998). This should be combined with the recent finding that tobacco plants may contain large numbers of copies of pararetroviral-like sequences, in some cases reaching copy numbers of about 10 000 (Jakowitsch et al., 1999). This study is quite striking as it was previously thought that plant viruses rarely integrate, if at all, into host genomes. Furthermore, such integrated viral genetic material is normally silenced via methylation, so that there could be a lot of dormant viral sequences in plants. Interestingly, the cauliflower mosaic virus promoter (CaMV 35) used in virtually all transgenic plants on the market is a pararetrovirus-derived sequence (i.e., CaMV is a pararetrovirus).

With methylation, the danger exists that the CaMV 35S promoter, being a very powerful 'on switch' that can have effects on thousands of base pairs upstream and downstream from an insertion point, could inadvertently 'turn on' a foreign gene that has previously been silent. Given the studies in the last couple of years that suggest that horizontal gene transfer may be more common than previously thought and that most such foreign DNA, if it survives and is able to incorporate itself in the host genome, is frequently 'silenced' via methylation, there is a potential risk that some nasty dormant genetic material is inadvertently turned on due to the presence of the CaMV promoter. *Thus, it becomes important to know the exact insertion site of any and all genetic constructs as well as knowing what the genetic sequence is for thousands of base pairs upstream and downstream from the insertions site, and do long-term toxicity tests with the whole engineered food.*

## What Data Should FDA Require?

For all the reasons stated above and because of the random nature of the genetic transformation process each random insertion of transgenic DNA will differ in location and in structure from all other inserts. It will be accompanied by a different pattern of unintended positional and pleiotropic effects due, respectively, to the location of the insert and the functional interaction of the insert with host genes. Thus, each transgenic line resulting from the same process, despite using the same vector system and plant materials under the same conditions, will be distinct, and must be treated as such. Consequently, we think FDA should require the companies to submit data for each separate transgenic line. For every line, FDA should require a complete molecular characterization of each line with respect to the identity, stability and unintended positional and pleiotropic effects. And based on the results of such characterization, the agency could decide on how much toxicity data to require.

The components of a complete molecular characterization for molecular identity would include, for each transgenic or transformed line:

- Total number of inserts of transgenic DNA.
- Location of each insert (organelle (chloroplast, mitochondria, etc.) or chromosomal).
- Exact chromosomal position of each insert.
- Structure of each insert (whether duplicated, deleted, rearranged, etc.).
- Complete genetic map of each insert including all elements (coding region, noncoding regions, marker gene, promoters, enhancers, introns, leader sequences, terminators, T-DNA borders, plasmid sequences, linkers, etc. including any truncated, incomplete sequences).
- Complete (nucleotide) base sequence of each insert.
- (Nucleotide) base sequence of at least 10 kbp (10 000 base pairs) of flanking host genome DNA on either side of the insert, including changes in methylation patterns.

To determine stability, the FDA needs data on both functional stability (level of expression remains constant over time and over successive generations) and structural stability (location in the genome and structural arrangement of the insert). For functional stability, FDA would need data on the level of expression of the transgene over time – throughout the lifetime of the plant as well as over a number of generations (say three to five generations). For structural stability, the FDA would need data on the physical location of the insert in the genome as well as the structure of the insert – throughout the lifetime of the plant as well as over successive generations (say three to five). In addition, the FDA would require appropriate molecular probes for each insert with flanking host genome (organelle sequence) sequences in order to monitor the structural stability of the insert.

To test for unintended positional effects, the FDA could look carefully at the methylation patterns of the genes in the flanking host genome DNA (data we suggest be required under molecular identity characterization). To look for pleiotropic (as well as positional effects), each transformed line must be identified in terms of total protein profile and metabolic profiles. The total protein profiles would help to monitor for unintended changes in the pattern of gene expression while the metabolic profile would help to monitor for unintended changes in metabolism. The use of mRNA fingerprinting and protein fingerprinting as part of the protein profiles would represent a better, finer screen for detecting novel biochemical, immunological or toxicological hazards. Some such tests have been suggested by a Dutch government team and should be more carefully considered by the FDA (Kuiper *et al.*, 1998). If any of these tests found differences, there would be more reasons to ask for more comprehensive toxicity testing.

## Public health/antibiotic marker genes

In 1991–1992, when FDA was developing its policy of GE plants, the conventional wisdom in the scientific community was that DNA was a very fragile molecule that would be readily broken down in the environment and would not survive digestion in the gut. We now know that both assumptions may not always be valid (Traavik, 1998). Even though DNases (molecules that break down DNA) are widely distributed in the environment, free DNA has been found in all ecosystems (marine, fresh water, sediments) studied (Lorenz and Wackernagel, 1994). Indeed, pooled data suggest that free DNA is present in significant amounts in the environment. Larger amounts of DNA are extracted from soil than can be extracted from the cells in the soil (Steffan *et al.*, 1988). Further studies have shown that this free DNA in the soil comes from microorganisms that no longer occur in that habitat (Spring *et al.*, 1992), thus demonstrating that DNA can out-survive the organism it came from and still be capable of being taken up and expressed by microorganisms. Finally, yet other studies have found that pollution (i.e., xenobiotics) can affect the survivability of DNA and the possibility of its transfer to other organisms (Traavik, 1998).

These data lead to serious concerns about the antibiotic resistance marker genes that are present in virtually all engineered plants presently on the market. These genes code for proteins that confer resistance to a given antibiotic. The possibility therefore exists that these genes for antibiotic resistance could be taken up by bacteria, thus exacerbating the already very serious problem of antibiotic resistance in disease-causing organisms.

In the mammalian system, the question is whether foreign DNA can survive digestion, be taken up through the epithelial surfaces of the gastrointestinal or respiratory tract or not, or be excreted in feces. Studies in the 1970s (Maturin and Curtiss, 1977) and 1980s (McAllan, 1982) on rats and ruminants, respectively, failed to find evidence that genetic material (e.g., DNA and RNA) survived digestion. Consequently, many scientists assumed that DNA was readily digested. However, the methods used to detect DNA were not very sensitive. In the mid-1990s, researchers in Germany re-investigated the issue, using far more sensitive methods (Schubbert *et al.*, 1994). Mice were fed DNA from the M13 bacteriophage either by pipette or by adding it to the feed pellets. Using sensitive hybridization methods and PCR (polymerase chain reaction) the authors found 2–4% of the M13 DNA in feces and 0.01–0.1% in the blood – both in serum and cell fraction. Sizeable DNA fragments (almost a quarter of the M13 genome) could be found up to seven hours after uptake.

If free DNA is not immediately digested in the gastrointestinal tract, the possibility also exists that it can be transferred to bacteria that live there. A recent study utilizing a simulated human gut demonstrated that naked DNA had a half-life of six minutes, more than enough time for such DNA to transform bacteria (MacKenzie, 1999).

In another experiment, a genetically engineered plasmid was found to survive (6 to 25%) up to an hour of exposure to human saliva (Mercer *et al.*, 1999). Partially degraded plasmid DNA also successfully transformed *Streptococcus gordonii*, a bacteria that normally lives in the human mouth and pharynx, although the frequency of transformation dropped exponentially with time. Transformation occurred with either filter-sterilized human saliva or unfiltered saliva. The study also found that human saliva contains factors that increase the ability of resident bacteria to become transformed by 'naked' DNA. Since transgenic DNA from food is highly unlikely to be completely broken down in the mouth, it may be able to transform resident bacteria. Of particular concern would be the uptake of transgenic DNA containing antibiotic resistance marker genes, which are found in the majority of GE crops presently on the market. It should be pointed out that the antibiotic marker gene present in Novartis' *Bt* corn, which codes for resistance to ampicillin, is under the control of a bacterial promoter rather than a plant promoter which would further increase the possibility of expression of the ampicillin resistance gene if it were taken up by bacteria.

In September 1998, the British Royal Society put out a report on genetic engineering that called for ending the use of antibiotic resistance marker genes in engineered food products (Anonymous, 1998). In May 1999, the British Medical Association released a report calling for a prohibition on the use of antibiotic resistance marker genes in genetically engineered plants (BMA, 1999).

We therefore urge FDA to prohibit use of antibiotic resistance marker genes as there is no consumer benefit for the presence of such genes in engineered foods and a significant potential risk.

## Allergenicity

In the United States, about a quarter of all people say they have an adverse reaction to some food (Sloan and Powers, 1986). Studies have shown that 2% of adults and 8% of children have true food allergies, mediated by immunoglobin E (IgE) (Bock, 1987; Sampson *et al.*, 1992). People with IgE mediated allergies have an immediate reaction to certain proteins that ranges from itching to potentially fatal anaphylactic shock. The most common allergies are to peanuts, other nuts and shellfish.

Allergens can be transferred from foods to which people know they are allergic, to foods that they think are safe, via genetic engineering. In March 1996, researchers at the University of Nebraska in the United States confirmed that an allergen from Brazil nuts had been transferred into soybeans. The Pioneer Hi-Bred International seed company had put a Brazil nut gene that codes for a seed protein into soybeans to improve their protein content for animal feed. In an *in vitro* and a skin prick test, the engineered soybeans reacted with the IgE of individuals with a Brazil nut allergy in a way that indicated that the individuals would have had an adverse, potentially fatal reaction to the soybeans (Nordlee *et al.*, 1996).

This case was resolved successfully. As Marion Nestle, the head of the Nutrition Department at New York University summarized in an editorial in the *New England Journal of Medicine*, 'In the special case of transgenic soybeans, the donor species was known to be allergenic, serum samples from persons allergic to the donor species were available for testing and the product was withdrawn' (Nestle, 1996, p. 726). However, for virtually every food, allergists will tell you, there is someone allergic to it. Proteins are what cause allergic reactions, and virtually every gene transfer in crops results in some protein production. Genetic engineering will bring proteins into food crops not just from known sources of common allergens, like peanuts, shellfish and dairy, but from plants of all kinds, bacteria and viruses, whose potential allergenicity is largely uncommon or unknown. Furthermore, there are no fool-proof ways to determine whether a given protein will be an allergen, short of tests involving serum from individuals allergic to the given protein. This point is strongly driven home in the case of the transgenic soybean containing a Brazil nut gene, where animal tests had suggested that the transferred Brazil nut seed storage protein was not an allergen (Nordlee *et al.*, 1996). Had the results of the animal tests been relied on and the soybeans approved, the results could have been disastrous.

Most biotechnology companies increasingly use microorganisms rather then food plants as gene donors, or are designing proteins themselves, even though the allergenic potential of these proteins is unpredictable and untestable. Consequently, Nestle continues, 'The next case could be less ideal, and the public less fortunate. It is in everyone's best interest to develop regulatory policies for transgenic foods that include premarketing notification and labeling' (Nestle, 1996, p. 727). Furthermore, Pioneer commendably withdrew its Brazil nut/soybean product before it was marketed. But this is not an FDA requirement; FDA should prohibit introduction of known allergens into new engineered foods.

In April 1994, the EPA, FDA, and USDA hosted a 'Conference on Scientific Issues Related to Potential Allergenicity in Transgenic Food Crops.' The conference revealed how little is actually known about the topic. Indeed, two conclusions/observations noted by the scientists at the meeting were that there are: (i) no direct methods to assess potential allergenicity of proteins from sources that are not known to produce food allergy, and (ii) although some assurance can be provided to minimize the likelihood that a new protein will cause an allergic reaction by evaluating its similarity with characteristics of known food allergens (i.e., whether the new protein has a similar protein sequence, is prevalent in food, is resistant to enzymatic and acid degradation, is heat stable, and is of the appropriate molecular size), no single factor is predictive. Since this meeting, FDA appears to have taken no significant steps to increase the scientific understanding of allergenicity or to develop a truly predictive methodology for assessing allergenicity of transgenic crops.

We urge FDA to take a leadership role in pushing for scientific research that could result in the development of a truly predictive test for allergenicity. Furthermore, at present, companies evaluate allergenicity by looking only at the similarity of the engineered proteins with characteristics of known food allergens. As pointed out at the April 1994 Interagency conference, such a rudimentary approach is not completely predictive. We think this approach is not stringent enough. We call on FDA to develop a stringent protocol for testing for allergenicity and to publish such a protocol for comment. Furthermore, since there is no fool-proof predictive methodology for testing for allergenicity, FDA must require labeling of all GE foods to facilitate the ability to detect the appearance of new allergies.

Introduction of genes that code for known allergens should be prohibited. But not all allergens are known. Genetically engineered foods present a qualitatively different risk of allergenicity than do conventional foods. If a consumer develops an allergy to a new food, they will always react to that food, thereby facilitating identification of the food causing the allergic reaction. With a genetically engineered food, the person will react only to the genetically engineered variety and not to all varieties of the same food. Without labeling, it will be exceedingly difficult to be able to determine what the offending food is.

## Future Products

The industry is discussing plans to develop products with altered nutritional content and/or the insertion of biologically active substances such as are now included in dietary supplements. Psychoactive substances, vitamins, and other substances may be engineered into food.

Such plants raise significant questions that we feel the FDA is not equipped to handle. The public health implications of wide-scale consumption of foods with significantly altered nutrient profiles could be enormous, yet FDA may not have the legal authority to address such issues. This underlines the need to label such food as genetically engineered, so that people know it is not part of a traditional diet. Research is also proceeding on engineering food plants to produce pharmaceuticals and industrial chemicals. FDA's current framework does not have mechanisms for addressing the unique public health risks these products pose.

## Public Information Issues

FDA currently requires labeling of genetically engineered food only if there are changes in nutrients or introduction of known allergens. Even in these cases

FDA would not require that the food be labeled 'genetically engineered' but rather only labeled as to the nutritional change or presence of the allergen.

The FDA should require mandatory labeling of all genetically engineered food. Consumer Reports tested foods purchased in the supermarket in 1999 as to whether it was genetically engineered and found that many foods contain genetically engineered ingredients and are not labeled as such. There are both health-based, and 'right-to-know' reasons for requiring labeling.

The current FDA policy does not serve the public well. A majority of consumers want labeling of all genetically engineered food, as numerous polls attest. A recent International Food Information Council poll shows that public comfort levels with genetic engineering are declining, and support for labeling is growing. An October 1999 Gallup poll found that 68% of US consumers want labeling even if it increased food costs.

Recently fifty members of Congress co-signed a letter with House Minority Leader David Bonior advising FDA that in their view, FDA currently has the authority to require mandatory labeling and should do so.

All such foods should be labeled including engineered whole food, processed food containing engineered ingredients, and food produced through genetic engineering, such as milk from cows treated with a genetically engineered drug (which contain residues of the drug). It should be labeled even if the food contains no 'foreign' DNA (from a source with which the crop would not cross normally in nature), as is the case with oils derived from engineered soy, corn, cotton, or canola. In part, this is because the process of genetic engineering, regardless of what is introduced, can induce unexpected effects.

The food should be labeled regardless of whether current testing technology is capable of verifying whether the food is engineered or not. Consumer Reports found that current test technology cannot detect engineered foods that are highly processed, like corn flakes, or that do not contain protein, like oils. However, the identity of such foods can be maintained through record keeping and certification procedures.

The terminology of labeling should be simple and straightforward, such as 'contains genetically engineered material.' The terminology should not be value laden or promotional, such as 'improved through modern biotechnology.'

Labeling should also be permitted which states that food does not contain genetically engineered material. We believe FDA should define 'not containing' as 'not detectable' where current test methodology is available. The current reliable limit of detection appears to be 0.01%; this is the benchmark Consumer Reports used in its testing. Foods containing detectable amounts of GE material should be required to label as such.

It will not be sufficient to merely allow voluntary labeling. Such a policy would lead to very few products being labeled, due to the extra effort involved in meeting standards for 'no GE material,' thus failing to facilitate consumer information and choice. Voluntary labeling also would not protect people with unusual food allergies or food sensitivities.

In any labeling proposals, we urge FDA not to suggest, propose or require any 'contextual statement' on any products labeled as to presence or absence of genetically engineered material. All product labeling required by FDA is 'shorthand' and can be misinterpreted. FDA has in the past, correctly and appropriately, generally taken the view that public education programs of various kinds, including company advertising, can dispel misconceptions. Explanatory statements along with labels have generally been deemed unnecessary and inappropriate.

The exception was with labeling of milk from cows not treated with rbGH, where the agency, in an unprecedented action, proposed a contextual statement. The FDA suggested that companies should indicate that FDA has found no significant difference between milk from cows treated with rbGH and milk from cows not treated with the drug. However, there are many similar contexts in which consumers could theoretically be misled in which FDA has not required explanatory statements – indeed there is an almost endless list of such potentially misleading situations. Consumers could for example possibly think that cheese 'made in Wisconsin' or 'made by the Amish' is somehow superior – indeed that is the goal of the labelers. Yet FDA has correctly not seen fit to suggest that labels should say 'FDA sees no significant difference between cheese made in Wisconsin or cheese made elsewhere' or sees 'no significant difference between cheese made by the Amish and other cheese.' We urge FDA to rescind the guidance suggesting a special contextual statement on milk from cows not treated with rbGH, and issue no further requirements for contextual statements on genetically engineered products.

There are both health reasons and basic 'right-to-know' reasons to require labeling of genetically engineered food. There are two health concerns that are relevant: allergens and unexpected effects.

Regarding allergens, while we urge FDA to prohibit introduction of known allergens into genetically engineered food, science is not yet at a point where it can successfully predict all allergens. Indeed, allergies are so varied, that for almost any food, there is probably someone who is allergic to it. Also, there are many food sensitivities (causing indigestion and the like) that do not have the same mechanism as the IgE-mediated food allergies that cause severe reactions. Finally, genetic engineering is introducing many proteins into the food supply that have not previously been eaten, or which have been eaten only in small quantities, such as *Bt* endotoxin. Though no one is currently allergic to them, allergies may develop as people are exposed.

The problems of uncommon allergens, food sensitivities, and unknown and new allergens can all best be addressed through mandatory labeling of all genetically engineered food. Mandatory labeling of GE food will allow individuals to identify and avoid foods that cause them difficulty. Without such labeling, it will be impossible to distinguish the problem food from its conventional look-alike counterpart.

As discussed, there may also be unexpected effects from genetically engineered food. At this point, we do not know if such effects will occur frequently, rarely or hardly ever, in a way that affects health. However, as discussed earlier in these comments, there is considerable new data showing unexpected effects of genetic transformations. It therefore would be prudent to require mandatory labeling so that if any effects that affected health did occur, they could be identified and their origin determined. For example, tracking a problem such as occurred with the genetically engineered L-tryptophan dietary supplement, which led to illness and death but whose exact cause has never been determined, could be facilitated if there were mandatory labeling.

FDA also has ample precedent for requiring mandatory labeling of genetically engineered food, because it is a 'material fact.' FDA has generally required labeling of all processes that are of interest to consumers, including whether food is frozen, irradiated, or from concentrate. It requires ingredients and additives to be indicated. It also has many standards of identity whose sole purpose is to facilitate consumer choice. FDA should also require a label if food is genetically engineered, as this is a 'material fact' to consumers.

# 25

# Ag Biotech: Our Past and Our Future

**Roger Krueger**
*Director, Environmental Technical Stewardship and Global Product Development, Monsanto Corporation*[1]

## The Advent of Modern Agricultural Biotechnology

The science of modern agricultural biotechnology emerges from the work of Gregor Mendel, an Austrian monk who began his studies of the principles of heredity in 1856. Mendel discovered that various traits of parent plants were evident in offspring in consistent ratios and he developed a set of rules to explain how these traits were passed from generation to generation. Walter Sutton built on this information and in 1903 concluded that hereditary information was located in chromosomes.

Nucleic acid, a component of deoxyribonucleic acid (DNA), was discovered in 1869 by Frederick Miescher, a Swiss chemist. This discovery went largely

---

[1] Roger Krueger is Technical/Environmental Stewardship Lead within the Global Product Management Group in Monsanto. Previously, Dr Krueger served as Director of Technology Development and part of the United States Business Team in US Markets. Dr Krueger has held a range of positions in the agriculture business including product development, biotechnology strategy, molecular genetics and trait development through genetic engineering and transposon tagging, while working at American Cyanamid, Yale University and Dekalb-Pfizer Genetics.

Dr Krueger earned a BSc in Plant Science from the University of New Hampshire, MSc in Plant Nutrition from the University of Rhode Island, and PhD in Biology from the University of Missouri-Columbia. Dr Krueger also worked as a Post Doctoral Associate in the Biochemistry Department at the University of Missouri prior to working with Dekalb-Pfizer.

---

Genetically Modified Organisms in Agriculture
ISBN 0-12-515422-4

ignored until the 1930s when it was being speculated that DNA was the critical element of genes. The basic chemistry of DNA was determined by the late 1920s, and by 1952, it was confirmed that DNA was responsible for transmitting genetic information.

A critical breakthrough in the science of genetics came in 1953, when James Watson, an American biochemist, and Francis Crick, a British biophysicist, identified the double helix-structure of DNA (often referred to as the 'Blueprint of Life'). In the early 1970s, researchers at Stanford University created the first recombinant DNA (rDNA) molecule by combining segments of DNA from different types of organisms. Shortly thereafter, researchers Stanley Cohen and Herbert Boyer transferred a recombinant molecule from one organism to another and, in so doing, created the first genetically engineered organism.

## Biotechnology Transforms and Improves Agriculture

The exponential growth in computing capacity, which speeds the evaluation of complex genetic data, is supporting exponential growth in mankind's knowledge of genes. Today, the library of genetic information is doubling every 12 to 24 months.

This better understanding of the genetic heritage of plants has prompted the development of techniques to improve crops. These techniques are superior to traditional methods of plant breeding, whereby many traits are moved unintentionally along with the desired one. With biotechnology and recombinant DNA techniques, a selected gene (or genes) can be inserted into the cells of a plant, thus providing a more directed and rapid way to improve crop plants.

The primary methods used for the majority of gene insertions are *Agrobacterium* and biolistic/gene gun. The *Agrobacterium* method transfers DNA to its plant host naturally using a plasmid (a genetic element occurring outside the nucleus that is present in the cytoplasm of some bacteria cells) as a carrier or vector of the engineered gene. The introduced genes then integrate into the DNA of the host plant. The biolistic/gene gun method uses DNA-coated metal projectiles, which are fired into target cells.

Much discussion has occurred as to whether these gene transfer methods are 'natural.' Professor Conrad Lichenstein of Queen Mary and Westfield College has found evidence that gene transfer can occur naturally. When introducing genes into tobacco plants to inoculate them against virus infection, Lichenstein discovered that similar, foreign viruses were already present. The same DNA was found in five species of *Nicotina* plants. Lichenstein also discovered that the implanted DNA came from a single event. He reported that, 'One fine morning in May, about five million years ago, a single plant, in a genetic accident, acquired these DNA characteristics' (Highfield, 1999).

### Enhancing crop plants

Crop plants have been designed to be insect, disease and herbicide resistant. These crops were first commercially planted in the United States in 1996. In 1999, an estimated 70 million acres of biotechnology crops were grown in the

United States, and almost 99 million were grown globally (National Research Council, 2000, p. xi).

Crops like corn, potatoes, and cotton have been enhanced to contain a gene from a natural insecticide, *Bacillus thuringiensis* (*Bt*), that home gardeners have been using for decades. *Bt* is a naturally occurring soil microbe that produces proteins that are lethal to various insects but harmless to humans, animals and nontarget insects.

Potatoes, squash, and papaya, among other important crop plants, have been enhanced to protect themselves against viral disease by introducing genes that produce viral coat proteins. Plants with these proteins resist viruses, thus, in effect, vaccinating themselves against disease.

Various crop plants have also been enhanced through biotechnology to be tolerant to the herbicidal ingredient in Roundup herbicide, glyphosate, which controls more than 100 weed species but has a low selective toxicity, is not a threat to humans or animals, and degrades quickly in the soil.

Insect-, disease-, and herbicide-resistant crop plants represent the first wave of biotechnology products. With these crops in place, the second wave of agricultural biotechnology has begun with the design of products that will help growers and processors create more value for consumers by enhancing the quality of the food and fiber produced by the plant. In the third wave, agricultural biotechnology will bring the creation of outputs that are only beginning to exist in plants today such as crops from which renewable industrial resources like lubricating oils and biodegradable plastics could be derived as well as other outputs that would be powerful weapons in the fight against cancer and heart disease.

## Creating benefits

Through agricultural biotechnology, crop yields can be increased to provide more food product for the world's hungry and malnourished, and the increases can help feed the burgeoning population in the developing world.

Higher crop yields are already proving to be a boon to US farmers, who are constantly battling rising costs. Increased yields helped create (through reduced chemical and other input costs) more than $830 million in captured value in 1997. Consumers benefited by more than $55 million in savings (Falck-Zepeda *et al.*, 2000b). The global market for biotech-based crops is expected to grow from $500 million in 1996 to $6 billion by 2005 and to $20 billion by 2010 (Falck-Zepeda *et al.*, 2000b).

Products of agricultural biotechnology are also promoting disease prevention. Plants damaged by insects are more receptive to disease and can carry disease. Mycotoxins can build up in damaged corn stalks and kernels, and some (fumonisins) can be fatal to animals and have been identified as carcinogenic for humans. *Bt* corn resists insect damage and this protects against disease.

Further, agricultural biotechnology promotes the sustainability of natural resources by reducing the use of energy as well as the use of pesticides, herbicides, fungicides, and other agricultural chemicals. And by easing crop damage and disease, improved pest controls allow for the more efficient use of existing farmland, which is critical to the preservation of natural habitat for biodiversity.

In addition to increasing crop yields, providing numerous economic benefits, fighting disease and promoting the sustainability of natural resources, agricultural biotechnology can enhance the nutritional value of food products and develop crop plants that contain medicinal qualities – often called pharmafoods. This is especially important to most parts of the developing world where the provision of health care is extremely difficult.

## Regulating Agricultural Biotechnology

Given the enormous resources, time, and reputation that are involved with the development of biotechnology crop plants, strict self-regulation is practiced in the biotech industry. Breeders of biotechnology crop plants are responsible for a number of factors during the developmental phase, including sensitivity to environmental stresses and disease. Food crops must also be evaluated for toxins, allergens, and nutritional content, as well as for the safety of any newly expressed protein(s).

The governmental regulatory process under which biotechnology crop plants are assessed today derives from guidelines first issued in the early 1970s by the National Institutes of Health (NIH) Recombinant DNA Advisory. These guidelines were replaced in 1986 by the Coordinated Framework for Regulation of Biotechnology, which stated that biotechnology would be regulated 'in essentially the same manner for safety and efficacy as products obtained by other techniques.'[2] The regulations governing the biotech industry in the United States are generally based on the concepts and guidance provided by international scientific organizations.

The federal regulatory system governing biotech products in the United States involves three agencies:

*   The *US Department of Agriculture (USDA)* regulates the plant and is responsible for the oversight of field trials and interstate movement involving plants modified through biotechnology. 'In no instance has a biotech plant approved for field testing by USDA created an environmental hazard or exhibited any unpredictable or unusual behavior compared to similar crops modified using conventional breeding methods.'[3]
*   The *Environmental Protection Agency (EPA)* is responsible for regulating plant protection traits, such as insect protection, and regulates labels for herbicide usage on herbicide-tolerant plants. 'EPA . . . has found no

---

[2] For a more complete early history of the regulatory process see the report of the Subcommittee on Basic Research of the Committee on Science, United States House of Representatives, *Seeds of Opportunity: An Assessment of the Benefits, Safety, and Oversight of Plant Genomics and Agricultural Biotechnology* (pp. 16–18) from which this information has been extracted and condensed.

[3] Subcommittee on Basic Research of the Committee on Science, United States House of Representatives, *Seeds of Opportunity: An Assessment of the Benefits, Safety, and Oversight of Plant Genomics and Agricultural Biotechnology*, April 13, 2000, p. 20.

documented case of environmental harm caused by a plant-pesticide produced through biotechnology.'[4]

• The *Food and Drug Administration (FDA)* regulates food and feed safety.
Food producers are not required to seek pre-market approval from the
FDA for a new food variety if it is substantially equivalent (defined below)
to a variety on the market. Without exception, however, biotech companies have consulted with FDA before bringing a new biotech food to
market. There are no examples of a biotech food causing human health
problems.[5]

   In May 2000, the FDA announced plans to refine its regulatory approach
regarding biotechnology foods in order to build consumer confidence.
FDA will publish a proposed rule mandating that developers of biotech
foods notify the agency when they intend to market such products and
that the information submitted along with the agency's conclusions be
posted on the FDA website for public viewing. FDA also announced plans
to draft labeling guidance to assist manufacturers who wish to voluntarily
label their foods.[6]

Countries around the world have also developed regulations that assure that
the foods derived from biotechnology crop plants are rigorously assessed for
their safety.

The concept of comparing the safety of the food from a biotechnology crop
plant to that of a food with an established history of safe use is referred to as
'substantial equivalence.' The process of substantial equivalence involves comparing the characteristics, including key nutrients and other components, of the
food derived from biotech crop plants to the food derived from conventionally
bred plants. When a food is shown to be substantially equivalent to a food with
a history of safe use, 'the food is regarded to be as safe as its conventional counterpart and no further safety consideration is needed.' (FAO/WHO, 1996, p. 22;
see also OECD, 1993; WHO, 1995).

## Ensuring the Safety of Products

Numerous national and international organizations have considered the safety
of products derived from biotech crop plants and have declared that the safety
considerations for these products are basically the same as for those derived
using other methods like traditional breeding. For example:

---

[4] Janet L. Anderson, *Testimony Before the US House of Representatives Subcommittee on
Basic Research hearing on 'Plant Genome Research: From the Lab to the Field to the Market,
Part III,'* Serial No, 106–60, Government Printing Office, Washington, October 19, 1999.
[5] Subcommittee on Basic Research of the Committee on Science, United States House of
Representatives, *Seeds of Opportunity: An Assessment of the Benefits, Safety, and Oversight
of Plant Genomics and Agricultural Biotechnology*, April 13, 2000, p. 23.
[6] Food and Drug Administration, Center for Food Safety and Applied Nutrition, May 3,
2000 (http://vm.cfsan.fda.gov/~lrd/biotechm.html).

- The American Medical Association (1991) has concluded that the use of agricultural biotechnology is a safe, useful tool with which to enhance food safety, quality and nutrition. The American Dietetic Association (1993) and the National Center for Nutrition and Dietetics (1996) have also judged biotech foods to be as safe as traditional foods.
- The National Research Council, which provides science, technology and health policy advice under a congressional charter, conducted three separate studies on agricultural biotechnology, one in 1987, another in 1989 and the most recent in the spring of 2000. All three studies reached essentially the same conclusion, that is: 'There is no evidence that unique hazards exist either in the use of rDNA techniques or in the movement of genes between unrelated organisms.' (National Research Council, 2000, p. 6).
- The Food and Agriculture Organization of the United Nations (FAO) and the World Health Organization (WHO) have conducted a number of studies on the safety of food products derived from biotech crop plants. Most recently, a joint consultation in 2000 by the two organizations found that, 'for genetically modified foods, the pre-marketing safety assessment already gives assurance that the food is as safe as its conventional counterpart. Accordingly it was considered that the possibility of long-term effects being specifically attributable to genetically modified foods would be highly unlikely.' (FAO/WHO, 2000, p. 20).
- The Organization for Economic Cooperation and Development (OECD) held a conference on the science and health aspects of biotech foods in February/March 2000 in Edinburgh, Scotland. Attendees concluded that: 'Worldwide, many people are eating GM foods (especially in North America and China) with no adverse affects on human health having been reported in the peer-reviewed scientific literature.' (OECD, 2000a, p. 3). This reconfirmed previous evaluations of biotech foods conducted by the OECD (OECD, 1993, 1996).
- The Australia New Zealand Food Authority (ANZFA) and the National Academies of Science of six nations (Brazil, China, England, India, Mexico, and the United States), along with the Third World Academy of Sciences, have also issued reports on agricultural biotechnology that support the safety aspects of biotech foods. The ANZFA assessment held that, 'the level of safety associated with GM foods is at least as high as that of all other available foods because the safety assessment process undertaken for GM foods is far more thorough than that undertaken for any other food. The safety assessment process ensures that GM foods provide all the benefits of conventional foods and no additional risks.' (ANZFA, 2000, p. 8). The National Academies reported that, 'To date, over 30 millions hectares of transgenic crops have been grown and no human health problems associated specifically with the ingestion of transgenic crops or their products have been identified.'[7]

---

[7] Brazilian Academy of Sciences, Chinese Academy of Sciences, Indian National Science Academy, Mexican Academy of Sciences, National Academy of Sciences of the USA, The Royal Society (UK) and the Third World Academy of Sciences, *Transgenic Plants and World Agriculture*, p. 12.

## Managing and eliminating risk

### Outcrossing

Concerns have been expressed that either the genetically modified plant will itself become a weed, or as a result of cross-pollination with compatible relatives (outcrossing), other weed problems will result.

Leading environmental scientists and international bodies of scientific experts, such as the National Academy of Sciences (1987) (National Research Council), have described as 'negligible' the risk of a domesticated crop plant accidentally reverting to a weedy condition.[8]

Of outcrossing, a report by the House Subcommittee on Basic Research, which held comprehensive hearings on agricultural biotechnology, concluded that:

> The risks that new plant varieties developed using agricultural biotechnology will become weedy or outcross are the same as those for similar varieties developed using classical breeding methods for introduced species.[9]

The Environmental Protection Agency (EPA) assessed each of the *Bt* plant pesticide registrations to determine the likelihood of transgene movement to weedy relatives and reported that, 'in almost all cases, the likelihood of occurrence of such movement is almost non-existent'.[10]

### Pest-resistant crops and pesticide-resistant insects

Another concern related to the potential impact of agricultural biotechnology on the environment involves the question of whether biotech crops could confer to insects a resistance to pesticides.

A critical component for the use of biotechnology crops is managing and minimizing the possibility that crops and insects will develop a resistance to *Bt*, the naturally occurring soil bacteria that are toxic to various insects. To accomplish this, insect resistance management (IRM) strategies have been devised, implemented and refined to minimize resistance development. The core element of any IRM plan is the planting of a 'refuge' of non-*Bt* crops to ensure that an adequate population of susceptible insects of the target species are available to mate with any resistant insects that survive the protected crop.

The EPA reported on the refugee strategy that:

> The high dose/structured refuge strategy has been widely endorsed by the scientific community. . . . EPA believes it is significant that, after four years

---

[8] See also: OECD, 1993; Ellstrand, 1991; Scheffler and Dale, 1994; Raybould and Gray, 1994; Jorgensen *et al.*, 1996; Snow and Jorgensen, 1999.

[9] Subcommittee on Basic Research of the Committee on Science, United States House of Representatives, *Seeds of Opportunity: An Assessment of the Benefits, Safety, and Oversight of Plant Genomics and Agricultural Biotechnology*, April 13, 2000, p. 38.

[10] US Environmental Protection Agency (EPA), 'Response of the Environmental Protection Agency to Petition for Rulemaking and Collateral Relief Concerning the Registration and Use of Genetically Engineered Plants Expressing *Bacillus Thuringiensis* Endotoxins,' April 20, 2000.

of full-scale commercialization of *Bt* crops, with approximately 17 million total acres of *Bt* corn, *Bt* potato and *Bt* cotton planted in 1998, EPA has received no confirmed evidence that field resistance to any *Bt* endotoxin expressed in these crops has occurred in any insect species.[11]

## Monarch butterfly

Researchers at Cornell University and Iowa State University conducted laboratory studies that reported milkweed dusted with pollen from *Bt* corn is harmful to the growth and survival of the nontarget monarch butterfly. These results were reported in a letter to the scientific journal *Nature* (Losey *et al.*, 1999), and on the website edition of the journal *Oecologia* (Obrycki and Hansen, 2000), respectively, and stoked fears about the potential environmental impact of agricultural biotechnology.

In neither study did the researchers match the conditions that would be present in a natural setting. The Cornell researchers dusted leaves with corn pollen and the larvae were given no choice but to feed on these leaves. The Iowa researchers gathered leaves from plants growing in and around cornfields that had been dusted with *Bt* corn pollen. Here again, the leaves were taken back to a laboratory and fed to caterpillars.

The findings of the Cornell study were not unexpected. The House Subcommittee on Basic Research heard that, '[T]he *Bt*/Monarch study has been heavily criticized in the scientific community because every entomologist knows that . . . if you feed Monarch butterfly larvae *Bt* toxin, whether it be in corn or whether it be on a spray, that insect will die.'[12]

A critique of the Iowa State study found that:

- Toxicity results were not consistent with those of other studies.
- The sample size of the larvae was small.
- The concentrations that caused toxicity were not representative of field concentrations that were reported.
- Larvae were exposed to very small sections of leaf and had no choice of diet (Sears and Shelton, 2000).

The researchers concluded that '[W]e believe this study does little to help understand potential risks of deploying *Bt* plants in the field' (Sears and Shelton, 2000).

## Allergens/toxins

A major safety concern raised with regard to agricultural biotechnology is the risk of introducing allergens and toxins into otherwise safe foods.

All proteins/genes used to enhance crop plants are carefully assessed for allergenic potential early in the development process using an internationally accepted testing approach. If there is any indication that the newly produced

---

[11] *Ibid.*

[12] Anthony M. Shelton, *Testimony Before the US House of Representatives Subcommittee on Basic Research hearing on 'Plant Genome Research: From the Lab to the Field to the Market, Part II,'* Serial No, 106-60, Government Printing Office, Washington, October 5, 1999.

protein may be an allergen, clinical allergy testing methods are applied. When proteins/genes known to be allergenic are used as donors of genetic material, the highest standard of proof of non-allergenicity is required for the resulting product.

In addition to analysis early in the developmental process, foods derived from biotechnology crop plants must meet stringent regulatory and food safety standards for allergenicity before approval by regulatory agencies. If a company does launch a new food product that contains an allergen, regulatory agencies around the world, including the FDA, require that the product be labeled.

Strict standards similar to those used to govern allergens are used to protect against the introduction of toxins into biotech crop plants.

## Antibiotic resistance

Antibiotic resistance genes are used to identify and trace the trait of interest that has been introduced into plant cells to ensure that the transfer was successful. Use of these markers has raised concerns that new antibiotic resistant strains of bacteria will emerge.

The safety of crops that contain antibiotic resistance markers has been thoroughly reviewed and confirmed by such organizations as the Food and Agriculture Organization of the United Nations (FAO) and the World Health Organization (WHO) as well as by scientific committees in the European Union (EU) and by government experts from Canada, Japan, Switzerland, United States, Australia, and New Zealand.

## Applying Agricultural Biotechnology in the Developing World

It has been estimated that the global population will grow by more than 30% to almost 8 billion people by 2020. According to a report prepared by the International Food Policy Research Institute (IFPRI), most of this growth will occur in the developing world. To meet the increased demand for food, the world's farmers will have to produce 40% more grain in 2020. Improvements in crop yields will be required to bring about the necessary production increases. The IFPRI study pointed to agricultural biotechnology as a potential valuable contributor to food production needs (Pinstrup-Andersen *et al.*, 1999, p. 5).

A recent study conducted under the auspices of the Royal Society of London also confirmed the value of agricultural biotechnology for world agriculture in the 21st century: 'Foods can be produced through the use of GM (genetic modification) technology that are more nutritious, stable in storage and in principle, health promoting – bringing benefits to consumers in both industrialised and developing nations.'[13]

There are a number of prominent challenges to the success of agricultural biotechnology in developing nations. Among the most prominent are the

---

[13] Brazilian Academy of Sciences, Chinese Academy of Sciences, Indian National Science Academy, Mexican Academy of Sciences, National Academy of Sciences of the USA, The Royal Society (UK) and the Third World Academy of Sciences, *Transgenic Plants and World Agriculture*, p. 6.

challenges that the private sector and the global agricultural research system are experiencing with the protection of intellectual property rights.[14] This is threatening the sale and use of enhanced seeds in developing nations.

Agricultural biotechnology has also been the subject of political considerations that affect its acceptance and establishment in developing countries. During its hearings on agricultural biotechnology, the House Subcommittee on Basic Research studied various political considerations as they relate to the acceptance of ag biotech. Among the aspects studied was how politics affects acceptance of the system in the developing world. The subcommittee reported: 'What is perhaps of greater concern is the impact European attitudes could have in other agricultural and food markets around the world, particularly in the developing world, where biotechnology can address so many health and environmental problems.'[15]

The promise of agricultural biotechnology for the developing world is great and the challenges to the full realization of this promise are formidable.

A number of projects are underway to promote agricultural biotechnology in developing nations.

## Individual projects

### Golden rice

This vitamin-fortified rice offers a powerful weapon against vitamin A deficiency, a leading cause of infant mortality in developing nations. There is also evidence that vitamin A can reduce or delay the progression of HIV to AIDS. In August 2000, the Monsanto Company became the first company to announce that it will donate its draft sequence of the 'golden rice' genome in order to accelerate research on the development of this biotech product.[16]

### Healthy Harvest Foundation

Monsanto Company has joined with a number of African nations and renowned scientists to establish a foundation to promote agricultural biotechnology in Africa.

## Projects by nation

A survey conducted by the International Service for National Agricultural Research (1999) found compelling examples of the practice of agricultural biotechnology in several developing nations.

---

[14] For more information on the subject of agricultural biotechnology and intellectual property rights, see the 'Intellectual Property' section of report on 'Transgenic Plants and World Agriculture' conducted under the auspices of the Brazilian Academy of Sciences, Chinese Academy of Sciences, Indian National Science Academy, Mexican Academy of Sciences, National Academy of Sciences of the USA, The Royal Society (UK) and the Third World Academy of Sciences, pp. 17–18.

[15] Subcommittee on Basic Research of the Committee on Science, United States House of Representatives, *Seeds of Opportunity: An Assessment of the Benefits, Safety, and Oversight of Plant Genomics and Agricultural Biotechnology*, April 13, 2000, p. 61.

[16] For more information on 'golden rice,' see 'Grains of Hope,' *TIME* Magazine, July 31, 2000, pp. 39–46.

## Mexico

National biotechnology research units were established here in the early 1980s and Mexico is one of the most advanced in agricultural biotechnology among the developing nations today.

Virus-resistant potatoes are being developed by private industry and the Center for Advanced Studies (CINVESTA) in Irapuat. The goal is to develop potato virus X (PVX) resistance in varieties of potatoes grown by Mexican farmers. The partnership is funded by the Rockefeller Foundation and facilitated by the International Service for the Acquisition of Agri-biotech Applications (ISAAA).

## Indonesia

A National Program Development of Biotechnology was formulated in 1983 as the national strategy and policy for biotechnology. The Inter-University Center for agricultural biotechnology was established in 1985 to train university faculty in biotechnology, as was the Research and Development Center for Biotechnology (RDCB), which was charged with enhancing the national capacity for biotechnology. Four years later, the Biotechnology Division of the Central Research Institute for Food Crops (CRIFC) of the Agency for Agricultural Research and Development (AARD) was established.

## Zimbabwe and Kenya

Established in 1982, the Kenya Research Institute (KARI) is the center of biotechnology activities in that country. In Zimbabwe, the Biotechnology Research Institute (BRI) at the Scientific and Industrial Research and Development Centre (SIRDC) is the nation's leader in biotechnology.

Biotechnology in both nations gained momentum in 1992 with the support of the Netherlands for the Special Program on Biotechnology. The Special Program introduced the first elements of biotechnology planning in Kenya and Zimbabwe. As a result, the Kenya Agricultural Biotechnology Platform and the Zimbabwe Biotechnology Advisory Committee were founded in 1996.

# Conclusions

Agricultural biotechnology offers exciting possibilities and is already accomplishing much as a safe, productive alternative to traditional farming methods. What is perhaps most important about this agricultural system is that it presents the possibility of a new life for hundreds of millions of people who, with the benefits offered by agricultural biotechnology, will have an opportunity to change their lives. They could be relieved of the burdens of hunger, malnutrition and disease and enjoy the prospects of improved yields of high nutrition food products, which would bring new economic opportunities as well. No doubt, Gregor Mendel would be extremely pleased with the progress his work has made possible.

# Part 3

## Special Topics

# 26

# A Short History of Agricultural Biotechnology[1]

**Julie Babinard**
*International Food Policy Research Institute, Washington DC, USA*

Throughout the history of agriculture, biotechnological processes have been used in the creation of agricultural products. Food and agriculture are biological by nature and are the result of biological applications. Early examples include bread making and the fermentation of fruits and grains to make wine and beer. Many conventionally bred crops could be considered transgenic as they contain genes or segments of chromosomes found in totally different crop species. For instance, many crop varieties of sugarcane, tomato, potato, rice, corn, oats, and other highly bred crops contain genes or chromosome segments derived from different wild relative species (Spillane, 1999).

Farmers have taken advantage of the natural processes of genetic exchange through sexual reproduction to produce varieties of organisms that display desired biological traits. In conventional breeding, the genes pre-existing within a species are brought together in new combinations by making sexual crosses, a process in which tens of thousands of genes are mixed together through the fusion of pollen and egg. Each parent contributes half of its genome (an organism's entire repertoire of genes) to the offspring, but the composition of that half varies in each parental sex cell and hence in each cross. In practice, many crosses are necessary before the 'right' recombination of genes occurs to create the desired crop. The genetic variation we experience today in agriculture comes for the most part from mutations that occurred long ago in natural populations or during cultivation (OECD, 1992).

Genetic improvement as we know it today is the result of a lengthy process of research and scientific discoveries that occurred throughout the 20th century. Though plant breeding existed for thousand of years, it became a scientific endeavor only after Gregor Mendel formulated his laws on inheritance in 1866. Mendel's basic discovery was that each heritable property in any living organism is determined by a physical factor contained within the cell of the organism.

---

[1] Based on Babinard (1999).

Today, these factors of heredity are referred to as genes (see Table 26.1 for a brief overview of the significant periods in the development of the science of genetics).

**Table 26.1**: Significant periods in the development of genetics

| | |
|---|---|
| 1866–1920 | Classical genetics: Mendel's basic principles accepted and the existence of genes demonstrated; location of genes on chromosomes demonstrated; linear arrangements and location of genes shown by genetic mapping. |
| 1944–1966 | Central dogma: genes shown to consist of a chemical, deoxyribonucleic acid (DNA); structure of DNA determined; genetic code deciphered. |
| 1971 to date | Genetic engineering: development of rDNA, which allows genes to be manipulated and transferred from one species to another, whether or not they are related; genetic engineering of bacteria, plants, and animals initiated. |

*Source:* Persley (1990). Beyond Mendel's Garden. Biotechnology in the Service of World Agriculture. Oxon, UK, C.A.B. International.

## From Traditional Methods to Revolutionary Productivity Gains

Although Mendel made his observations about inheritance patterns in peas in the mid-1800s, they were lost from the scientific community until botanists rediscovered them in 1900. The understanding of inheritance and the underlying genetic factors eventually contributed to substantial increases in productivity of crops and animals. The constraints that characterize traditional agriculture were overcome as biologists began to understand and exploit hybrid vigor and as advances in other fields, such as techniques to produce cheap nitrogenous fertilizers, resulted in an upsurge in the development of superior hybrids. The first widely grown crop developed by scientists applying and expanding on Mendel's principles was hybrid corn in the 1920s (Wortman and Cummings, 1978). For example, the average US corn yield increased from 25 bushels per acre (1.5 metric tons per hectare) in the 1930s to 100 bushels per acre (6.2 metric tons per hectare) by 1980 (http://www.usda.gov/nass/pubs/trackrec/track00a.htm#corn).

In the developing world, the agricultural revolution that took place had its greatest successes with rice and wheat. Combined with increased use of chemical fertilizers and irrigation, worldwide efforts for desirable genetic traits led to significant improvements in yields in developing countries. With the Green Revolution, rice yields in the Philippines increased from 2 metric tons per hectare in 1970 to over 3 metric tons per hectare in the early 1980s. Between 1965 and 1980, wheat and rice production increased by about 75% in the developing world as a whole (Nottingham, 1998).

### The Next Agricultural Revolution?

Past improvements in agricultural productivity have relied on the application of modern technology and on sexual exchange of genetic material. Whereas genetic improvement research behind the earlier biotech originated in university and other public sector laboratories, private sector research expenditures have become increasingly important to recombinant DNA technologies. Modern biotechnology was first used commercially in the United States where several start-up companies were created to exploit its application. In Europe, relatively few companies that focus exclusively on biotechnology were established. Instead, major European pharmaceutical and chemical companies, especially those operating globally from France, Germany, Switzerland, and the UK, expanded their biotechnology research and product development activities. These multinationals began investing significantly in biotechnology research, and created life science departments to consolidate all research efforts in rDNA technologies, including medical and agricultural applications (James and Persley, 1990; Doyle and Persley, 1998). The first microorganism patent granted in the US and the granting of the Cohen–Boyer process patent for their genetic transfer technique in 1980 generated a rapidly growing interest in biotechnology commercial applications. Combined, these two events provided the necessary protection and incentive for commercial development of biotechnology products. Another incentive for the industry to focus research on increasing commercial applications in agriculture and food processing was the growing awareness of environmental and regulatory issues related to farm chemicals and the incentive to generate high-value agricultural products (James and Persley, 1990).

The first wave of agricultural biotechnology products initiated in the early 1990s has benefited farmers and producers by providing agronomic traits that make it easier to grow crops while reducing production costs. The products are primarily modified to include pest or herbicide resistance genes. Biotechnology is also being applied with some success in the livestock sector.

Single-gene products such as the growth hormones of cattle and pigs can be produced in commercial quantities. Biotechnology has also been used to produce experimental 'transgenic' animals, in which the genetic material was deliberately modified to produce 'clones'; animals are reproduced artificially but the DNA is not modified. Transgenic animals have mainly been developed for 'niche' markets such as production of high value pharmaceutical proteins where transgenic plants cannot fulfill the same functions (Spillane, 1999).

While genetically modified plants have been modified only for a single trait, such as herbicide tolerance or pest resistance, ongoing genome sequencing efforts should lead the way to identifying the function of more genes; breeding for more complex traits such as drought tolerance, which is controlled by many genes, should then become common (Persley and Doyle, 1999). Genetic transfer will also give crops increased resistance to other environmental and biological stresses such as heat, nutrient deficiencies, insects, and diseases (Arntzen, 1984; National Academy of Sciences, 1982). Through these developments, the benefits of agricultural biotechnology are expected to go well beyond increases in crop yields. The second wave of agricultural biotechnology products is scheduled to reach the market in 2002–2005 and will target output traits, promising to enhance the value of crops from the farmer to the consumer. A third wave of agro-biotechnology, intended to focus on the development of plants as nutrient factories to supply food, feed, and fiber, is also anticipated after 2005.

As a vast range of approaches for the improvement of agronomic traits is either under study or in early development phases, one of the major keys in capturing the promising benefits of crop biotechnology is the development of genomics. Genomics is concerned with the mapping and sequencing of genes (structural genomics) and with the assessment of gene function on the basis of the information thus obtained (functional genomics) (Kern, 2000). Genomic research will enable identification of new genes and gene functions, allowing the discovery and validation of new traits for seed improvements. Since December, 1999, major developments have occurred in genomics, when biologists decoded a large part of the DNA of a plant, affording the first glimpse of the genetic makeup of the plant world. By completing the first genetic sequence of the *Arabidopsis thaliana* plant, which is informally known as the 'weed,' biologists are greatly reducing the time for isolating genes and speeding up genetic discoveries ranging from more healthful soybean oil to a protein that may lead to faster growing crops (Wade, 1999). As it is cheaper and easier to determine the DNA sequence of species with smaller genomes, a species like the 'weed' can emerge as an 'anchor genome' acting as a 'Rosetta Stone' for understanding the larger genome's hieroglyphics of other related species.

A number of major multinational efforts are underway to determine the sequence and structure of major cereal crop genomes, which will provide valuable information for understanding the genomes of other cereals. Examples include the International Grass Genome Initiative, the International Rice Genome Initiative and the National Corn Genome Initiative (Spillane, 1999).

Biotechnology research for the next waves of biotechnology products is currently identifying and isolating a vast range of genes related to pest and disease resistance as well as protein and micronutrient deficiencies. A team of Brazilian scientific researchers deciphered for the first time the complete DNA sequence of an organism that causes a plant disease. The plant pathogen, *Xylella fastidiosa*, is an insect-borne bacterium that is a serious disease of citrus fruit. Certain strains of *X. fastidiosa* also affect other important produce like coffee, nuts, and other fruits (Bevan, 2000). Exploring the bacterium's biology will provide new perspectives on host–pathogen interactions, eventually leading to the development of tolerance and resistance in crop plants. A range of transgenic approaches is now being developed to nutritionally improve the amino acid profile of crop protein and also to increase the vitamin A and iron content in crops to facilitate dietary intake. A variety of genetically modified rice, given the name golden rice for its golden hue, was recently enriched in beta-carotene, the building block of vitamin A, ensuring an adequate supply of that vital nutrient.

The impending completion of several genome sequences adds excitement to the possibilities in agricultural genomics and biotechnology. Both are still in their infancy but further technological improvements present significant opportunities for agriculture, including reduced reliance on chemical use, increased flexibility in crops planted, and yield increases in some cases. For farmers, the application of biotechnology means potential increased net returns through savings in production cost. The importance of biotechnology for future agricultural productivity can also be considerable. The steady increases in agricultural productivity that occurred over the past several decades are beginning to level off as the possibilities of present methods of production are exhausted. In addition, it is unlikely that significant production expansion will be possible, or desirable, from the opening of new lands. Finally, agricultural biotechnology offers tools to tackle current or emerging problems in food production and human nutrition.

# Plant Genetic Modification Technologies

27

**Jack M. Widholm**
*University of Illinois, Urbana, Illinois, USA*

## Introduction

This chapter describes the materials and techniques available for plant genetic manipulation to both the plant breeder (conventional plant breeding methods) and biotechnologist (genetic engineering methods). It should be pointed out, however, that these are not mutually exclusive technological approaches to crop improvement. Plants altered by genetic engineering methods become an additional source of genetic material for the plant breeder using conventional approaches to produce commercially acceptable cultivars with the altered trait.

## Conventional Plant Breeding Methods

*Farmers* have been doing genetic manipulation for thousands of years (Duvick, 1996). However, these manipulations are limited by sexual compatibility, i.e., the transfer of genes can only take place between plants that can pollinate one another. This is most common within the same species like *Glycine max* (soybeans). Sometimes closely related species (for soybeans these would be *Glycine soja* or wild perennial species like *Glycine tomentella*) can be crossed to obtain unique traits not already present in commercial crop lines. Even where some sexual compatibility with relatives exists, there is clearly a limit to the type of gene that can be transferred, i.e., the plant must contain it. Furthermore, since a large amount of other DNA will be introduced as well, the breeder must backcross many times to the desired line to obtain the desired plants carrying the gene of interest and only a small amount of other DNA from the donor parent. Conner and Jacobs (1999) state 'In many respects, the precise manner in which

Genetically Modified Organisms in Agriculture
ISBN 0-12-515422-4

genetic engineering can control the nature and expression of the transferred DNA offers greater confidence for producing the desired outcome compared with traditional plant breeding.'

Plant breeders generally work with natural variation in plant populations and do large-scale crossings and assessments of progeny to find desirable lines. They have, for example, been very good at increasing yield and finding disease resistance genes to use. Yield is measured directly by harvesting grain in field plots and disease resistance usually is tested by inoculating the plants with the pathogen or growing the plants in an infested soil. Due to the environmental variation usually found in the field, pathogen inoculations are often carried out under more controlled conditions in the greenhouse or growth chamber. One long-term selection experiment that was initiated in 1896 has produced corn lines with very high and very low levels of oil and protein simply by selecting the highs and lows at each generation and growing these again (Dudley and Lambert, 1992). The high oil corn varieties that are now being grown commercially are a result of this kind of selection.

## Molecular markers

Plant breeders use a number of different molecular markers in their work that actually identify and follow specific DNA sequences on the plant chromosome. These markers make it possible to map genes, determine genetic variability and to do marker-assisted selection to follow traits when backcrossing (Phillips and Vasil, 1994). These molecular marker systems have various designations such as RFLP, SNP, AFLP, RAPD, SCAR and microsatellites, but they all simply identify a specific location on the chromosome using different molecular biological methods thus allowing breeders and geneticists to identify and follow these sites in the plant material they are using. While these techniques use isolated DNA and molecular biology methodology, they are only tools to expedite genetics and breeding and do not change the genes in the lines being studied.

## Induced mutations

The plant breeder can also work with traits resulting from induced mutations, i.e., changes in the DNA sequences that compose the genes (Van Harten, 1998). Mutations can be induced by treatments including radiation and certain chemicals. These treatments alter the DNA by changing the bases, thus changing the code, or by breaking the DNA chains, which can cause rearrangements, deletions, or duplications. The mutations result in an altered gene product, change in expression levels or timing of expression or no gene product. Mutations are random and are usually detrimental. Mutations that affect any single specific gene are rare so one usually needs to examine a large number of individuals to find any desired change. Thus development of an efficient screening method is necessary. Mutation breeding has been used by DuPont to identify some commercially useful traits such as the STS soybeans that are more resistant than normal to certain herbicides and low raffinose and stachyose soybean lines with low flatulence factor content.

## Plant tissue culture

Plant tissue cultures are plant tissues or cells that are grown under microorganism-free conditions in or on a medium containing sugar as a carbon and energy source, minerals to serve as nutrients similar to those needed by whole plants and growth regulators to control the growth. Such cultures have widespread use for commercial plant propagation, germplasm storage, experimentation, plant regeneration and compound production. Since plants can often be regenerated from the cells, the cultures can be used to select mutants or for inserting genes to then produce plants with altered characteristics as described below.

## Somaclonal variation

Mutations can occur at a rather high frequency in plants regenerated from tissue cultured cells (Widholm, 1996). This phenomenon is called somaclonal variation. The frequency of mutations induced in this way varies with the species, time in tissue culture and tissue culture method. In the case of maize (*Zea mays*, corn) the frequency of heritable phenotypic changes that can be seen by eye has been as high as 1.2 per regenerated plant. Soybean plants, on the other hand, can be regenerated that have no detrimental agronomic traits (Stephens *et al.*, 1991). Thus many plants have genetic changes but not all have damage that can be observed. Very little practical use of somaclonal variation has been made even though industry has evaluated a large amount of germplasm produced from regenerated plants.

## Tissue culture selection

Tissue culture selection is another way to produce plants with an altered trait. The challenge is to design an appropriate selection method. Many successful selections for resistance to compounds that are toxic to plant cells have been accomplished, especially herbicides (Newhouse *et al.*, 1991). Usually the resistant mutant cells are found at a low frequency, probably due to spontaneous mutations that occur naturally. However, since billions of cells can be readily used in the selection it is possible to find them due to their ability to grow in the presence of the toxic selection agent. It is also possible to apply mutagens to the cultures to increase the mutation frequency.

An example of a commercially used product resulting from tissue culture selection is the imidazolinone herbicide-resistant maize that has been sold in the US since the early 1990s. These hybrids are now called Clearfield Corn. The selection method produced cell lines with an altered target enzyme, acetolactate synthase, that was no longer very sensitive to inhibition by the herbicide (Newhouse *et al.*, 1991). Plants were regenerated from the resistant cells and these plants were also resistant to the herbicide. These were then used to produce inbred lines by backcrossing to make the commercial hybrids.

## Anther culture

Another technique that plant breeders can use with certain crops is called anther culture since the anthers that contain the developing pollen grains are removed from the plant and are placed in culture (Wan and Widholm, 1993). The haploid microspores (immature pollen grains) in the anthers can develop abnormally to form callus (unorganized growing cells) or embryos. Both callus or embryos can then regenerate into plants depending upon the species. Since the microspores are haploid, i.e., have half the normal plant chromosome number, either spontaneous or induced chromosome doubling will form completely homozygous plants (inbreds). Since these plants can be obtained within months rather than the years required by conventional self-pollination methods, inbred lines can be produced more quickly. This technique speeds up the breeding process but does not usually affect the genes themselves. Anther culture is presently being used in canola (rape) and rice breeding programs since the procedure is very efficient with these species.

## Protoplast fusion

Protoplast fusion is a genetic manipulation method that can produce new hybrids or transfer genes between sexually incompatible species. Protoplasts are prepared from plant cells by using enzymes that digest the rigid cell walls that normally encase the protoplasm. These protoplasts can then be fused together using several methods to produce hybrid cells containing the genomes of both cell types used (symmetrical fusion). It is also possible to transfer a portion of the genome from the donor protoplast, that is irradiated to break up the chromosomes and to prevent growth, into a recipient cell (asymmetrical fusion, Skarzhinskaya *et al.*, 1998). Most of the experimentation done so far has been with model species since generally it is difficult to regenerate plants from the hybrids. In the cases where regeneration was possible with symmetrical hybrids, the plants were quite abnormal. An example is the case of a tomato and potato fusion hybrid plant that did not produce either fruit or tubers (Melchers *et al.*, 1978).

# Biotechnological Methods for Producing GMOs

Many terms have been used to describe the techniques involved in producing GMOs. These include transformation, biotechnology, genetic engineering, and transgenic. All involve the insertion of foreign DNA or genes into the genome of the recipient (Goodman *et al.*, 1987; Gasser and Fraley, 1992; Conner and Jacobs, 1999).

## Genes and promoters

In order to produce a GMO one first needs to have an isolated gene with a promoter DNA attached to control the expression of the gene. Genes consist of a DNA chain made up of a four letter alphabet that can be copied exactly by an enzyme called RNA polymerase to make an RNA molecule (messenger RNA). When read by ribosomes, the messenger RNA molecule makes a specific chain of amino acids, called a protein, using a three letter code. Genes with many different functions can now be isolated from almost any organism (plant, animal, microbe, virus). Most are enzymes that carry out metabolic reactions. An example is the gene for glyphosate (Roundup) herbicide resistance. This gene encodes an enzyme in the pathway that synthesizes the amino acids phenylalanine, tryptophan, and tyrosine, which are necessary for plant growth. Roundup inhibits the activity of the normal plant enzyme and stops plant growth. Researchers at the Monsanto Company found a bacterium (*Agrobacterium tumefaciens*) with an enzyme that is resistant to the herbicide. The gene for this enzyme was inserted into soybean to produce the glyphosate-resistant commercial lines 'Roundup Ready.' Another example is *Bt* crops that include a gene from one of several candidate subspecies of the bacterium *Bacillus thuringiensis* (*Bt*). The gene encodes a protein that is not an enzyme, but is toxic to certain families of insects.

The expression of any gene is controlled by a segment of DNA usually adjacent to it in the DNA chain called a promoter. Gene expression is controlled by proteins that bind to the promoter region to regulate the binding of the RNA polymerase enzyme that makes messenger RNA from the gene. These proteins are found in varying levels in different tissues or conditions to thus regulate expression of the gene. Clearly it is important to control the expression of each of the many genes (about 50 000) in the plant cells since this is what makes each tissue what it is. Roots, for instance, must express different genes than leaves in order to carry out their unique functions. There are several general kinds of promoters that have been studied including constitutive (for expression in all tissues), tissue specific (for expression in certain plant tissues such as seeds), and inducible (that turns the gene on only under certain conditions such as pathogen attack, heat stress, drought). The biotechnologist must choose the correct promoter to control the gene of interest so that it is expressed at the right place at the right time at the right level.

## Selection markers

A selectable marker gene that confers resistance, usually to an antibiotic or herbicide, must be included with the useful gene during the transformation process since the number of cells that successfully integrate the novel DNA into their genome is small. The cells expressing the selectable marker gene are identified using toxic concentrations of the selection agent that inhibits the growth of normal cells, but not those expressing the selectable marker gene. The expression of the selectable marker gene must also be controlled by a promoter region attached to it. Usually high level expression is needed to be able to successfully select out the cells expressing the gene.

## Gene insertion methods

While there are a number of techniques available for inserting genes into plants, most of the GMOs being made today use one of two methods (Gasser and Fraley, 1992). The first of these methods involves shooting about 1 µm diameter gold or tungsten particles coated with the DNA at high speed into the plant cells. This method, called 'biolistics,' was first described by Sanford (1988). Initially the projectiles were propelled by gunpowder. Later versions of the 'gene gun' have used compressed helium gas or electro-volatilized water for propulsion. The particles are propelled into a chamber under a vacuum since air molecules slow the particles. Plant cells are usually about 30–50 μm in diameter. The DNA coated particles enter the plant cells and some of the DNA is able to integrate into the plant chromosomal DNA apparently randomly.

The other commonly used method utilizes the natural genetic engineer, a bacterium, *Agrobacterium tumefaciens*, which can transfer a defined piece of its own DNA into plant cells at wound sites. Normally the DNA placed in the plant cells by *A. tumefaciens* carries genes to make plant growth hormones, causing abnormal tissue growth to produce galls, resulting in the so-called crown gall disease. *A. tumefaciens* also inserts genes that produce compounds called opines. The *A. tumefaciens* cells can use the opines as carbon and nitrogen sources while most other organisms cannot, thus giving them a selective advantage over other microbes. Researchers replace the growth hormone and opine-producing genes with the desired ones. It was initially assumed that *A. tumefaciens* could only transfer DNA to dicotyledonous plants such as tobacco and grapes since crown gall disease was not found on monocotyledonous crops such as cereals. However, recent research has shown that *A. tumefaciens* can insert genes into cereals such as rice (Hiei *et al.*, 1994).

The use of *A. tumefaciens* has an important advantage over biolistics. The bombardment procedure often inserts multiple gene copies that can be rearranged in undesirable ways while the *A. tumefaciens* system is more likely to result in the insertion of one copy of the correct, full-length DNA fragment since the bacterium has the ability to direct the specific fragment to the plant nucleus.

As described above, a selectable marker gene is usually used in order to select out the cells that are actually carrying the inserted genes. Once enough resistant tissue is produced plants are regenerated. These plants will carry the selectable marker gene and in most cases the useful gene. Analysis for the presence and expression of the useful gene must be done and progeny obtained by self-pollination to produce homozygous lines.

Since the transformation process produces plants with a variety of expression levels one must usually regenerate plants from many independent transformation events to obtain the desired result. Thus like the plant breeder, the biotechnologist must work with relatively large numbers so that lines with poor characteristics can be eliminated. In addition, in the case of most commercial applications, the transgene would be backcrossed into elite lines that are sold commercially or inbreds used to produce hybrids. This process takes several years.

## Turning off genes – antisense and silencing

As mentioned above, specific DNA segments called promoters are attached to the gene one wants to express and this usually results in high-level expression of the desired gene. However, it is also possible to decrease gene expression using transformation. The first method involves expressing a gene already present in the plant as antisense, i.e., expressing the complement of the coding sequence so that the messenger RNA produced can bind to the natural message. Somehow the plant recognizes this as being abnormal and degrades the complex thus decreasing the levels of the encoded protein. In the early 1990s another method was discovered to turn off gene expression. This method results from high level expression of a gene that is identical or very similar to one already being expressed by the plant. Somehow the plant recognizes the 'overexpression' of the novel gene and shuts off both genes so instead of high gene product levels there is none. This process has been called gene silencing.

## GMO Challenges

### Transfer to wild relatives

Some crop plants like sorghum (*Sorghum vulgare*) and canola (*Brassica napus*) can cross with weed relatives of the same respective genus like Johnson grass (*Sorghum halepense*) and field mustard (*Brassica campestris*). Thus it would appear that herbicide resistance should not be placed in these crops when grown in areas that have the corresponding weeds since cross species pollination would lead to weeds resistant to the specific herbicide used.

One possible strategy to prevent spread of genes through pollen to weeds is to place the genes in the genome of the chloroplast, rather than the cell itself. The chloroplast, where photosynthesis is carried out within a cell, has its own chromosome that carries a relatively small number of genes. Most plants transmit this chromosome to their progeny only through the egg of the mother and not through pollen from the father. Thus if, for example, a herbicide resistance gene was carried by the chloroplast DNA, in most cases it could not be transmitted to sexually compatible weeds by pollen. Chloroplast DNA transformation is possible (Maliga *et al.*, 1993). Thus far success has been accomplished with only a very small number of species (tobacco, *Arabidopsis thaliana*, rice), but further success seems likely with continued research effort.

### Side effects from selection markers

Due to the debate about the use of certain selectable marker genes, like antibiotic and herbicide resistance, research is underway to develop less controversial selectable markers or markers that can be lost from the plants. An example of both possibilities involves the use of the gene *ipt* (isopentenyltransferase), taken from *Agrobacterium tumefaciens*. The expression product of this gene in a tissue cultured cell is a plant hormone (a cytokinin) that can induce shoots with an abnormal compact appearance, thus easing the process of identifying cells carrying the gene. Sugita *et al.* (1999) used this gene with tobacco (a good model system) to select transformed shoots with the abnormal appearance and then looked for normal shoots that formed from the original compact shoots. They used a special gene system from a fungus that can remove specific DNA sequences in these experiments. Many normal appearing shoots formed that had lost the selectable marker gene. Kunkel *et al.* (1999) used the same *ipt* gene but added a promoter that required the presence of a synthetic steroid not found in plants. About half the tobacco and lettuce shoots formed in the presence of the synthetic steroid inducer were transformed. The transformed plants had normal appearance if the inducer was not present; i.e., the marker gene was still present but was turned off. It should also be possible to use selectable marker genes that are not expressed in the plants by using a promoter that is 'tissue culture specific.' This approach would allow expression for selection in the tissue culture process, but would be turned off in the regenerated plants (Song *et al.*, 1998).

Newly developed selectable marker genes impart resistance to some compound or allow growth with a new energy source in ways that should also be less objectionable. These new marker genes include a feedback altered form of the tryptophan control enzyme, anthranilate synthase, that can provide resistance to the toxic tryptophan analog, 5-methyltryptophan (Song *et al.*, 1998). Another new marker gene is mannose-6-phosphate isomerase that converts mannose-6-phosphate to fructose-6-phosphate, thus allowing cells to use mannose as a carbon and energy source (Joersbo *et al.*, 1999). These and other new markers have not been widely used at this time but their availability would seem to open the possibility of producing crop plants without objectionable genes.

Another approach is to use a gene whose expression can be seen (a 'reporter' gene) so that selectable marker genes are not necessary. The most commonly used reporter genes are a bacterial β-glucuronidase (*gus*) gene, a firefly luciferase (*luc*) gene and a jellyfish green fluorescent protein (*gfp*) gene (Blumenthal *et al.*, 1999). The tissue turns blue but is killed when *gus* enzyme activity is measured upon adding a substrate for the enzyme reaction. The *luc* gene expression also requires an added substrate and a sensitive light measuring system but the tissue is not killed by the assay. The *gfp* gene product (GFP) fluorescence can be seen just by shining light on the tissue, but chlorophyll can interfere with the easy detection of GFP in regenerated plants. Some transformed plants are now being produced using GFP visual selection so this approach may have utility (Kaeppler *et al.*, 2000).

## Conclusion

While there is debate about the acceptability of GMOs in parts of the world, it would be unfortunate if the technology cannot be put to work to help solve problems that have no other solution, to make food safer, more nutritious and cheap, to help clean up the environment, and to help Third World and developing countries especially in their quest for a better life.

# 28

# *Bt* Corn and the Monarch Butterfly: Research Update[1]

## Richard L. Hellmich

*United States Department of Agriculture, Agricultural Research Service, and Iowa State University, Ames, Iowa, USA*

## Blair D. Siegfried

*University of Nebraska, Lincoln, Nebraska, USA*

## Introduction

The effects of *Bacillus thuringiensis* (*Bt*) corn, *Zea mays*, on nontarget organisms have received much attention since Losey *et al.* (1999) suggested in a correspondence to *Nature* that pollen from *Bt* corn could be hazardous to the larvae of the monarch butterfly, *Danaus plexippus*. In Losey's laboratory investigation, young monarch larvae given no choice but to feed on milkweed, *Asclepias curassavica*, leaves dusted with pollen from a *Bt* corn hybrid ate less, grew more slowly, and had a significantly higher mortality rate than larvae feeding on leaves dusted with nontransgenic pollen. The authors questioned the environmental safety of *Bt* corn and called for scientific investigations. These preliminary findings were largely misrepresented by mainstream media before the potential impact of *Bt* corn pollen on monarch populations could be adequately assessed. Such reports have heightened public awareness, increased scrutiny of transgenic plants in terms of potential environmental impact, and

[1] The use of trade, firm, or corporation names in this article is for the information and convenience of the reader. Such use does not constitute an official endorsement or approval by the United States Department of Agriculture or the Agricultural Research Service of any product or service to the exclusion of others that may be suitable.

Genetically Modified Organisms in Agriculture
ISBN 0-12-515422-4

intensified one of the most controversial and polarizing issues to face agricultural scientists in recent memory.

The corn hybrids in question were genetically modified to express an insecticidal protein derived from the bacterium *B. thuringiensis*. *Bt* provides yield protection from pest species such as the European corn borer, *Ostrinia nubilalis*, and some protection from other Lepidoptera (Pilcher *et al.*, 1997) without the use of traditional insecticides or other management practices. The first transgenic plants were planted on a large scale in the United States in 1996 and have quickly been adopted by growers. Nearly 25 million acres of *Bt* field corn were planted in 1999, representing approximately 30% of total corn. Preliminary estimates for 2000 suggest this amount has deceased to 20 million acres, about 25% of total corn.[2] More than a year has passed since initial concerns were expressed in the *Nature* correspondence. In response to these concerns, several researchers have begun detailed studies to evaluate the effects of *Bt* pollen on monarch larvae. This chapter provides an overview of these investigations and significant related events.

## *Bt* Corn Registration and Nontarget Insects

The Environmental Protection Agency (EPA) has required all *Bt* corn registrants to provide information on possible effects of *Bt* corn on honey bees and other beneficial nontarget insects, such as ladybird beetles and green lacewings. Examples of tests required by registrants for approval of *Bt* corn are listed at the EPA website.[3] Additionally, EPA has provided fact sheets for all *Bt* corn events that have been registered and for those that are seeking registration.[4] In the EPA-required nontarget insect tests, insects were fed high concentrations of *Bt* protein and no negative effects were observed. The toxicity of *Bt* proteins expressed by transgenic corn to larval stages of butterflies and moths is well known (MacIntosh *et al.*, 1989). Many studies, particularly those conducted on the extensive use of *Bt* sprays in forests for gypsy moth control, have shown that some *Bt* Cry proteins can adversely affect nontarget Lepidoptera (Miller, 1990; Johnson *et al.*, 1995). Field data from these studies indicated a temporary reduction in lepidopteran populations during prolonged *Bt* use, although widespread irreversible harm was not apparent (Hall *et al.*, 1999). Based on such information, the EPA made the assumption that *B. thuringiensis* is a hazard to all Lepidoptera, but that exposure from agricultural uses of *Bt* was not expected to be as high as in forest spraying. *Bt* corn also was not expected to impact nontarget butterflies and moths.[5]

---

[2] Released June 30, 2000, by the National Agricultural Statistics Service (NASS), Agricultural Statistics Board, US Department of Agriculture, posted at http://usda.mannlib.cornell.edu/reports/nassr/field/pcp-bba/acrg0600.txt.

[3] Environmental testing data protocol sites posted at www.epa.gov/docs/OPPTS_Harmonized/850_Ecological_Effects_Test_Guidelines/Drafts/.

[4] *Bt* corn fact sheets are posted at www.epa.gov/pesticides/biopesticides/.

[5] Testimony of Janet L. Andersen, Director of Biopesticides and Pollution Prevention Division, US EPA, to Committee on Science Subcommittee on Basic Research, US House of Representatives, October 19, 1999; posted at www.epa.gov/pesticides/biopesticides/otherdocs/testimony-whouse.htm.

## Toxicity Studies

The amount of pollen dusted onto the milkweed leaves in the Losey *et al.* (1999) study was not quantified, and as a result, the dose of *Bt* protein consumed by the larvae could not be quantified. To formulate a quantitative risk assessment, the level of toxicity must first be determined. Generally dose–response studies are conducted to determine estimates of the $LC_{50}$, or lethal concentration that kills 50% of tested insects. Dose–response relationships of four *Bt* proteins were conducted by Blair Siegfried (University of Nebraska) with monarch neonates (newly hatched larvae). Neonates were exposed for seven days to purified *Bt* toxins incorporated into an artificial diet. All toxins currently available in *Bt* corn (Cry1Ab, Cry1Ac, and Cry9C) and one under development (Cry1F) were tested. Results of these studies indicate that monarch larvae are highly sensitive to certain *Bt* toxins, whereas others are relatively nontoxic. Monarch neonates were most sensitive to Cry1Ab and Cry1Ac, even more so than European corn borer neonates, although direct comparisons have yet to be conducted. In contrast, Cry9C and Cry1F were considerably less toxic; therefore, risks associated with plants expressing one or the other of these proteins are likely to be reduced relative to the Cry1Ab and Cry1Ac events. The commercially available Cry1Ac event, DBT418, is in the process of being phased out,[6] and has received little

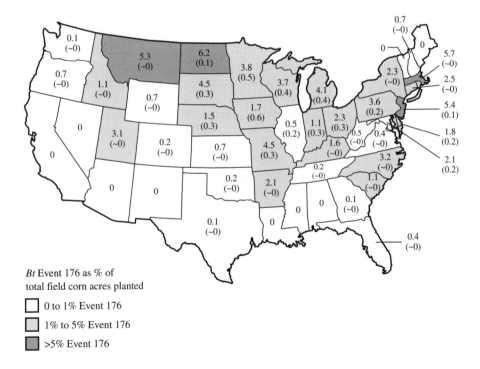

*Bt* Event 176 as % of
total field corn acres planted

☐ 0 to 1% Event 176

▨ 1% to 5% Event 176

■ >5% Event 176

**Figure 28.1**: Event 176 use: % of total corn (% of total land area), 1999

---

[6] DBT418 fact sheet posted at www.epa.gov/pesticides/biopesticides/factsheets/fs006463t.htm.

further attention. Consequently, most of the exposure questions have focused on the Cry1Ab events: 176, *Bt*11, and MON810.

Bioassays conducted by Richard Hellmich (USDA–ARS) and Mark Sears (University of Guelph) with monarch neonates fed milkweed leaves treated with pollen collected from *Bt* corn plants generally support Siegfried's findings. It should be noted, however, that susceptibility to *Bt* pollen varied among events known to express the same toxin. Events 176 and MON810 both express the Cry1Ab toxin, but 176 expresses at higher levels due to a pollen-specific gene promoter (Koziel *et al.*, 1993). Event 176 pollen affected neonates by increasing mortality and reducing body weight at pollen concentrations substantially lower than that of event MON810. Wraight *et al.* (2000) found that black swallowtail larvae, *Papilio polyxenes*, also were more sensitive to 176 pollen compared with MON810 pollen, based on field exposures.

Preliminary data suggest that pollen from event 176, compared with all the other *Bt* events, may pose the highest risk to monarch larvae. However, the potential risks to monarch populations are likely to be minimal, because sales of 176 corn hybrids in 1999 represented only approximately 2% of total corn. Fields of 176 corn represented very small portions of the total land area of each state (Figure 28.1).

## Exposure Studies

Although toxicity of *Bt* pollen and the toxins expressed by pollen can be clearly documented and could represent a hazard, the risk to monarch populations from exposure to *Bt* corn pollen is also a function of exposure. If exposure to the toxin under field conditions is minimal, then even highly toxic materials pose little risk. For monarchs to be exposed and potentially affected by corn pollen, larval development must coincide with corn anthesis (pollen shed). Throughout much of the Midwest, corn anthesis occurs during a short period (3–7 days) in July. Exposure to the pollen, however, may be longer depending on the duration of pollen availability on milkweed and the rate of *Bt* protein degradation. In 1999, anthesis overlapped with larval development in Iowa, but according to observations made by John Foster (University of Nebraska) in Nebraska and Galen Dively (University of Maryland) in Maryland, the overlap was of short duration. Dennis Calvin (Pennsylvania State University), Joe Russo (ZedX, Inc.), and Orley Taylor (University of Kansas) are developing models to identify areas where corn anthesis and monarch larval development are likely to occur simultaneously. During the 2000 summer, monarch observations from corn-growing areas will provide information to validate these predictions.

In areas where monarch larval stages overlap with corn anthesis there are a number of questions associated with pollen deposition and movement and their joint effect on exposure. Determining how much pollen is shed and how far the pollen drifts is necessary for an accurate assessment of exposure. Experiments were conducted in the summer of 1999 to measure deposition of corn pollen on milkweed plants in and near cornfields. Results from four independent laboratories (John Foster; Galen Dively; Mark Sears, University of Guelph; and John Pleasants, Iowa State University) indicate that pollen deposition sharply decreases at short distances from cornfields. Most pollen deposition occurs within the cornfield, and pollen quantities found on milkweed leaves and the proportion of leaves with pollen decreased rapidly 2 to 3 m from the fields. Similar pollen deposition results were reported by Raynor *et al.* (1972) and Wraight *et al.* (2000). This steep deposition gradient is influenced by the large size of corn pollen, typically 90 to 100 micrometers in diameter. Pleasants *et al.* (1999) examined pollen deposition within and at various distances from cornfields in Iowa by using glass slides coated with glycerin to trap pollen deposits, and pairs of milkweed leaves termed 'boutonnieres.' The concentration of pollen deposited on milkweed leaves followed the same pattern with distance and wind direction as the concentration on glycerin-coated slides, although the total amount deposited on leaves was much less than on slides. On average, milkweed leaves retained only 30% of the pollen available at a given sampling location. Also, a rain event during one of the experiments removed 90% of the pollen from milkweed leaves.

Another important factor in assessing exposure of monarch larvae to *Bt* pollen is milkweed distribution. Milkweeds in or adjacent to cornfields are likely to have more corn pollen on them than milkweeds in other habitats. Thus, it is important to assess the relative numbers of milkweeds in different habitats to fully assess the potential impact of *Bt* corn pollen on monarch populations. In the summer of 1999 Doug Buhler (USDA–ARS) and Robert Hartzler (Iowa State University) conducted a comprehensive survey of common milkweed plants across Iowa. Common milkweed was found in 71% of the roadsides and approximately 50% of the corn and soybean, *Glycine max*, fields (Hartzler and Buhler, 2000). Corn and soybean fields had 85% fewer patches than roadsides. Conservation reserve program (CRP) fields had the greatest average area infested. While common milkweed was frequently found in corn and soybean fields, average frequency and patch size were much greater in noncrop areas. This survey will be modified in the summer of 2000 to evaluate the stability of common milkweed populations and to better document common milkweed populations in CRP fields. There are several important questions that remain regarding the distribution of milkweed in different habitats across the Midwest and the relative importance of milkweed in these habitats as hosts for developing monarch larvae.

Behavior of monarch larvae and adults also can influence exposure to *Bt* pollen. Larval preference for milkweed leaves with pollen would increase exposure to *Bt* protein, but an avoidance of such leaves would have the opposite effect. Likewise, *Bt* exposure would be influenced if females had an oviposition preference for or against milkweed plants in or near cornfields.

Preliminary choice tests conducted by Hellmich and Leslie Lewis (USDA–ARS) suggest that monarch larvae are influenced by the presence of pollen, although their behavior changes with pollen concentrations. When presented with a choice of leaf discs without pollen and leaf discs with high amounts of pollen, more larvae were found on the discs without pollen. However, when presented with leaf discs without pollen and leaf discs with small amounts of pollen, more larvae were found on the leaves with pollen. Studies are being conducted to determine the importance of larval choice under field conditions and to determine the role, if any, that feeding cessation has when larvae encounter *Bt* protein.

Corn hybrids are taller than milkweed plants when corn anthesis occurs, but it is not known whether female oviposition is influenced by corn plant height.

Coordinated field studies are being conducted in Minnesota (Karen Oberhauser and Michelle Prysby), Iowa (Pleasants and Hellmich), New York (John Losey), Maryland (Dively), and Ontario (Sears) to determine the distribution of monarch life stages found in cornfields, other crop fields, roadsides, pasture, and set-aside land. These studies should determine whether female oviposition preference is influenced by habitat and the importance of milkweed plants growing in cornfields to monarch development.

## Nontarget Insect Meetings

Preliminary results from the 1999 laboratory and field studies were first reported in Chicago, Illinois at the November 1999 Monarch Conference. This conference served as an important forum for researchers, regulators, environmental advocacy groups, and industry. It was clear, however, that the results presented at this meeting were preliminary and incomplete. As a result of this meeting, a consortium of researchers was formed to identify gaps and overlaps in the data, promote an open exchange of information, and provide a coherent research agenda. The Chicago conference was followed in February 2000 by an USDA-sponsored Monarch Research Workshop in Kansas City, Missouri. A steering committee, including Adrianna Hewings (USDA–ARS), Eldon Ortman (Purdue University), Mark Scriber (Michigan State University), Eric Sachs (Monsanto), and Margaret Mellon (Union of Concerned Scientists), was formed to provide guidance for the workshop and subsequent consortium activities. The goal of the workshop was to identify research priorities regarding *Bt* corn and monarch butterflies and establish cooperation among researchers. Approximately 40 government, academic, and industry scientists participated in the workshop. Attendees identified short- and long-term research priorities, which were summarized by the steering committee. Short-term research objectives included the following: (1) determine importance of cornfields in monarch population production; (2) continue laboratory bioassays to further define dose–response relationships and sublethal effects; (3) expand land-use surveys to determine milkweed distribution, and abundance; (4) determine monarch distribution, abundance and survival in *Bt* and non-Bt corn production systems; and (5) collect data to field verify models that predict co-occurrence and dose–mortality response of monarch larvae and corn anthesis. A request for proposals based on these priorities was announced April 7, 2000. USDA–ARS and industry, through an unrestricted gift from the Agricultural Biotechnology Stewardship Technical Committee, each made $100 000 available for research projects outlined by the consortium.

In December 1999, a Scientific Advisory Panel was convened to provide advice, information, and recommendations to EPA regarding characterization and nontarget organism data requirements for protein plant-pesticides. The panel recommended that selection of nontarget insects should be made on a case-by-case basis and that species that have an ecological association with the crop plant or target insect should be considered. These species could include nontarget relatives of the target pest (e.g., nontarget Lepidoptera). The EPA is taking steps to address issues raised by the panel to more adequately address issues surrounding effects of nontarget organisms for current and future registrations of transgenic plants.

## Summary

The ability to transform crop plants to express the insecticidal toxins from *B. thuringiensis* is likely to have profound effects on the future of pest management. The benefits of the technology in terms of yield protection and reduced environmental disruption relative to synthetic insecticides must be balanced in terms of the uncertainty associated with risk to nontarget organisms, such as the monarch butterfly. The issue has drawn attention to an important aspect of genetic engineering. More attention needs to be directed at clarifying risk assessment and communicating that information, particularly to non-scientific audiences.

The consortium of researchers, industry, and environmental advocacy groups that has been assembled to address concerns related to the monarch butterfly and *Bt* corn represents an unparalleled level of cooperation and is indicative of the product stewardship that is essential for the full benefits of the technology to be realized. One of the goals of the consortium is to develop high quality research data so that decision making is based on sound science.

# 29

# The Beef Hormone Dispute Between the United States and the EU[1]

**Timothy Josling**
*Institute for International Studies, Stanford University, Stanford, California, USA*

**Donna Roberts**
*Economic Research Service, USDA, US Mission to the WTO, Geneva, Switzerland*

## Background

The beef hormone conflict has become one of the longest-running and most intransigent trade disputes. In the 1970s, European consumers became alarmed about the human health effects of hormones in livestock production. A ban on the use of certain growth hormones in livestock followed in the 1980s, and this ban was eventually extended to include imports of meat from animals that had been given hormones. Protests from the United States and other exporters were to no avail. By the 1990s, the issue became a test case for the new SPS Agreement (Agreement on the Application of Sanitary and Phytosanitary Measures), and eventually a test (in the eyes of many) of the willingness of the European Union to abide by an adverse ruling under the strengthened legal architecture of the Dispute Settlement Understanding of the WTO. The amount of trade originally involved was only about $100 million, a small fraction of the billions of dollars of trade that flow across the Atlantic in each direction. Compared to agricultural trade flows, the sum is very small, but the conflict

---

[1] Based on Josling *et al.* (1999).

Genetically Modified Organisms in Agriculture
ISBN 0-12-515422-4

over hormone-treated beef has had a major impact on trade relations far beyond the confines of the beef sector.

The events leading up to the ban on EU use of hormones in cattle raising and on imports of hormone-treated beef are important in explaining the political longevity of the issue in Europe. In many ways, the story begins with the emergence of non-governmental institutions such as consumer and environmental groups, together with the rise of the European Parliament, each cutting their political teeth on issues that appeared to resonate with public opinion. The beef hormone controversy was an ideal issue for these organizations. Trade concerns were not dominant in the early years, and the disciplines applied by trade rules were in any case weak. European livestock producers were searching for ways to stimulate growth in cattle and took eagerly to the use of hormones, sometimes with inadequate knowledge of the consequences of misuse of such chemicals. Regulatory control sometimes slipped between the cracks, as coordination and harmonization of national regulations progressed haltingly in the European Union. And, as the GMO debate later demonstrated, the media could not resist a good story of commercial greed, administrative incompetence, and consumer vulnerability.

## The DES Scare

The European ban on the use of hormones arose out of the DES scare of the 1970s. The illegal use of diethylstilbestrol (DES) in veal production in France was, at least in the public mind, linked to incidents of adolescents in Italy reportedly displaying hormonal irregularities. In addition, DES was found in baby food made from veal, and cases of children born with birth defects due apparently to exposure to DES were reported from other places in Europe. European consumers became alarmed over the possible negative health effects of using hormones in livestock production. European consumer organizations called for a boycott of veal. This had a significant adverse effect on the market and incidentally on the administration of the agricultural market policy which, at that time, supported veal as well as beef prices.

The DES scare created a consumer climate in Europe suspicious of hormones in livestock production and fearful of the potentially harmful health effects of these practices. Many of the same suspicions resurfaced with the outbreak of bovine spongiform encephalopathy (BSE) 15 years later and the more recent introduction of GMOs into the food chain. But the immediate legacy of the DES scare was the ban on the use of six specific hormones in the production of beef.

## The EC Ban on Hormones in Livestock Production

The DES scare eventually led the European Community Council of (Agriculture) Ministers in September 1980 to adopt a declaration favoring a ban on the use of estrogen and to support measures ensuring harmonized legislation of veterinary medicines and animal rearing. This was followed by more regulatory activity. In October 1980, the European Commission proposed a ban on the use of all hormones in livestock production unless the hormones were administered for therapeutic purposes (COM (80) 614). The expanded proposals of the Commission, COM (80) 920 and COM (80) 922, presented January 6, 1981, allowed for controlled use of three natural hormone products for therapeutic and zootechnical purposes. These proposals also introduced several control measures for the production and handling of these three hormone products and proposals for animal testing for the presence of these hormones. On February 13, 1981, the European Parliament adopted the Nielsen Report, which approved the Commission's proposals, and the EC Economic and Social Committee endorsed the proposals in February 1981. Three member states (Belgium, Ireland, and the United Kingdom) questioned the need for a total ban on the use of hormones in livestock production and sought to have the three natural hormones available for therapeutic and growth-production purposes. Ireland and the United Kingdom also argued that two of the synthetic hormones should also be retained for these purposes. The positions of these three member states on hormone use were echoed by non-European producers, including Argentina, Australia, Canada, New Zealand, South Africa, and the United States, who feared exclusion from the EU market.

On July 31, 1981, the EC Council of Ministers adopted the first directive banning hormones in livestock production (81/602/EEC). This directive addressed the use of five of the six hormones at issue, and the Council directed the Commission to provide a report on the scientific assessment of the harmful health effects of these five hormones when used to fatten animals. A scientific group was formed to investigate the issue and present its results to the Council. The Lamming Report concluded that most of the hormones would not present any harmful health effects when used under appropriate conditions as growth promoters in animals, but that control programs and monitoring systems for the appropriate use of these hormones were essential, and that additional scientific investigations were necessary to assess the health effects of some of the hormones. The 'Lamming Report' was followed by further action to establish programs to monitor and control the use of these hormones. The Commission proposed a directive to allow the controlled use of the three natural hormones for growth-promotion purposes and proposed reexamination of two of the synthetic hormones once further scientific investigations had been completed. This proposal was rejected by the European Parliament, the EC Economic and Social Committee, and eventually by the EC Council of Ministers.

The EC Commission, in deference to this opposition, amended its proposal and resubmitted it to the Council, which adopted it as Directive 85/649/EEC on December 31, 1985. This directive banned the use of all of the six hormones for growth-promotion purposes and established provisions for the use of these hormones for therapeutic purposes. This directive was challenged in the European Court of Justice and was annulled by the Court on procedural grounds. Finally, the Directive was reintroduced by the EC Commission and readopted by the EC Council as Directive 88/146/EEC on March 16, 1988. The ban applied to imported meat and products produced with the hormones in question, to the chagrin of the United States and other exporters.

## A Test for the WTO's New Safety Rules

This case was the first big test of the WTO's new food safety rules, the SPS Agreement.[2] The SPS rules were instituted to reduce conflict over these issues, but the complexities of scientific standards, interpretation of the rules provisions, and implications for trade, health, and consumer protection issues are widespread. The SPS rules do not articulate how scientific risk assessments should be conducted, how the findings of such assessments should be interpreted, or how much risk justifies trade restrictions by a member country.

Some trade experts believe that for the WTO rules to work, the dispute panels will have to define the application of the rules more precisely. This hormone case was the first major test of the SPS rules. It has had broad precedent-setting power for other health-related and biotechnology trade disputes governed by the SPS rules and for how the WTO's rules will be interpreted and applied from an institutional perspective.

## Outcome

For the United States, three objectives were achieved. First, the right of exporters to sell without trade restrictions posing as sanitary measures, but not scientifically based, was upheld. Second, the panel maintained the balance between the right to adopt higher levels of protection than are recognized internationally and the WTO requirement that sanitary measures be based on scientific principles and not be more restrictive than they need be. Third, the panel ruled that sanitary measures predating the SPS Agreement are not exempt from its requirements. The EU also found some comfort in the result. The ruling affirmed its right to establish a level of consumer protection higher than the level set by international health standards, provided it is backed by an objective risk assessment.

Not all individuals and groups go along with the panel in giving the hormones a clean bill of health. Some ecologists and health scientists cite studies that show a strong link between the rise in reproductive cancer incidents in the United States and contaminants such as at least one of the hormones at issue. Others worry more about the impact that the ruling has for local control of regulations. They feel that the WTO decision favors corporate interests and see it as a blow to national and subnational legislation to protect the environment and public health. They cite the fact that trade experts on the panel and trade lawyers advocating positions are not scientific experts, and that the rules do not allow for non-governmental organizations to present evidence at the hearings. Therefore they believe that the WTO rulings are fundamentally flawed and place corporate interests above health or environmental concerns.

---

[2] The SPS Agreement has since been tested in other disputes, including a Canadian complaint against Australian salmon import restrictions and a US complaint about Japanese testing of fruit varieties.

# Agricultural Biotech Glossary[1]

***Bacillus thuringiensis (Bt):*** A naturally occurring soil bacterium that occurs worldwide; produces a toxin specific to certain insects (e.g., moths, beetles, blackflies, mosquitoes).

**Biotechnology:** The science and art of genetically modifying an organism's DNA, such that the transformed individuals can express new traits that enhance survival (e.g., insect or disease resistance, herbicide resistance) or modify quality (e.g., oil, amino acids).

**Cry proteins:** Any of several proteins that comprise the crystal found in spores of *Bacillus thuringiensis*. Activated by enzymes in the insect's midgut, these proteins attack the cells lining the gut, cause gut paralysis, and subsequently kill the insect.

**Deoxyribonucleic acid (DNA):** Double-stranded molecule consisting of paired nucleotide units grouped into genes and associated regulatory sequences. These genes serve as blueprints for protein construction from amino-acid building blocks.

**Event:** Successful transformation of an organism by insertion of exotic genetic material (DNA). Events vary in the genetic package inserted into the organism and the particular place where this genetic package is inserted into the host DNA.

**Expression:** Production of the desired trait (e.g., protein concentration) in a transgenic plant. Expression varies with the gene, its promoter, and its insertion point in the host DNA.

**Gene:** The basic unit of inheritance and diversity; a section of DNA that codes for a specific product (e.g., protein) or trait.

**High-dose strategy:** An approach for minimizing the rapid selection for resistance to transgenic plants by using plants that produce Cry proteins at a concentration sufficient to kill all but the most resistant insects.

**Host plant resistance:** Ability of a plant to avoid insect attack, kill attacking insects, or tolerate their damage.

---

[1] Source: Ostlie *et al.*, 1997, used with permission.

**Integrated pest management (IPM):** A management approach that integrates multiple, complementary control tactics (e.g., biological control, crop rotation, host plant resistance, insecticides) to manage pests in a profitable yet environmentally sound manner.

**Lethal concentration (LC):** Concentration at which a toxin kills a given percentage of the insect test group, e.g., $LC_{50}$ refers to the concentration necessary to kill 50% of the insect test group.

**Marker:** A genetic flag or trait used to verify successful transformation and to indirectly measure expression of inserted genes. For example, a gene used as a marker in *Bt*11 confers tolerance to the herbicide Liberty.

**Mode of action:** Mechanism by which a toxin kills an insect. For example, the mode of action of *Bt* is ingestion and disruption of cells lining the midgut.

**Promoter:** A DNA sequence that regulates where, when, and to what degree an associated gene is expressed.

**Refuge:** An area planted to nontransgenic plants (e.g., non-*Bt* corn or alternative hosts for European corn borer), where susceptible pests can survive and produce a local population capable of mating with any possible resistant survivors from *Bt* corn.

**Registration:** Legal approval by EPA of pesticides and transgenic crops for use in the United States, after extensive review of toxicology to mammals, birds, fish, and other nontarget organisms, environmental fate, and health and safety issues, and precautions.

**Resistance:** The capacity of an organism to survive exposure to a toxin.

**Resistance management:** A proactive process of limiting or delaying resistance development in a pest population with a focus on preserving susceptible genes (individuals).

**Resistance ratio:** A measure of an insect population's resistance to a toxin, typically calculated by dividing the $LC_{50}$ of the resistant population by the $LC_{50}$ of a susceptible population.

**Selection:** A natural or artificial process that results in survival and better reproductive success of some individuals over others. Selection results in genetic shifts if survivors are more likely to have particular inherited traits.

**Stacking:** Introduction of two or more sets of novel DNA into an organism. For example, combining a *Bt* toxin gene with glyphosate resistance.

**Transgenic:** An organism genetically altered by addition of foreign genetic material (DNA) from another organism into its own DNA.

# GMO Field Trials in Europe

This chapter reproduces two tables developed by the Joint Research Center of the European Commission about the development of GMO crops in the EU. Table 1 reports the number of GMO crops broken down by country and species for which permission to do field tests has been approved.

Table 2 provides a correspondence between the common name and the scientific name for important crops.

**Table 1**: Plant species released per country, as of January 10, 2000

| Common name<br>Species | AT | BE | DE | DK | ES | FI | FR | GB | GR | IE | IT | NL | PT | SE | Total |
|---|---|---|---|---|---|---|---|---|---|---|---|---|---|---|---|
| African violet<br>*Saintpaulia ionantha* | · | · | · | · | · | · | · | · | · | · | · | 1 | · | · | **1** |
| Alfalfa<br>*Medicago sativa* | · | 1 | · | · | 1 | · | · | · | · | · | · | · | · | · | **2** |
| Apple<br>*Malus domestica* | · | · | · | · | · | · | · | · | · | · | · | 1 | · | 1 | **2** |
| Barley<br>*Hordeum vulgare* | · | · | · | · | · | 2 | · | 1 | · | · | · | · | · | · | **3** |
| Beet<br>*Beta vulgaris* | · | · | · | · | 6 | · | 1 | 6 | 1 | · | · | 1 | · | 2 | **17** |
| Broccoli<br>*Brassica oleracea* | · | · | · | · | · | 1 | · | · | · | · | · | · | · | · | **1** |
| Cabbage<br>*Brassica oleracea* | · | · | · | · | · | 1 | · | · | · | · | · | 1 | · | · | **2** |
| Cantaloupe<br>*Cucumis melo* | · | · | · | · | 1 | · | · | · | · | · | · | · | · | · | **1** |
| Carnation<br>*Dianthus caryophyllus* | · | · | · | · | · | · | · | · | · | · | · | 8 | · | · | **8** |
| Carrot<br>*Daucus carota* | · | · | · | · | · | · | · | · | · | · | · | 2 | · | · | **2** |
| Cauliflower<br>*Brassica oleracea* | · | 5 | · | · | · | 1 | · | · | · | · | · | · | · | · | **6** |
| Chicory<br>*Chicorium intybus* | · | 12 | · | · | · | · | 4 | · | · | · | 3 | 7 | · | · | **26** |
| Chrysanthemum<br>*Dendranthema indicum* | · | · | · | · | · | · | · | · | · | · | · | 1 | · | · | **1** |
| Cotton<br>*Gossypium hirsutum* | · | · | · | · | 14 | · | 1 | · | 10 | · | · | · | · | · | **25** |
| Eggplant (aubergine)<br>*Solanum melongena* | · | · | · | · | · | · | · | · | · | · | 7 | · | · | · | **7** |
| Eucalyptus<br>*Eucalyptus grandis* | · | · | · | · | 1 | · | · | 1 | · | · | · | · | · | · | **2** |

**Table 1**: *cont.*

| Common name<br>    Species | AT | BE | DE | DK | ES | FI | FR | GB | GR | IE | IT | NL | PT | SE | Total |
|---|---|---|---|---|---|---|---|---|---|---|---|---|---|---|---|
| European aspen<br>    *Populus tremula* | · | · | 1 | · | · | · | · | · | · | · | · | · | · | · | **1** |
| European aspen (alba × tremula)<br>    *Populus alba × Populus tremula* | · | · | · | · | · | · | 1 | · | · | · | · | · | · | · | **1** |
| European plum<br>    *Prunus domestica* | · | · | · | 1 | · | · | · | · | · | · | · | · | · | · | **1** |
| Fodder beet<br>    *Beta vulgaris* | · | 2 | · | 9 | 3 | · | 5 | 8 | · | · | 2 | · | · | · | **29** |
| Grape<br>    *Vitis vinifera* | · | · | 1 | · | · | · | 2 | · | · | · | 1 | · | · | · | **4** |
| Grape (berlandieri × riparia)<br>    *Vitis berlandieri × Vitis riparia* | · | · | · | · | · | · | 1 | · | · | · | · | · | · | · | **1** |
| Grape (berlandieri × rupestris)<br>    *Vitis berlandieri × vitis rupestris* | · | · | · | · | · | · | 1 | · | · | · | · | · | · | · | **1** |
| Green hearted chicory<br>    *Chicorium intybus* | · | · | · | · | · | · | · | 2 | · | · | · | 7 | · | · | **9** |
| Kiwi<br>    *Actinidia deliciosa* | · | · | · | · | · | · | · | · | · | · | 3 | · | · | · | **3** |
| Lettuce<br>    *Lactuca sativa* | · | · | · | · | · | 6 | · | · | · | · | · | · | · | · | **6** |
| Maize (corn)<br>    *Zea mays* | 1 | 23 | 18 | 1 | 59 | · | 188 | 7 | 6 | · | 89 | 13 | 5 | · | **410** |
| Marigold<br>    *Osteospermum ecklonis* | · | · | · | · | · | · | · | · | · | · | 15 | · | · | · | **15** |
| Melon<br>    *Cucumis melo* | · | · | · | · | 4 | · | 3 | · | · | · | 1 | · | · | · | **8** |
| Norway spruce<br>    *Picea abies* | · | · | · | · | · | 2 | · | · | · | · | · | · | · | · | **2** |
| Oilseed rape<br>    *Brassica napus* | · | 45 | 36 | 4 | 3 | 2 | 106 | 86 | · | · | 2 | 10 | · | 22 | **316** |
| Olive<br>    *Olea europea* | · | · | · | · | · | · | · | · | · | · | 2 | · | · | · | **2** |
| Paradise apple<br>    *Malus pumila* | · | · | · | · | · | · | · | 1 | · | · | · | · | · | · | **1** |
| Pea<br>    *Pisum sativum* | · | · | 1 | · | · | · | · | · | · | · | · | · | · | · | **1** |
| Petunia<br>    *Petunia petunia × Petunia hybrida* | · | · | 2 | · | · | · | · | · | · | · | · | · | · | · | **2** |
| Poplar<br>    *Populus deltoides* | · | · | · | · | · | · | 2 | 1 | · | · | · | · | · | · | **3** |
| Poplar (alba × tremula)<br>    *Populus alba × Populus tremula* | · | · | · | · | 1 | · | 4 | 1 | · | · | · | · | · | · | **6** |
| Potato<br>    *Solanum tuberosum* | 2 | 1 | 33 | 10 | 7 | 2 | 8 | 28 | · | · | 6 | 43 | 4 | 20 | **164** |
| Raspberry<br>    *Rubus idaeus* | · | · | · | · | · | · | · | · | · | · | 1 | · | · | · | **1** |
| Red hearted chicory<br>    *Chicorium intybus* | · | · | · | · | · | · | · | 2 | · | · | 6 | · | · | · | **8** |
| Rice<br>    *Oryza sativa* | · | · | · | · | 1 | · | 1 | · | · | · | 3 | · | · | · | **5** |
| Robusta (coffee)<br>    *Coffea canephora* | · | · | · | · | · | · | 1 | · | · | · | · | · | · | · | **1** |
| Rose gum<br>    *Eucalyptus grandis* | · | · | · | · | · | · | 1 | · | · | · | · | · | · | · | **1** |

**Table 1**: *cont.*

| Common name<br>Species | AT | BE | DE | DK | ES | FI | FR | GB | GR | IE | IT | NL | PT | SE | Total |
|---|---|---|---|---|---|---|---|---|---|---|---|---|---|---|---|
| Sand grape<br>*Vitis rupestris* | · | · | · | · | · | · | 1 | · | · | · | · | · | · | · | **1** |
| Scented pelargonium<br>*Pelargonium odoratissimum* | · | · | · | · | · | · | · | · | · | · | 1 | · | · | · | **1** |
| Scotch pine<br>*Pinus sylvestris* | · | · | · | 2 | · | · | · | · | · | · | · | · | · | · | **2** |
| Sea beet<br>*Beta vulgaris* | · | 1 | · | · | · | · | · | · | · | · | · | · | · | · | **1** |
| Silver birch<br>*Betula pendula* | · | · | · | 2 | · | · | · | · | · | · | · | · | · | · | **2** |
| Soybean (soya)<br>*Glycine max* | · | · | · | · | 4 | · | 6 | · | · | · | 4 | · | · | · | **14** |
| Spinach beet<br>*Beta vulgaris* | · | 1 | · | · | · | · | · | · | · | · | · | · | · | · | **1** |
| Spring turnip rape<br>*Brassica rapa* | · | · | · | · | · | 1 | · | · | · | · | · | · | · | 2 | **3** |
| Squash<br>*Cucurbita pepo* | · | · | · | · | 2 | · | 1 | · | · | · | 3 | · | · | · | **6** |
| Strawberry (adj. ananassa)<br>*Fragaria fragaria × Fragaria ananassa* | · | · | · | · | 2 | · | · | 1 | · | · | 4 | · | · | · | **7** |
| Sugar beet<br>*Beta vulgaris* | · | 8 | 22 | 21 | 12 | 5 | 61 | 24 | 1 | 4 | 39 | 21 | · | 7 | **225** |
| Sunflower<br>*Helianthus annuus* | · | · | · | · | 3 | · | 7 | · | · | · | · | 2 | · | · | **12** |
| Sweet cherry<br>*Prunus avium* | · | · | · | · | · | · | · | · | · | · | 3 | · | · | · | **3** |
| Sweet orange<br>*Citrus sinensis* | · | · | · | · | 1 | · | · | · | · | · | · | · | · | · | **1** |
| Tasmanian blue gum<br>*Eucalyptus globulus* | · | · | · | · | · | · | · | · | · | · | · | · | 1 | · | **1** |
| Thale cress<br>*Arabidopsis thaliana* | · | · | · | · | · | · | · | · | · | · | · | · | · | 1 | **1** |
| Tobacco<br>*Nicotiana tabacum* | · | · | 1 | · | 5 | 1 | 39 | 7 | · | · | 1 | · | · | · | **54** |
| Tomato<br>*Lycopersicon esculentum* | · | · | · | 16 | · | · | 5 | 1 | 1 | · | 43 | 2 | 2 | · | **70** |
| Watermelon<br>*Citrullus lanatus* | · | · | · | · | · | · | · | · | · | · | 1 | · | · | · | **1** |
| Wheat<br>*Triticum aestivum* | · | 2 | · | · | 3 | · | · | 8 | · | · | 3 | · | · | · | **16** |
| Wine grape<br>*Vitis vinifera* | · | · | · | · | · | · | 1 | · | · | · | · | · | · | · | **1** |

*Source*: Joint Research Center, European Commission, *http://food.jrc.it/gmo/gmo.asp*, accessed September 23, 2000.

Country code: Austria (AT), Belgium (BE), Germany (DE), Denmark (DK), Spain (ES), Finland (FI), France (FR), Great Britain (GB), Greece (GR), Ireland (IE), Italy (IT), Netherlands (NL), Portugal (PT), Sweden (SE).

**Table 2**: Taxonomy table of released crops

| Common name | Family name | Genus | Species | Subspecies |
|---|---|---|---|---|
| African marigold | Compositae | *Osteospermum* | *Osteospermum ecklonis* | |
| African violet | Gesneriaceae | *Saintpaulia* | *Saintpaulia ionantha* | |
| Alfalfa | Fabaceae | *Medicago* | *Medicago sativa* | *sativa* |
| Apple | Rosaceae | *Malus* | *Malus domestica* | |
| Barley | Poaceae | *Hordeum* | *Hordeum vulgare* | *vulgare* |
| Beet | Chenopodiaceae | *Beta* | *Beta vulgaris* | *vulgaris* |
| Broccoli | Brassicaceae | *Brassica* | *Brassica oleracea* | *botrytis* |
| Cabbage | Brassicaceae | *Brassica* | *Brassica oleracea* | |
| Canola | Brassicaceae | *Brassica* | *Brassica napus* | *napus* (syn. *oleifera*) |
| Cantaloupe | Cucurbitaceae | *Cucumis* | *Cucumis melo* | |
| Cape marigold | Compositae | *Osteospermum* | *Osteospermum ecklonis* | |
| Carnation | Caryophyllaceae | *Dianthus* | *Dianthus caryophyllus* | |
| Carrot | Apiaceae | *Daucus* | *Daucus carota* | *sativus* |
| Cauliflower | Brassicaceae | *Brassica* | *Brassica oleracea* | *botrytis* |
| Chicory | Asteraceae | *Cichorium* | *Chicorium intybus* | |
| Chrysanthemum | Asteraceae | *Dendranthema* | *Dendranthema indicum* | |
| Cotton | Malvaceae | *Gossypium* | *Gossypium hirsutum* | |
| Durum wheat | Poaceae | *Triticum* | *Triticum durum* | |
| Eggplant (aubergine) | Solanaceae | *Solanum* | *Solanum melongena* | |
| Eucalyptus | Myrtaceae | *Eucalyptus* | *Eucalyptus grandis* | |
| European aspen | Salicaceae | *Populus* | *Populus tremula* | *Populus tremola* × *Populus tremuloides* |
| European aspen (alba × tremula) | Salicaceae | *Populus* | *Populus alba* × *Populus tremula* | |
| European plum | Rosaceae | *Prunus* | *Prunus domestica* | *domestica* |
| Fodder beet | Chenopodiaceae | *Beta* | *Beta vulgaris* | *vulgaris var. crassa* |
| Grape | Vitaceae | *Vitis* | *Vitis vinifera* | |
| Grape (berlandieri × riparia) | Vitaceae | *Vitis* | *Vitis berlandieri* × *vitis riparia* | |
| Grape (berlandieri × rupestris) | Vitaceae | *Vitis* | *Vitis berlandieri* × *vitis rupestris* | |
| Grape (vinifera × berlandieri) | Vitaceae | *Vitis* | *Vitis vinifera* × *vitis berlandieri* | |
| Grapevine | Vitaceae | *Vitis* | *Vitis vinifera* | |
| Green chicory | Asteraceae | *Cichorium* | *Chicorium intybus* | |
| Green hearted chicory | Asteraceae | *Cichorium* | *Chicorium intybus* | |
| Kiwi | Actinidaceae | *Actinidia* | *Actinidia deliciosa* | |
| Lettuce | Asteraceae | *Lactuca* | *Lactuca sativa* | |
| Maize | Poaceae | *Zea* | *Zea mays* | *mays* |
| Marigold | Compositae | *Osteospermum* | *Osteospermum ecklonis* | |
| Melon | Cucurbitaceae | *Cucumis* | *Cucumis melo* | |
| Norway spruce | Pinaceae | *Picea* | *Picea abies* | *Abies* |
| Oilseed rape | Brassicaceae | *Brassica* | *Brassica napus* | *Napus* (syn. *oleifera*) |
| Olive | Oleaceae | *Olea* | *Olea europea* | |
| Paradise apple | Rosaceae | *Malus* | *Malus pumila* | |
| Petunia | Solanaceae | *Petunia* | *Petunia petunia* × *petunia hybrida* | |
| Poplar | Salicaceae | *Populus* | *Populus deltoides* | |

**Table 2**: *cont.*

| Common name | Family name | Genus | Species | Subspecies |
|---|---|---|---|---|
| Poplar (alba × tremula) | *Salicaceae* | *Populus* | *Populus alba × populus tremula* | |
| Potato | Solanaceae | *Solanum* | *Solanum tuberosum* | *tuberosum* |
| Rape | Brassicaceae | *Brassica* | *Brassica napus* | *napus* (syn. *oleifera*) |
| Rapeseed | Brassicaceae | *Brassica* | *Brassica napus* | *napus* (syn. *oleifera*) |
| Raspberry | Rosaceae | *Rubus* | *Rubus idaeus* | *rubus idaeus vulgatus* |
| Red chicory | Asteraceae | *Cichorium* | *Chicorium intybus* | |
| Red hearted chicory | Asteraceae | *Cichorium* | *Chicorium intybus* | |
| Rice | Poaceae | *Oryza* | *Oryza sativa* | |
| Robusta (coffee) | Rubiacae | *Coffea* | *Canephora* | |
| Rose gum | Myrtaceae | *Eucalyptus* | *Eucalyptus grandis* | |
| Sand grape | Vitaceae | *Vitis* | *Vitis rupestris* | |
| Scented pelargonium | Geraniaceae | *Pelargonium* | *Odoratissimum* | |
| Scotch pine | Pinaceae | *Pinus* | *Pinus sylvestris* | |
| Sea beet | Chenopodiaceae | *Beta* | *Beta vulgaris* | *vulgaris* var. *maritima* |
| Silver birch | Betulaceae | *Betula* | *Betula pendula* | |
| Soybean (soya) | Fabaceae | *Glycine* | *Glycine max* | |
| Spinach beet | Chenopodiaceae | *Beta* | *Beta vulgaris* | *vulgaris* var. *cicla* |
| Spring barley | Poaceae | *Hordeum* | *Hordeum vulgare* | *vulgare* |
| Spring oilseed rape | Brassicaceae | *Brassica* | *Brassica napus* | *Napus* (syn. *oleifera*) |
| Spring turnip rape | Brassicaceae | *Brassica* | *Brassica rapa* | *Oleifera* |
| Spring wheat | Poaceae | *Triticum* | *Triticum aestivum* | |
| Squash | Cucurbitaceae | *Cucurbita* | *Cucurbita pepo* | |
| Strawberry (adj. ananassa) | Rosaceae | *Fragaria* | *Fragaria fragaria × fragaria ananassa* | |
| Strawberry (virginiana × chiloensis) | Rosaceae | *Fragaria* | *Fragaria virginia × fragaria chiloensis* | |
| Sugar beet | Chenopodiaceae | *Beta* | *Beta vulgaris* | *vulgaris* var. *saccharifera* |
| Sunflower | Asteraceae | *Helianthus* | *Helianthus annuus* | |
| Swede | Brassicaceae | *Brassica* | *Brassica napus* | *napus* (syn. *oleifera*) |
| Sweet cherry | Rosaceae | *Prunus* | *Prunus avium* | |
| Sweet orange | Rutaceae | *Citrus* | *Citrus sinensis* | |
| Tasmanian blue gum | Myrtaceae | *Eucalyptus* | *Eucalyptus globulus* | *globulus* |
| Thale cress | Brassicaceae | *Arabidopsis* | *Arabidopsis thaliana* | |
| Tobacco | Solanaceae | *Nicotiana* | *Nicotiana tabacum* | *tabacum* |
| Tomato | Solanaceae | *Lycopersicon* | *Lycopersicon esculentum* | *esculentum* |
| Upland cotton | Malvaceae | *Gossypium* | *Gossypium hirsutum* | |
| Watermelon | Cucurbitaceae | *Citrullus* | *Citrullus lanatus* | *lanatus* |
| Wheat | Poaceae | *Triticum* | *Triticum aestivum* | |
| Wild strawberry | Rosaceae | *Fragaria* | *Fragaria vesca* | *Fragaria vesca f. semperflorens* |
| Wine grape | Vitaceae | *Vitis* | *Vitis vinifera* | |
| Winter oilseed rape | Brassicaceae | *Brassica* | *Brassica napus* | *napus* (syn. *oleifera*) |

*Source*: Joint Research Center, European Commission, *http://food.jrc.it/gmo/gmo.asp*, accessed September 23, 2000.

# References and Further Reading

ACC/SCN (United Nations Administrative Committee on Coordination/Subcommittee on Nutrition) (1997). *Third Report on the World Nutrition Situation*. Geneva: ACC/SCN.

ACC/SCN and IFPRI (International Food Policy Research Institute) (2000). *Fourth Report on the World Nutrition Situation*. Geneva: ACC/SCN and Washington: IFPRI.

Acharya, R. (1992). 'International Property, Biotechnology and Trade: The Impact of the Uruguay Round on Biodiversity.' Biopolicy International. Nairobi, Kenya: ACTS Press.

Alam, M.F., Datta, K., Abrigo, E., Oliva, N., Tu, J., Virmani, S.S. and Datta, S.K. (1999). 'Transgenic Insect Resistant Maintainer Line (IR68899B) for Improvement of Hybrid Rice.' *Plant Cell Reports* 18: 672–5.

Alston, J.M., Norton, G.W. and Parley, P.G. (1995). *Science Under Scarcity: Principles and Practice for Agricultural Research Evaluation and Priority Setting*. Ithaca, NY: Cornell University Press.

Alston, J.M., Sexton, R.J. and Zhang, M. (1997). 'The Effect of Imperfect Competition on the Size and Distribution of Research Benefits.' *American Journal of Agricultural Economics* 79 (November): 1252–65.

Alston, J.M., Marra, M.C., Pardey, P.G. and Wyatt, T.J. (2000). 'Research Returns Redux: A Meta-analysis of the Returns to Agricultural R&D.' *Australian Journal of Agricultural and Resource Economics* 44: 2 (June).

Altieri, M., Rosset, P. and Thrupp, L.A. (1998). 'The Potential of Agroecology to Combat Hunger in the Developing World.' *Food First Policy Brief*.

American Dietetic Association (1993). 'Position of the American Dietetic Association. Biotechnology and the Future of Food.' *Journal of the American Dietetic Association* 93(2): 189.

American Farm Bureau Federation (AFBF). (2000) 'Farm Bureau Policies for 2000. Resolutions on National Issues Adopted by Elected Voting Delegates of the Member State Farm Bureaus to the 81st Annual Meeting of the American Farm Bureau Federation.' Houston, Texas.

American Medical Association, Council on Scientific Affairs (1991). 'Biotechnology and the American Agricultural Industry.' *Journal of the American Medical Association* 266(3): 363.

Animal Plant Health Inspection Service, APHIS (2000). 'Agricultural Biotechnology Web Site.' www.usda.aphis.gov/biotech

Anonymous (1988). 'La Nueva Biotecnologia en Agricultura y Salud.' Serie 7 de Documentos de Programas. IICA (Instituto Interamericano de Cooperacion para la Agricultura.

Anonymous (1994). 'Foodborne Pathogens: Risks and Consequences.' Task Force Report. Ames, IA: Council for Agricultural Science and Technology, September.

Anonymous (1995a). 'At a Glance: RAFI's US Plant Utility Patent Database.' *RAFI Communiqué* 8.

Anonymous (1995b). 'Ground Cover Forum: Herbicide-Resistant Crops.' *CRDC Ground Cover* 11: 10.

Anonymous (1996). 'Another Study Warns of Transgenic Superweeds.' *Biotech Reporter* 13(11): 2.

Anonymous (1997). 'Bioseguridad Agropecuaria: Hacia la Comercializacion de vegetales y vacunas geneticamente modificados.' Secretaria de Agricultura, Ganaderia, Pesca y Alimentacion, Septiembre.

Anonymous (1998a). 'Monsanto Apologizes Over GM Soya Bean.' *Farmers Guardian* May 22.

Anonymous (1998b). 'Safeway Confirms Numbers of 170g GM Tomato Puree Cans Sold at 29p.' Available at http://www.biotechknowledge.com/showlib.php3?536, June 4.

Anonymous (1998c). 'Call for UK Genetic Food Watchdog.' *Nature* online service. Sept. 3.

Anonymous (1998d). 'Bt Plant-Pesticide Resistance Management.' EPA Publication. US Biopesticides and Pollution Prevention Division (7511C). Available at http://www.epa.gov/pesticides/biopesticides/white_bt.pdf.

Anonymous (1999a). 'Cattle Hormones: Hot Flushes.' *The Economist*, p. 94, May 15.

Anonymous (1999b). 'Sticky Labels.' *The Economist* May 1.

Anonymous (1999c). 'Tesco to Respond to Customer Views on GM Food.' Tesco News Center. Available at http://www.tesco.co.uk/, April 27.

Anonymous (1999d). 'Sachs on Development: Helping the World's Poorest.' *The Economist*. August 14, pp. 17–20.

Anonymous (2000). *Transgenic Plants and World Agriculture*. Washington DC: National Academy Press.

Arntzen, C.J. (1984). *Cutting Edge Technologies*. Washington DC: National Academy of Engineering, pp. 52–61.

Australia New Zealand Food Authority (2000). *GM Foods and the Consumer: ANZFA's Safety Assessment Process for Genetically Modified Foods*. June, p. 8.

Avery, A. (1995). 'Infantile Methemoglobinemia: Re-Examining the Role of Drinking Water.' *Environmental Health Perspectives* 107(7): 583–6.

Avery, D.T. (1997). 'Saving Nature's Legacy Through Better Farming.' *Issues in Science and Technology*, National Academy of Sciences and the University of Texas/Dallas, Fall.

Avery, D. (2000). 'Cattle Not to Blame For Antibiotic Resistance in Humans.' *Bridge News Forum, BridgeNews* May 12; opinion@bridge.com.

Baarda, J.R. (1999). 'A Transgenic Theory of the Firm.' Paper presented at the NE-165 Conference Transitions in Agbiotech: Economics of Strategy and Policy, Washington DC, June 24–25.

Babinard, J. (1999). 'The Role of Government in the Approval and Regulation of Genetically Modified Crops.' *International Policy Studies*, Stanford University, CA.

Bacheler, J.S. (1999). '1999 Bollgard Cotton Performance Expectations for North Carolina Producers.' *Carolina Cotton Notes* May: 99-5-A.

Bacheler, J.S., Mott, D.W. and Morrison, D.E. (1998). 'Large Scale Evaluation of Bollgard Resistance to Multiple Pests in North Carolina Under Grower Conditions.' *1998 Proceedings Beltwide Cotton Conferences* (eds. P. Dugger and D. Richter), pp. 961–4. Memphis, TN: National Cotton Council of America.

Bailey, J. (2000). USDA/FAS Mexico City Office, personal communication.

Bains, W. (1998). *Biotechnology from A to Z* (2nd edn). Oxford: Oxford University Press.

Ballenger, N., Bohman, M. and Gehlhar, M. (2000). 'Biotechnology: Implications for US Corn and Soybean Trade.' *Agricultural Outlook*. Economic Research Service, US Department of Agriculture, April.

Barboza, D. (2000). 'Modified Foods Put Companies in a Quandary.' *New York Times* June 4, p. 27. Available at www.nytimes.com archive.

Barinaga, M. (1997). 'Making Plants Aluminum Tolerant.' *Science* 276(June 6): 1497.

Barton, J., Crandon, J., Kennedy, D. and Miller, H. 'A Model Protocol to Assess the Risks of Agricultural Introductions.' *Nature Biotechnology* 15(9): 845–8.

Bazett, M. (2000). *Long-term Changes in the Location and Structure of Forest Industries*. World Bank/WWF Project, Global Vision 2050 for Forestry, January.

Bedsworth, J. (2000) 'Big Bang for Your Beef.' *Beef Today* February, p. 22.

Benbrook, C. (1999). 'Evidence of the Magnitude of the Roundup Ready Soybean Yield Drag from University-Based Varietal Trials in 1998.' Ag BioTech InfoNet Technical Paper No. 1, July 13. Available at http://www.biotech-info.net/RR_yield_drag_98.pdf. Accessed July 2000.

Benbrook, C.M. and Hansen, M. (1997). 'Return to the Stone Age of Pest Management.' Conference on Plant Pesticides Resistance Management, Washington DC, May 21.

Bender, K., Hill, L., Wenzel, B. and Hornbaker, R. (1999). 'Alternative Market Channels for Specialty Corn and Soybeans.' Agric. Exp. Sta. Res. Bull. No. AE-4726, February. University of Illinois, Urbana-Champaign.

Benedict, J.H. (1996). '*Bt* Cotton: Opportunities and Challenges.' *1996 Proceedings Beltwide Cotton Production Conferences* (eds. P. Dugger and D. Richter), pp. 25–9. Memphis, TN: National Cotton Council of America.

Benfey, P. and Chua, N.H. (1989). 'Regulated Genes in Transgenic Plants.' *Science* 244: 174–81.

Bergelson, J., Purrington, C.B. and Wichmann, G. (1998). 'Promiscuity in Transgenic Plants.' *Nature* 395: 25.

Berkes, F., Feeny, D., McCay, B.J. and Acheson, J.M. (1989). 'The Benefits of the Commons.' *Nature* 340: 91–3.

Bessin, R. (1996). *Bt Corn*. University of Kentucky Department of Entomology Entfacts. May. Available at http://www.uky.edu/Agriculture/Entomology/entfacts/flcrops/ef118.htm.

Bevan, M. (2000). 'The Bugs from Brazil.' *Nature* 406(July 13): 140–1.

Bieir, F.K., Crespo, R.S. and Straus, J. (1985). *Biotechnology and Patent Protection: An International Review*. Geneva: Organization for Economic Cooperation and Development.

Birch, A.N.E., Geoghegan, I. E., Majerus, M.E.N., Hackett, C. and Allen, J. (1997). 'Interactions Between Plant Resistance Genes, Pest Aphid Populations and Beneficial Aphid Predators.' *Annual Report*. Invergowrie Scotland: Scottish Crop Research Institute.

Blumenthal, A., Kuznetzova, L., Edelbaum, O., Raskin, V., Levy, M. and Sela, I. (1999). 'Measurement of Green Fluorescent Protein in Plants, Quantification, Correlation to Expression, Rapid Screening and Differential Gene Expression.' *Plant Science* 142: 93–9.

BMA (British Medical Association) (1999). 'The Impact of Genetic Modification on Agriculture, Food and Health. An Interim Statement.' May, 1999. Board of Science and Education, British Medical Association.

Bock, S.A. (1987). 'Prospective Appraisal of Complaints of Adverse Reactions to Foods in Children During the First 3 Years of Life.' *Pediatrics* 79: 683–8.

Bode, W.M. and Calvin, D.D. (1990). 'Yield-Loss Relationships and Economic Injury Levels for European Corn Borer Populations (Lepidoptera: Pyralidae) Infesting Pennsylvania Field Corn.' *Journal of Economic Entomology* 83: 1595–603.

Bohnert, H.J. (1996). 'How Do Plants Tolerate Water Deficit – How Can We Use Such Knowledge for Plant Engineering.' Workshop on Transgenic Plants: Biology and Applications, Tuskegee University.

Bowen, D.J. and Ensign, J.C. (1998). 'Purification and Characterization of a High-Molecular-Weight Insecticidal Protein Complex Produced by the Entomopathogenic Bacterium *Photorhabdus luminescens.*' *Applied and Environmental Microbiology* 64: 3029–35.

Bowen, D., *et al.* (1998). 'Insecticidal Toxins from Bacterium *Photorhabdus luminescens.*' *Science* 280: 2129–32.

Briggs, S.P., and Guse, C.A. (1986). 'Forty Years of European Corn Borer: What Have We Learned?' In: *Thirty-eighth Illinois Custom Spray Operators Training Manual.* University of Illinois Cooperative Extension Service, pp. 169–173.

Bryant, K.J., Robertson, W.C. and Lorenz, G.M. (1998). 'Economic Evaluation of Bollgard Cotton in Arkansas.' *1998 Proceedings Beltwide Cotton Conferences* (eds. P. Dugger and D. Richter), pp. 388–9. Memphis, TN: National Cotton Council of America.

Buckwell, A., Brookes, G. and Bradley, D. (1999). 'Economic of Identity Preservation for Genetically Modified Crops.' Food Biotechnology Communications Initiative.

Bull, D.L. and Menn, J.J. (1990). 'Strategies for Managing Resistance to Insecticides in Heliothis Pests of Cotton.' *American Chemical Society (ACS) Symposium Series* 421: 118–33.

Bullock, D.S., Desquilbet, M. and Nitsi, E. (2000). 'The Economics of Non-GMO Segregation and Identity Preservation.' Unpublished manuscript, University of Illinois, Department of Agricultural and Consumer Economics. Available online at: http://w3.aces.uiuc.edu/ACE/faculty/bullockd.html

Bureau, J.C., Marette, S. and Schiavina, A. (1997). 'Non-Tariff Trade Barriers and Consumers' Information: The Case of EU-US Trade Disputes on Food Products.' Institut National de la Recherche Agronomique, Department d'Economie, Unité de Grignon, July.

Burges, H.D. (1998). *Formulation of Microbial Biopesticides: Beneficial Microorganisms, Nematodes, and Seed Treatments.* Kluwer Academic Publishers.

Butler, L.J. and Marion, B.W. (1985). *The Impacts of Patent Protection on the US Seed Industry and Public Plant Breeding.* Madison, WI: University of Wisconsin Press.

Buzby, J.C. and Roberts, T. (1997). 'Economic Costs and Trade Impacts of Microbial Foodborne Illness.' *World Health Statistics Quarterly* 50.

Buzby, J.C., Roberts, T., Lin, C.T.J. and MacDonald, J. (1996). 'Bacterial Foodborne Disease Medical Costs and Productivity Losses.' Economic Research Service Report, US Department of Agriculture, Washington DC: GPO, August.

Calvin, D.D. (1995). 'The Economic Benefits of Transgenic Corn Hybrids for European Corn Borer Management in the United States.' Report to Monsanto Company, October 18.

Carlson, G.A. and Sappie, G. (1995). 'Economic Evaluation of Insect Eradication: The Case of Boll Weevils in the Southeast.' *Cotton Insects and Mites: Characterization and Management* (ed. J. Brown). Memphis, TN: The Cotton Foundation.

Carlson, G.A., Marra, M.C. and Hubbell, B.J. (1997). 'Yield, Insecticide Use, and Profit Changes from Adoption of *Bt* Cotton in the Southeast.' *1998 Proceedings Beltwide Cotton Conferences* (eds. P. Dugger and D. Richter), pp. 973–4. Memphis, TN: National Cotton Council of America.

Carlson, J.B. and Lersten, N.R. (1987). 'Reproductive morphology.' *Soybeans: Improvement, Production, and Uses* (ed J.R. Wilcox), 2nd edn, pp. 95–134. Madison, WI: American Society of Agronomy.

Carpenter, J. and Gianessi, L. (2000). 'Herbicide Use on Roundup Ready Crops.' *Science*, February 4.

Carson, D. (2000). Managing Director, J.P. Morgan Mortgage Capital, Inc., personal communication.

Casela, C., Renfro, B. and Krattiger, A. (1998). 'Diagnosing Maize Diseases in Latin America.' *ISAAA Briefs*. Ithaca, NY: International Service for the Acquisition of Agri-Biotech Applications.

Caswell, J.A. (1999). 'An Evaluation of Risk Analysis as Applied to Agricultural Biotechnology (With a Case Study of GMO Labeling).' Paper presented at the NE-165 Conference, Transitions in Agbiotech: Economics of Strategy and Policy, Washington DC, June 24–25.

Caswell, J. and Padberg, D. (1992). 'Toward a More Comprehensive Theory of Food Labels.' *American Journal of Agricultural Economics* 74: 460–8.

Caswell, J., Bredahl, M.E. and Hooker, N.H. (1998). 'How Quality Management Metasystems Are Affecting the Food Industry.' *Review of Agricultural Economics* 20: 547–57.

Caviness, C.E. (1970). 'Cross-Pollination in the Soybean.' *The Indispensable Pollinators*, pp. 33–6. Arkansas Agricultural Extension Service.

'Chasing Butterflies' (2000). *The Furrow*, Spring. Peoria, IL: John Deere.

Chèvre, A.M., Eber, F., Jenczweski, E., Darmency, H. and Renard, M. (2000). 'Gene Flow From Rapeseed.' *The Biosafety of Genetically Modified Organism* (eds. C. Fairbairn, G. Scoles and A. McHughen), pp. 45–50. Proceedings of the 6th International Symposium, Saskatoon, Canada, July.

Clark, J.S. and Carlson, G.A. (1990). 'Testing for Common Versus Private Property: The Case of Pesticide Resistance.' *Journal of Environmental Economics and Management* 19: 45–60.

CNN.com (1999). 'Researchers Find Bio-Engineered Corn Harms Butterflies.' CNN.com, May 20. Available at http: //www.cnn.com/NATURE/9905/20/ butterfly.killers. Accessed June 26, 2000.

Coghlan, A. (1999). 'Splitting Headache: Monsanto's Modified Soya Beans Are Cracking Up in the Heat.' *New Scientist* November 20.

Coghlan, A. and Fox, B. (1999). 'Keep that Spray: Crops Made Resistant to Pests Still do Better With Chemicals.' *New Scientist* December 18.

Cohen, M.J. (1994). 'Powerlessness and Politics.' *Causes of Hunger: Hunger 1995* (ed. M.J. Cohen). Silver Spring, MD: Bread for the World Institute.

Cohen, P. (1998). "Strange Fruit' Living in a GM World.' *New Scientist*, October 31. Available at http: //www.newscientist.com/nsplus/insight/gmworld/ gmfood/fruit.html.

Compeerapap, J. (1997). 'The Thai Debate on Biotechnology and Regulations.' *Biology and Development Monitor* 32: 13–15.

Connell, J.H. and Orias, E. (1964). 'The ecological regulation of species diversity.' *American Naturalist* 98: 399–414.

Conner, A.J. and Jacobs, J.M.E. (1999). 'Genetic Engineering of Crops as Potential Source of Genetic Hazard in the Human Diet.' *Mutation Research* 443: 223–34.

*Consumer Reports*, (1999). 'Seeds of Change.' September.

Cooke, F.T. and Freeland, T.B. (1998). 'Some Economic Considerations For *Bt* Cotton Planting in The Yazoo-Mississippi Delta.' *1998 Proceedings Beltwide Cotton Conferences* (eds. P. Dugger and D. Richter), pp. 383–4. Memphis, TN: National Cotton Council of America.

Csordas, A. (1990). 'On the Biological Role of Histone Acetylation.' *Biochemistry Journal* 265: 23–38.

Cunningham, C.J. (2000). 'Implications of Biotechnology for the Future of US Grain Markets: A Qualitative Analysis.' Unpublished MS Thesis, Department of Agricultural and Consumer Economics, University of Illinois.

Datta, K., Vasquez, A., Tu, J., Torrizo, L., Alam, M.F., Oliva, N., Abrigo, E., Khush, G.S. and Datta, S.K. (1998). 'Constitutive and Tissue-Specific Differential Expression of CryIa(b) Gene in Transgenic Rice Plants Conferring Resistance to Rice Insect Pests.' *Theoretical and Applied Genetics* 97: 31–6.

Davis, M.K., Layton, M.B., Verner, J.D. and Little, G. (1995). 'Field Evaluation of *Bt*-Transgenic Cotton in the Mississippi Delta.' *1995 Proceedings Beltwide Cotton Conferences* (eds. P. Dugger and D. Richter), pp. 771–5. Memphis, TN: National Cotton Council of America.

Dawkins, R. (1987). *The Selfish Gene.* Oxford: Oxford University Press.

Deaton, A. (1997). *The Analysis of Household Surveys: A Microeconomic Approach to Development Policy.* Baltimore and London: Johns Hopkins University Press for The World Bank.

de Castro, F.S. (1993) 'Agricultura, Biotecnologia y Propiedad Intellectual.' Instituto Interamericano de Cooperacion para la Agricultura (IICA).

Delgado, C.L., Hopkins, J., Kelly V., with Hazell, P., McKenna, A.A., Gruhn, P., Hojjati, B., Sil, J. and Courbois, C. (1998). 'Agricultural Growth Linkages in Sub-Saharan Africa.' *Research Report* No. 107. Washington DC: IFPRI.

Delta and Pine Land Company. (1997). *1997 Annual Report.* Scott, MS: Delta and Pine Land Company.

Demaske, C. (1996). 'Managing TBW Without *Bt* Cotton.' *Cotton Grower* 32: 24a.

Department of Agriculture Animal and Plant Health Inspection Service (1993). Genetically Engineered Organisms and Products; Notification Procedures for the Introduction of Certain Regulated Articles; and Petition for Nonregulated Status 58(60), March 31.

Department of Agriculture Animal and Plant Health Inspection Service (1997). Genetically Engineered Organisms and Products; Simplification of Requirements and Procedures for Genetically Engineered Organisms 62(85), May 2.

Doerfler, W., *et al.* (1997). Integration of Foreign DNA and its Consequences in Mammalian Systems. *Trends in Biotechnology,* 312: 401–6.

Donegan, K.K. and Seidler, R.J. (1999). 'Effects of Transgenic Plants on Soil and Plant Microorganisms.' *Recent Research and Developments in Microbiology* 3: 415–24.

Doyle, J. (1985). *Altered Harvest.* New York: Viking Press.

Doyle, J.J. and Persley, G.J. (1998). 'New Biotechnologies: an International Perspective.' *Investment Strategies for Agriculture and Natural Resources* (ed. G.J. Persley). Wallingford, UK: CAB International.

Dudley, J.W. and Lambert, R.J. (1992). 'Ninety Generations of Selection for Oil and Protein in Maize.' *Maydica* 37: 81–7.

Duffy, P.A., Wohlgenant, M.K. and Richardson, J.W. (1990). 'The Elasticity of Export Demand for US Cotton.' *American Journal of Agricultural Economics* 72: 468–74.

Durant, J. (1992). *Biotechnology in Public: A Review of Recent Research.* Chippenham, England: Science Museum for the European Federation of Biotechnology.

Duvick, D.N. (1996). 'Plant Breeding, An Evolutionary Concept.' *Crop Science* 36: 539–48.

Ebora, R.V., Ebora, M.M. and Sticklen, M.B. (1994). 'Transgenic Potato Expressing the *Bacillus thuringiensis* CryIA(c) Gene Effects on the Survival and Food Consumption of *Phthorimea operculella* (Lepidoptera: Gelechiidae) and *Ostrinia nubilalis* (Lepidoptera: Noctuidae).' *Journal of Economic Entomology* 87(4): 1122–7.

Edens, E.R., Slinsky, S., Larson, J.A., Roberts, R.K. and Lentz, G.L. (1998). 'Economic Analysis of Genetically Engineered *Bt* Cotton for Tobacco Budworm and Bollworm Control.' *1998 Proceedings Beltwide Cotton Conferences* (eds. P. Dugger and D. Richter), pp. 380–3. Memphis, TN: National Cotton Council of America.

Ellstrand, N.C. (1991). 'Gene Flow by Pollen: Implications for Plant Conservation Genetics.' *Oiko 63s* January, pp. 77–86.

Ellstrand, N. (2000). Transgene Escape Into Wild Populations. Unpublished, on file with author.

Emerich, M. (1996). 'Industry Growth: 22.6%.' *Natural Food Merchandiser* June: 1–39.

Ervin, D. (1999). 'Agricultural Biotechnology Is a Double-Edged Environmental Sword.' Remarks at Annual Meeting of American Agricultural Journalists, Washington DC, April 19.

EU Commission (1997a). 'Communication from the Commission: Consumer Health and Food Safety.' April 30.

EU Commission (1997b). 'Joint European Parliament and Commission Conference on Food Law and Food Policy.' November 4.

EU Commission (1997c). 'Report on United States Barriers to Trade and Investment.'

EU Commission (1997d). 'The Commission Proposes Rules for the Labeling of Food Made from Genetically Modified Soya Beans and Maize.' December 3.

EU Commission (1997e). 'The Consumer Committee's Comments on the Commission's Green Paper on the General Principles of Food Law in the European Union.' COM (97), September 18.

EU Commission (1997f). 'The General Principles of Food Law in the European Union.' Commission Green Paper, COM (97).

EU Commission (2000). 'Communication from the Commission on the Precautionary Principle.' COM (2000) 1, February 2, Brussels.

EU Regulation 49/2000, Available at http://www.europa.eu.int.

Ewen, S.W.B. and Pusztai, A. (1999). 'Effect of Diets Containing Genetically Modified Potatoes Expressing *Galvanthus nivalis* Lectin on Rat Small Intestine.' *The Lancet* 354(9187): 1353–4.

Falck-Zepeda, J.B., Traxler, G. and Nelson, R.G. (1999). 'Rent Creation and Distribution from the First Three Years of Planting *Bt* Cotton,' *ISAAA Brief* 14.

Falck-Zepeda, J.B., Traxler, G. and Nelson, R.G. (2000a). 'Surplus Distribution from the Introduction of a Biotechnology Innovation.' *American Journal of Agricultural Economics* 82(2): 360–9.

Falck-Zepeda, J.B., Traxler, G. and Nelson, R.G. (2000b). 'Rent Creation and Distribution from Biotechnology Innovations: The Case of *Bt* Cotton and Herbicide-Tolerant Soybeans in 1997.' *Agribusiness* 16: 1(Winter): 21–32.

FAO (Food and Agriculture Organization of the United Nations) (1996a). *Food, Agriculture, and Food Security: Developments Since the World Food Conference and Prospects, World Food Summit Technical Background Document Number 1*. Rome: FAO.

FAO (Food and Agriculture Organization of the United Nations) (1996b). *Investment in Agriculture: Evolution and Prospects, World Food Summit Technical Background Document Number 10*. Rome: FAO.

FAO (Food and Agriculture Organization of the United Nations) (1998). *State of Food and Agriculture 1998*. Rome: FAO.

FAO (Food and Agriculture Organization of the United Nations) (1999). *The State of Food Insecurity in the World 1999*. Rome: FAO.

FAO (Food and Agriculture Organization of the United Nations) (2000). 'Public Assistance and Agricultural Development in Africa.' Posted at http://www.fao.org/docrep/meeting/x3977e.htm#2. Accessed July 11.

Farrington, J. (1989). *Agricultural Biotechnology: Prospects for the Third World*. Boulder, CO: Westview Press.

Feeny, D., Hanna, S. and McEvoy, A.F. (1996). 'Questioning the Assumptions of the Tragedy of the Commons' Model of Fisheries.' *Land Economics* 72: 187–205.

Ferber, D. (1999). 'Biotech Critics Watch The Watchdogs.' *Science* 286(5445), November 26.

Fernandez-Cornejo, J., Klotz-Ingram, C. and Jans, S. (1999). 'Farm-Level Effects of Adopting Genetically Engineered Crops in the USA.' Paper presented at the NE-165 Conference, Transitions in Agbiotech: Economics of Strategy and Policy, Washington DC, June 24–25.

Fincham, J.R. and Ravetz, J. (1990). *Genetically Engineered Organisms: Benefits and Risks*. Toronto: University of Toronto Press.

Finnegan, H. and McElroy. (1994). 'Transgene Inactivation: Plants Fight Back!' *Bio/Technology* (12) 883–8.

Fiskel, J. and Covello, V.T. (eds.). (1987). *Safety Assurance for Environmental Introductions of Genetically-Engineered Organisms*. Berlin: Springer-Verlag.

Fitzgerald, D. (1990). *The Business of Breeding: Hybrid Corn in Illinois, 1890 -1940*. Ithaca, NY: Cornell University Press.

Florkowski, W.J. and Hill, L.D. (1985). 'Expected Commercial Application of Biotechnology in Crop Production: Results of the Survey.' Dept. Agric. Econ., Agric. Exp. Sta., University of Illinois at Urbana-Champaign, December.

Foltz, J., Barham, B. and Kim, K. (1999). 'Universities, Agricultural Biotechnology Patents, and Local Spillovers.' Paper presented at the NE-165 Conference, Transitions in Agbiotech: Economics of Strategy and Policy, Washington DC, June 24–25.

Food and Agriculture Organization of the United Nations/World Health Organization (1996). *Biotechnology and Food Safety: Report of a Joint FAO/WHO Consultation*, October 4.

Food and Agriculture Organization of the United Nations/World Health Organization (2000). *Safety Aspects of Genetically Modified Foods of Plant Origin/Report of a Joint FAO/WHO Expert Consultation on Foods Derived from Biotechnology*. Geneva, May 29–June 2.

Fowle III, J.R. (ed.) (1987). *Application of Biotechnology*. AAAS Selected Symposium 106. Boulder, CO: Westview Press.

Fox, J.L. (1997). 'Farmers Say Monsanto's Engineered Cotton Drops Bolls.' *Nature Biotechnology* 15: 1233.

Frost, E.L. (1997). *Transatlantic Trade: A Strategic Agenda*. Washington DC: Institute for International Economics.

Gachet, E., *et al.* (1999). 'Detection of Genetically Modified Organisms by PCR: A Brief Review of Methodologies Available.' Available at http://www.eurofins.com/gmo/PCR_methodologies.asp. Accessed October 19.

Galinat, W.C. (1988). 'The Origin of Corn.' *Corn and Corn Improvement* (3rd edn), (eds. G.F. Sprague and J.W. Dudley), pp. 1–31. Madison, WI: American Society of Agronomy, Crop Science Society of America, and Soil Science Society of America.

Gaskell, G., Bauer, M.W., Durant, J. and Allum, N.C. (1999). 'Worlds Apart? The Reception of Genetically Modified Foods in Europe and the US.' *Science* 285(5426): 384–7.

Gasser, C.S. and Fraley, R.T. (1992). 'Transgenic Crops.' *Scientific American* June, pp. 62–9.

Gendel, S.M., Line, A.D., Warren, D.M. and Yates, F. (1990). *Agricultural Bioethics: Implications of Agricultural Biotechnology*. Ames, IA: Iowa State University Press.

General Accounting Office. (1988). Biotechnology: Managing the Risks of Field Testing Genetically Engineered Organisms. GAO/RCED-88-27, June.

Genetic ID. (1999). *GMF Market Intelligence Newsletter*. No. 32. Available at http://www.genetic-id.com, May 1.

Gianessi, L.P. and Carpenter, J.E. (1999). 'Agricultural Biotechnology: Insect Control Benefits.' National Center for Food and Agricultural Policy. Available at http://www.bio.org/food&ag/bioins01.html. Accessed July.

Gianessi, L.P. and Carpenter, J.E. (2000). *Agricultural Biotechnology: Benefits of Transgenic Soybeans*. National Center for Food and Agriculture Policy. Washington DC. April. http://www.ncfap.org/soy85.pdf.

Gibbs, J., Cooper, I. and Mackler, B. (1987). *Biotechnology and the Environment: International Regulation*. New York: Stockton Press.

Gibson, J.W., Laughlin, D., Luttrell, R.G. Parker, D., Reed, J. and Harris, A. (1997). 'Comparison of Costs and Returns Associated with *Heliothis* Resistant *Bt* Cotton to Non-Resistant Varieties.' *1997 Proceedings Beltwide Cotton Conferences* (eds. P. Dugger and D. Richter), pp. 244–7. Memphis, TN: National Cotton Council of America.

Giroux, S., Cote, J.C., Vincent, C., Martel, P. and Coderre, D. (1999). 'Bacteriological Insecticide M-ONE Effects on Predation Efficiency and Morality of Adult *Coleomegilla maculata lengi* (Coleoptera: Coccinellidae).' *Journal of Economic Entomology* 87: 39–43.

Glickman, D. (2000). 'Remarks to National Academy of Sciences 1st Meeting of Standing Committee on Biotechnology Food and Fiber Production, and The Environment Washington, D.C. – May 4, 2000.' http://www.usda.gov/news/releases/2000/05/0146

Goklany, I. (1995). 'Richer is Cleaner: Long-term Trends in Global Air Quality.' *The True State of the Planet* (ed. R. Bailey), pp. 339–77. New York: The Free Press.

Goklany, I. (2000). 'Richer is More Resilient: Dealing With Climate Change and More Urgent Environmental Problems.' *Earth Report 2000* (ed. R. Bailey), pp. 155–87. New York: McGraw-Hill.

Golan, E. and Kuchler, F. (2000). 'Labeling Biotech Foods: Implications for Consumer Welfare and Trade.' Paper presented at Global Food Trade and Consumer Demand for Quality Conference. Montreal, Canada, June 25–26.

Golan, E., Kuchler, F., Mitchell, L., with Greene C. and Jessup, A. (2000). 'The

Economics of Food Labeling.' AER 793 Economic Research Service, USDA (in press, December).

Good, D., Bender, K. and Hill, L. (2000). 'Marketing of Specialty Corn and Soybean Crops.' AE-4733, Department of Agricultural and Consumer Economics, University of Illinois, March.

Goodman, R.M., Hauptli, H., Crossway, A. and Knauf, V.C. (1987). 'Gene Transfer in Crop Improvement.' *Science* 236: 48–54.

Gould, F.W. (1968). *Grass Systematics.* New York: McGraw Hill.

Gould, F.W. (1991). 'The Evolutionary Potential of Crop Pests.' *American Scientist* 79: 496–507.

Grain Inspection, Packers and Stockyards Administration, GIPSA (2000). 'Agricultural Biotechnology Web Site.' http://www.gipsa.usda.gov/biotech/biotech.htm

Grainnet. (1999). 'ADM Joins Staley in Rejecting Some Transgenic Corn.' Available at http: //www.grainnet.com/BreakingNews/articles.html?ID=3528, April 14.

Grant, W. (1999). 'Biotechnology: A Source of Tensions in US–EU Trade Relations.' Paper presented at conference, Liberalizing Agricultural Trade, University of Washington, May.

Gray, R., Malla, S. and Phillips, P.W.B. (1999). 'The Public and Not-for-Profit Sectors in a Biotechnology-Based, Privatizing World: The Canola Case.' Paper presented at the NE-165 Conference, Transitions in Agbiotech: Economics of Strategy and Policy, Washington DC, June 24–25.

Greene, A.E., and Alison, R.F. (1994). 'Recombination Between Viral RNA and Transgenic Plant Transcripts.' *Science* 263: 1423–5.

Gressel, J. (1996). 'Fewer Constraints Than Proclaimed to the Evolution of Glyphosate-Resistant-r-4 Weeds.' *Resistant Pest Management* 8: 2–5.

Gupta, A. (2000). 'Governing Trade in Genetically Modified Organisms: the Cartagena Protocol on Biosafety.' *Environment* 42(4): 23–33.

Gura, T. (1999). 'New Genes Boost Rice Nutrients.' *Science* 285(August 13): 994–5.

Hall, S.P., Sullivan, J.B. and Schweitzer, D.F. (1999). 'Assessment of Risk to Non-target Macro-moths after BTK Application to Asian Gypsy Moth in the Cape Fear Region of North Carolina.' *USDA Bulletin* No. FHTET-98-16.

Halweil, B. (1999). Pesticide-Resistant Species Flourish. In *Vital Signs 1999: The Environmental Trends That Are Shaping Our Future* (ed. L. Starke). New York: W.W. Norton.

Hansen, M.K. (2000). 'Genetic Engineering Is Not An Extension Of Conventional Plant Breeding: How Genetic Engineering Differs from Conventional Breeding, Hybridization, Wide Crosses and Horizontal Gene Transfer.' *Consumer Policy Institute/Consumers Union*, January.

Harbour G.C., Garlick, R.L., Lyse, S.B., Crow, F.W., Tobins, R.H. and Hoogerheide, J.G. (1992). 'N-Ə-acetylation Can Occur at Lysine Residues 157, 167, 171 and 180 of Recombinant Bovine Somatotropin.' *Techniques in Protein Chemistry III*: 487–95.

Hardell, L. and Eriksson, M. (1999). 'A Case-Control Study of Non-Hodgkin Lymphoma and Exposure to Pesticides.' *Cancer* 85(6): 1353–60.

Hartshorn, G. (1992). 'Biological Diversity in Tropical Forests.' *Global Warming and Biodiversity*, p. 140. New Haven, CT: Yale University Press.

Hartzler, R.G. and Buhler, D.D. (2000). 'Occurrence of Common Milkweed (*Asclepias syriaca*) in Cropland and Adjacent Areas.' *Crop Protection* 19: 363–6.

Hassanein, N. and Kloppenberg, J.J.R. (1995). 'Where the Grass Grows Again:

Knowledge Exchange in the Sustainable Agriculture Movement.' *Rural Sociology* 60: 721–40.

Hauptli, H., Katz, D., Thomas, B.R. and Goodman, R.M. (1990). 'Biotechnology and Crop Breeding for Sustainable Agriculture.' *Sustainable Agricultural Systems* (eds. C.A. Edwards, R. Lal, P. Madden, R.H. Miller and G. House), pp. 141–56. Akeny, OH: Soil and Water Conservation Society.

Hay, M.M.F. and Corman-Weinblatt, A. (eds.) (1985). *Catalogue of Cell Lines and Hybridomas,* (5th edn). Rockville, MD: American Type Culture Collection.

Hayenga, M.L. (1998). 'Biotechnology in the Food and Agricultural Sector: Issues and Implications for the 1990s.' Agricultural Issues Center Paper, University of California, Davis, September.

Hazell, P., Jagger, P. and Knox, A. (2000). 'Technology, Natural Resources Management, and the Poor.' Paper Prepared for the International Fund for Agricultural Development by IFPRI. Draft, March 10.

Hazell, P.B.R. and Ramasamy, C. (eds.) (1991). *The Green Revolution Reconsidered.* Baltimore, MD: The Johns Hopkins University Press for IFPRI.

Hefle, S. and Taylor, S. (1999). 'Beyond Substantial Equivalence.' *Nature*, October 7.

Heywood, V.H. (1995). *Global Biodiversity Assessment*, p. 724. UNEP, Cambridge University Press.

Hiei, Y., Ohta, S., Komari, T. and Kumashiro, T. (1994). 'Efficient Transformation of Rice (*Oryza sativa* L.) Mediated by *Agrobacterium* and Sequence Analysis of the Boundaries of the T-DNA.' *Plant Journal* 6: 271–82.

Highfield, R. (1999). 'Come and Quiz the GM Experts.' *The Daily Telegraph* July 16.

Hilbeck, A., Baumgartner, M., Fried, P.M. and Bigler, F. (1998a). 'Effects of Transgenic *Bacillus thuringiensis* Corn-Fed Prey on Mortality and Development Time of Immature *Chrysoperla carnea* (Neuroptera: Chrysopidae).' *Environmental Entomology* 27: 480–487.

Hilbeck A., Moar, W.J., Pusztai-Carey, M., Filippini, A. and Bigler, F. (1998b). 'Toxicity of *Bacillus thuringiensis* Cry1Ab toxin to the Predator *Chrysoperla Carnea* (Neuroptera: Chrysopidae).' *Environmental Entomology* 27: 1255–63.

Hilbeck A., Moar, W.J., Pusztai-Carey, M., Filippini, A. and Bigler, F. (1999). 'Prey-Mediated Effects of Cry1Ab Toxin and Protoxin and Cry2A Protoxin on the Predator, *Chrysoperla carnea.*' *Entomologia Experimentalis et Applicata* 91(2): 305–16.

Hileman, B. (1995). 'Views Differ Sharply Over Benefits, Risks of Agricultural Biotechnology.' *Chemical and Engineering News* August 21.

Hill, A.F., Desbruslais, M., Joiner, S., Sidle, K., Gowland, I., Collinge, J., Doey, L.J. and P. Lantos. (1997). 'The Same Prion Strain Causes vCJD and BSE.' *Nature* 389: 498–501.

Hill, D.S. (1987). *Agricultural Insect Pests of Temperate Regions and their Control.* Cambridge: Cambridge University Press.

Hill, L., Stockdale, J., Briemyer, H. and Klonglan, G. (1986). 'Economic and Social Consequences of Biological Nitrogen Fixation in Corn Production.' Dept. Agric. Econ., Agric. Exp. Sta., University of Illinois at Urbana-Champaign, November.

Hill, R.E., Chiang, H.C., Keaster, A.J., Showers, W.B. and Reed, G.L. (1973). 'Seasonal Abundance of the European Corn Borer *Ostrinia nubilalis* (Hbn.) Within the North Central United States.' NCR-216 North Central Regional Publication, Iowa State University.

Ho, M.W. (1998). *Genetic Engineering: Dream or Nightmare?* Bath, UK: Gateway Books.

Ho, M.W., Ryan, A. and Cummins, J. (1999). 'The Cauliflower Mosaic Virus Promoter – a Recipe for Disaster? *Microbial Ecology in Health and Disease* (in press).

Ho, M.-W., Ryan, A. and Cummins, J. (in press). 'Hazards of Transgenic Plants Containing the Cauliflower Mosaic Viral Promoter: Authors' reply to critiques of 'The Cauliflower Mosaic Viral Promoter – a Recipe for Disaster?' *Microbial Ecology in Health and Disease.*

Holm, L., Plunknett, D.L., Poncho, J.V. and Herberger, J.P. (1977). *The World's Worst Weeds: Distribution and Biology.* Honolulu: University Press of Hawaii.

Holzman, D. (1999). 'Agricultural Biotechnology: Report leads to Debate on Benefits of Transgenic Corn and Soya Bean Crops.' *Genetic Engineering News* 19(8): April 15.

Horbulyk, T.M. (1999). 'Strategy and Incentives in the Compulsory Licensing of Intellectual Property in Agriculture.' Paper presented at the NE-165 Conference, Transitions in Agbiotech: Economics of Strategy and Policy. Washington DC, June 24–25.

Hoy, C.W., Feldman, J., Gould, F., Kennedy, G.G., Reed, G. and Wyman, J.A. (1998). 'Naturally Occurring Biological Controls in Genetically Engineered Crops.' *Conservation Biological Control* (ed. P. Barbosa), pp. 185–205. New York: Academic Press.

Huang, F., Higgins, R.A. and Bushman, LL. (1997). 'Baseline Susceptibility and Changes in Susceptibility to *Bacillus thuringiensis* sub Sp. *kurstaki* Under Selection Pressure in European Corn Borer (Lepidoptera: *Pyralidae).*' *Journal of Economic Entomology* 90: 1137–43.

Huang, F., Buschman, L.L., Higgins, R.A. and McGaughey, W.H. (1999). 'Inheritance to *Bacillus thuringiensis* Toxin (Dipel ES) in European Corn Borer.' *Science* 284: 965–7.

Hubbell, B. and Welsh, R. (1988). 'Transgenic Crops: Engineering a More Sustainable Agriculture.' *Agriculture and Human Values* 15: 43–56.

Hubbell, B., Marra, M. and Carlson, G. (forthcoming). 'Estimating the Demand for a New Technology: *Bt* Cotton and Insecticide Policies.' *American Journal of Agricultural Economics.*

Hurburgh, C. (1999). 'Testing and Handling Procedures for Segregation of Non-GMO Grain.' Available at http: //www.extension.iastate.edu/Pages/grain/news/gmo /99915crh/index.htm. Accessed October 19.

Huston, M. (1994). *Biological Diversity.* Cambridge: Cambridge University Press.

Hyde, J., Martin, M.A., Preckel, P.V. and Edwards, C.R. (1999). 'The Economics of *Bt* Corn: Valuing the Protection from the European Corn Borer.' *Review of Agricultural Economics* 21(Fall/Winter).

Hymowitz, T. and Singh, R.J. (1987). 'Taxonomy and Speciation.' *Soybeans: Improvement, Production, and Uses,* 2nd edn (ed. J.R. Wilcox), pp. 23–48. Madison, WI: American Society of Agronomy.

Inose, T. and Murata, K. (1995). 'Enhanced Accumulation of Toxic Compound in Yeast Cells Having High Glycolytic Activity: a Case Study on the Safety of Genetically Engineered Yeast.' *International Journal of Food Science and Technology* 30: 141–6.

International Food Policy Research Institute (1999). *Are We Ready for a Meat Revolution?* March. Washington DC: IFPRI.

International Food Policy Research Institute (IFPRI) (2000). 'Global Study Reveals New Warning Signals: Degraded Agricultural Lands Threaten World's Food Production Capacity.' Press Release, May 21. Posted at http://www.cgiar.org/ifpri. Accessed June 23.

International Policy Council on Agriculture, Food, and Trade (1998). 'Plant Biotechnology and Global Food Production: Trade Implications.' IPS position paper.

International Service for National Agricultural Research (1999). *Agricultural Biotechnology Research Capacity in Four Developing Nations*, Briefing Paper 42, December.

Iowa State University (1999). 'Corn Borers Should Increase in 1999.' *Integrated Crop Management* April 4.

Isaac, G. and Phillips, P. (1999). 'The BioSafety Protocol and International Trade in Transgenic Canola: An Economic Assessment of the Impact on Canada.' Paper presented at the NE-165 Conference, Transitions in Agbiotech: Economics of Strategy and Policy. Washington DC, June 24–25.

Jackson, J.H. (1998). *The World Trade Organization: Constitution and Jurisprudence.* London: The Royal Institute of International Affairs.

Jakowitsch, J., Mette, M.F., van der Winden, J., Matzke, M.A. and Matzke, A.J.M. (1999). Integrated pararetroviral sequences define a unique class of dispersed repetitive DNA in plants. *Proceedings of the National Academy of Sciences* 96(23): 13241–6.

James, C. (1997). 'Global Status of Transgenic Crops in 1997.' *ISAAA Briefs*. Ithaca, NY: International Service for the Acquisition of Agri-Biotech Applications.

James, C. (1998). 'Global Review of Commercialized Transgenic Crops: 1998.' *ISAAA Briefs* No. 8. Ithaca, NY: International Service for the Acquisition of Agri-Biotech Applications.

James, C. (1999). 'Global Status of Commercialized Transgenic Crops: 1999.' *ISAAA Briefs* No. 12: Preview. Ithaca, NY: International Service for the Acquisition of Agri-Biotech Applications.

James, C. (2000). Global Status of Commercialized Transgenic Crops: 2000. *ISAAA Briefs* No. 21: Preview. ISAAA: Ithaca, NY.

James, C. and Krattiger, A.F. (1996). 'Global Review of the Field Testing and Commercialization of Transgenic Plants: 1986 to 1995.' *ISAAA Briefs* No. 1. Ithaca, NY: International Service for the Acquisition of Agri-Biotech Applications.

James, C. and Persley, G.J. (1990). *Agricultural Biotechnology: Opportunities for International Development* (ed. G.J. Persley). Wallingford, UK: CAB International.

Jasanoff, S. (2000). 'Technological Risk and Cultures of Rationality'. *Incorporating Science, Economics, and Sociology in Developing Sanitary and Phytosanitary Standards in International Trade: Proceedings of a Conference* (ed. National Academy of Sciences), pp. 23–30. Washington DC: National Academy Press.

Jasieniuk, M. (1995). 'Constraints on the Evolution of Glyphosate Resistance in Weeds.' *Resistant Pest Management* 7(2). Available at http://www.msstate.edu/entomology/v7n2/rpmv7n2.html, Winter.

Joersbo, M., Petersen, S.G. and Okkels, F.T. (1999). 'Parameters Interacting with Mannose Selection Employed for the Production of Transgenic Sugar Beet.' *Physiologia Plantarum* 105: 109–15.

Johnson, K.S., Scriber, J.M., Nitao, J.K. and Smitley, D.R. (1995). 'Toxicity of *Bacillus thuringiensis* var. *kurstaki* to Three Nontarget Lepidoptera in Field Studies.' *Environmental Entomology* 24(2): 288–97.

Jorgensen, R. and Andersen, B. (1995). 'Spontaneous Hybridization Between Oilseed Rape (*Brassica napus*) and Weed (*Brassica campestris*): A Risk of Growing Genetically Engineered Modified Oilseed Rape.' *American Journal of Botany* 81: 1620–6.

Jorgensen, R.B., Hauser, T., Mikkelsen, T.R. and Ostergard, H. (1996). 'Transfer of Engineered Genes from Crop to Wild Plants.' *Trends in Plant Science* 1: 356–8.

Josling, T. (1996). 'Agriculture in a Transatlantic Economic Area.' *Future Visions for US Trade Policy* (ed. B. Stokes). New York: Council on Foreign Relations.

Josling, T. (1999). 'Who's Afraid of the GMOs? EU–US Trade Disputes Over Food Safety and Biotechnology.' USC European Center, March.

Josling, T., Tangermann, S. and Warley, T.K. (1996). *Agriculture in the GATT*. Macmillan, Houndmills, UK.

Josling, T., Roberts, D. and Hassan, A. (1999). 'The Beef Hormone Dispute and its Implications for Trade Policy.' European Forum Working Paper No. 101, Institute for International Studies, Stanford University, August.

Juma, C. (1989). *The Gene Hunters: Biotechnology and the Scramble for Seeds*. Princeton, NJ: Princeton University Press.

Juma, C. and Aarti G. (1999). 'Safe Use of Biotechnology.' *Biotechnology for Developing Countries: Problems and Opportunities, 2020 Vision Focus 2, Brief 6 of 10*. Washington DC: IFPRI.

Just, R.E. and Hueth, D.L. (1993). 'Multimarket Exploitation: The Case of Biotechnology and Chemicals.' *American Journal of Agricultural Economics* 75: 936.

Kaeppler, H.F., Menon, G.K., Skadsen, R.W., Nuutila, A.M. and Carlson, A.R. (2000). 'Transgenic Oat Plants Via Visual Selection of Cells Expressing Green Fluorescent Protein.' *Plant Cell Reports* 19: 661–6.

Kalaitzondanakes, N. (2000). 'Agrobiotechnology and Competitiveness.' *American Journal of Agricultural Economics*, 82: in press.

Kennedy, G.G., and Whalon, M.E. (1995). 'Managing Pest Resistance to *Bacillus thuringiensis* Endoxtoxins: Constraints and Incentives To Implementation.' *Journal of Economic Entomology* 88: 454–60.

Kenney, M. (1986). *Biotechnology: The University-Industrial Complex*. New Haven, CT: Yale University Press.

Kern, M. (2000). *Future of Agriculture*. Paper prepared for the conference 'Global Dialogue EXPO 2000. The Role of the Village in the 21st Century: Crops, Jobs, and Livelihood' Hanover, Germany. August 15–17.

Kessler, D.A., Taylor, M.R., Maryanski, J.H., Flamm, E.L. and Kahl, L.S. (1992). 'The Safety of Foods Developed by Biotechnology.' *Science* 256: 1747.

Kinnucan, H.W. and Miao, Y. (1999). 'Optimal Generic Advertising Intensities for US Cotton in a Free Market.' Unpublished paper, Department of Agricultural Economics and Rural Sociology, Auburn University.

Kling, J. (1996). 'Could Transgenic Supercrops One Day Breed Superweeds?' *Science* 274: 180–1.

Kloppenburg, J.R. (1988). *First the Seed: The Political Economy of Plant Biotechnology, 1492–2000*. Cambridge, UK: Cambridge University Press.

Kloppenberg, J., Hassanein, N. and Burrows, B. (1996). 'Does Technology Know Where It's Going?' Edmonds Institute Occasional Paper. Edmonds, WA: The Edmonds Institute.

Koskella, J. and Stotzky, G. (1997). 'Microbial Utilization of Free and Clay-Bound Insecticidal Toxins for *Bacillus thuringiensis* and Their Retention of Insecticidal Activity after Incubation with Microbes.' *Applied and Environmental Microbiology* 63(9): 3561–8.

Kota, M. *et al.* (1999). 'Overexpression of the *Bacillus thuringiensis* (*Bt*) CryA2Aa2 Protein in Chloroplasts Confers Resistance to Plants Against Susceptible and *Bt*-Resistant Insects.' *Proceedings of the National Academy of Sciences* 96 (March): 1840–45.

Koziel, M.G., Beland, G.L., Bowman, C., Carozzi, N.B., Crenshaw, R., Crossland, L., Dawson, J., Desai, N., Hill, M. and Kadwell, S. (1993). 'Field Performance of Elite

Transgenic Maize Plants Expressing an Insecticidal Protein Derived from *Bacillus thuringiensis*.' *Bio/Technology* 11: 194–200.

Krattiger, A. (1997). 'Insect Resistance in Crops: A Case Study of *Bacillus thuringiensis* (Bt) and its Transfer to Developing Countries.' *ISAAA Briefs*. Ithaca, NY: International Service for the Acquisition of Agri-Biotech Applications.

Krimsky, S. (1991). *Biotechnics and Society: The Rise of Industrial Genetics*. New York: Praeger.

Krimsky, S. and Rubel, R. (1996). *Agricultural Biotechnology and the Environment*. Urbana and Chicago, IL: University of Illinois Press.

Kuiper, H.A., *et al.* (1998). *Food Safety Evaluation of Genetically Modified Foods as a Basis for Market Introduction*. Ministry of Economic Affairs, The Hague, Netherlands.

Kunkel, T., Niu, Q.-W., Chan, Y.-S. and Chua, N.-H. (1999). 'Inducible Isopentenyltransferase as a High-efficiency Marker for Plant Transformation.' *Nature Biotechnology* 17: 916–19.

Kunreuther, H. and Slavic, P. (1996). 'Science, Values, and Risk.' *Annals of the American Academy of Political and Social Science* 545: 116–25.

Lambrecht, B. (1998). 'Many Farmers Finding Altered Cotton Lacking.' *St. Louis Post-Dispatch*, April 12.

Lappe, M. and Bailey, B. (1999). *Against the Grain: The Genetic Transformation of Global Agriculture*. London: Earthscan Publications.

Lavoie, B.F. and Sheldon, I.M. (1999). 'The Source of Comparative Advantage in the Biotechnology Industry: A Real Options Approach.' Paper presented at the NE-165 Conference, Transitions in Agbiotech: Economics of Strategy and Policy, Washington DC, June 24–25.

Layton, B. (1996). 'Scouting *Bt* Fields.' *Cotton Grower* 32: 42d.

Layton, M. B., Stewart, S.D., Williams, M.R. and Long, J.L. (1997). 'Performance of *Bt* Cotton in Mississippi, 1998.' *1999 Proceedings Beltwide Cotton Conferences* (eds. P. Dugger and D. Richter), pp. 942–5; 970–3. Memphis, TN: National Cotton Council of America.

Leisinger, K.M. (1999). 'Disentangling Risk Issues.' *Biotechnology for Developing Countries: Problems and Opportunities, 2020 Vision Focus 2, Brief 5 of 10*. Washington DC: IFPRI.

Lesser, W. (1997). 'The Role of Intellectual Property Rights in Biotechnology Transfer Under the Convention on Biological Diversity.' *ISAAA Briefs*. Ithaca, NY: International Service for the Acquisition of Agri-Biotech Applications.

Lesser, W. (1999). ''Holding Up' the Public Agbiotech Research Sector over Component Technologies.' Paper presented at the NE-165 Conference, Transitions in Agbiotech: Economics of Strategy and Policy, Washington DC, June 24–25.

Lesser, W., Bernard, J. and Zillah, K. (1999). 'Methodologies for *Ex Ante* Projections of Adoption Rates for Agbiotech Products: Lessons Learned from rBST.' *Agribusiness* 15(2), Spring.

Lin, W., Anuratha, C.S., Datta, K., Potrykus, I., Muthukrishnan, S. and Datta, S.K. (1995). 'Genetic Engineering of Rice for Resistance to Sheath Blight.' *Bio/Technology* 13: 686–91.

Lin, W.W., Chambers, W. and Harwood, J. (2000). 'Biotechnology: US Grain Handlers Look Ahead.' *Agricultural Outlook*. Economic Research Service, US Department of Agriculture, April.

Linder, C.R. and Schmidt, J. (1995). 'Potential Persistence of Escaped Transgenes: Performance of Transgenic, Oil-Modified *Brassica* Seeds and Seedlings.' *Ecological Applications* 5: 1000–1068.

Lipton, M., with Longhurst, R. (1989). *New Seeds and Poor People*. London: Unwin Hyman.

Liu, Y.B., Tabashnik, B.E., Dennehy, T.J., Patin, A.L. and Bartlett, A.C. (1999). 'Development Time and Resistance to Bt Crops.' *Nature* 400: 519.

Lohr, L., Carter, H.O. and Logan, S.H. (1986). 'Agricultural Biotechnology Research: An Overview.' Agric. Issues Ctr. (AIC) working paper, University of California, Davis.

Lorenz, M.G. and Wackernagel, W. (1994). 'Bacterial Gene Transfer by Natural Genetic Transformation in the Environment.' *Microbial Reviews* 58: 563–602.

Losey, J.E., Rayor, L.S. and Carter, M.E. (1999). 'Transgenic Pollen Harm Monarch Larvae.' *Nature* 399: 214.

Lowe, P.T., Marsden and Watmore, S. (eds.) (1990). *Technological Change and the Rural Environment*. London: David Fulton Publishers.

MacArthur, M. (2000). 'Triple-Resistant Canola Weeds Found in Alta.' *The Western Producer*, February 10.

MacIntosh, S.C., Stone, T.B., Sims, S.R., Hunst, P.L., Greenplate, J.T., Marrone, P.G., Perlak, F.J., Fischhoff, D.A. and Fuchs, R.L. (1989). 'Specificity and Efficacy of Purified *Bacillus thuringiensis* Proteins Against Agronomically Important Insects. *Journal of Invertebrate Pathology* 56: 258–66.

Mack, D. (1998). 'Food for All Living in a GM World.' *New Scientist*. Available at http://www.newscientist.com/nsplus/insight/gmworld/gmfood/develop.html, October 31.

MacKenzie, D. (1999). 'Gut Reaction. Could a Mechanical Gourmet Help Us Digest a GM Future?' *New Scientist* January 30.

Magaña, J.E., García, J. Gonzalez, Rodríguez, A.J. O. and Garcia, J.M.O. (1999). 'Comparative Analysis of Producing Transgenic Cotton Varieties Versus No Transgenic Variety in Delicias, Chihuahua, Mexico'. *1999 Proceedings Beltwide Cotton Conferences* (eds. P. Dugger and D. Richter), pp. 255–6. Memphis, TN: National Cotton Council of America.

Magat, W. and Viscusi, W.K. (1992). *Informational Approaches to Regulation*. Cambridge, MA: MIT Press.

Maliga, P., Carrer, H., Kanevski, I., Staub, J. and Svab, Z. (1993). 'Plastid Engineering in Land Plants: A Conservative Genome is Open to Change.' *Philosophical Transactions of the Royal Society of London B* 342: 203–8.

Mann, C.C. (1999). 'Crop Scientists Seek a New Revolution.' *Science* 283(January 15): 310–16.

Marra, M., Carlson, G. and Hubbell, B. (1996). 'Economic Impact of the First Biotechnologies.' Raleigh, NC: North Carolina Agricultural Research Service, North Carolina State University. Available at http://www.cals.ncsu.edu/ag_rec/faculty/marra/FirstCrop/sld001.htm

Maryanski, J.H. (1995). 'FDA's Policy for Foods Developed by Biotechnology.' *Genetically Modified Foods: Safety Issues* (eds. Engel, Takeoka and Teranishi), pp. 12–22. American Chemical Society (ACS) Symposium Series No. 605.

Mason, C.E., Rice, M.E., Calvin, D.D., Van Duyn, J.W., Showers, W.B., Hutchison, W.D., Witkowski, J.F., Higgins, R.A., Onstad, D.W. and Dively, G.P. (1996). 'European Corn Borer: Ecology and Management.' NCR-327, North Central Regional Extension Publication, Iowa State University, July.

Matten, S.R. (1997). 'Pesticide Resistance Management Activities by the U.S. Environmental Protection Agency.' *Resistant Pest Management* 9: 3–5.

Matten, S. R., Lewis, P. I., Tomimatsu, G., Sutherland, D.W.S., Anderson, N. and Colvin-Snyder, T.L. (1996). 'The US Environmental Protection Agency's Role in Pesticide

Resistance Management.' *Molecular Genetics and Evolution of Pesticide Resistance* (ed. T.M. Brown), p. 645, American Chemical Society Symposium Series. Washington DC: American Chemical Society.

Maturin, L. and Curtiss, R. (1977). 'Degradation of DNA by Nucleases in Intestinal Tract of Rats.' *Science* 196: 216–18.

Mayeno, A.N. and Gleich, G.J. (1994). '*Eosinophilia myalgia* Syndrome and Tryptophan Production: A Cautionary Tale.' *TIBTECH* 12: 346–52.

Mayer, H. and Furtan, W.H. 'Economics of Transgenic Herbicide-Tolerant Canola. The Case of Western Canada.' *Food Policy* 24(4): 431–42.

McAllan, A.B. (1982). 'The Fate of Nucleic Acids in Ruminants.' *Progress in Nutrional Science* 41: 309–17.

McCalla, A.F. and Ayres, W.S. (1997). *Rural Development: From Vision to Action.* Washington DC: The World Bank.

McCormick, A.A. (1999). 'Rapid Production of Specific Vaccines for Lymphoma by Expression of the Tumor-Derived Single-Chain Fv Epitopes in Tobacco Plants.' *Proceedings of the National Academy of Sciences* 96 (January): 703–8.

McGoughlin, M. (2000). 'Biotech Crops: Rely on the Science.' *Washington Post* editorial page, June 14.

McGregor, S.E. (1976). *Insect Pollination of Cultivated Crop Plants.* Agriculture Handbook 496. Washington DC: United States Department of Agriculture.

Meikle, J. (2000). 'Soya Gene Find Fuels Doubts on GM Crops.' *Guardian*, May 31.

Melchers, G., Sacristan, M.D. and Holder, A.A. (1978). 'Somatic Hybrid Plants of Potato Regenerated from Fused Protoplasts.' *Carlsberg Research Communications* 43: 203–18.

Mellon, M. and Rissler, J. (1995). 'Transgenic Crops: USDA Data on Small-Scale Tests Contribute Little to Commercial Risk Assessment.' *Bio/Technology* 13: January 13.

Mercer, D.K., Scott, K.P., Bruce-Johnson, W.A., Glover, L.A. and Flint, H.J. (1999). 'Fate of Free DNA and Transformation of the Oral Bacterium *Streptococcus gordonii* DL1 by plasmid DNA in Human Saliva.' *Applied and Environmental Microbiology* 65: 6–10.

Meyer, P., Linn, F., Heidmann, I., Heiner Meyer, Z.A., Niedenhof, I. and Saddler, H. (1992). 'Endogenous and Environmental Factors Influence 35S Promoter Methylation of a Maize A1 Construct in Transgenic Petunia and Its Colour Phenotype.' *Molecular Genes and Genetics* 231: 345–52.

Michel, J.H. (1999). *Development Co-operation, 1998.* Paris: Organisation for Economic Co-operation and Development.

Mikkelsen, T.R., Andersen, B. and Bagger, J.R. (1996). 'The Risk of Crop Transgene Spread.' *Nature* 380: 31.

Miller, H.I. (1997). *Policy Controversy in Biotechnology: An Insider's View.* Austin, TX: R.G. Landes.

Miller, H.I. (1999). 'Baby Food for Thought.' *San Jose /Mercury News*, August 23.

Miller, J.C. (1990). 'Field Assessment of the Effects of a Microbial Pest Control Agent on Nontarget Lepidoptera.' *American Entomology* 36(2): 135–9.

Millstone, E., Brunner, E. and Mayer, S. (1999). 'Beyond 'Substantial Equivalence.'' *Nature* 401: October 7.

Minister of Agriculture, Fisheries and Food, United Kingdom. (1998). 'The Food Standards Agency: A Force for Change.' CM, The Stationery Office, January.

Molnar, J.J. and Kinnucan, H. (eds.) (1989). *Biotechnology and the New Agricultural Revolution.* American Association for the Advancement of Science (AAAS) Selected Symposium 108. Boulder, CO: Westview Press.

Monsanto (1997a). *1997 Annual Report Form 10-K*. St. Louis, MO: Monsanto Corporation.

Monsanto (1997b). 'Bollgard® Cotton Update 1996.' St. Louis, MO: Monsanto Corporation. Available at http: //www.monsanto.com/monsanto/mediacenter/background/97mar20_Bollgard.html.

Monsanto Company (1998). 'YieldGard Insect-Protected Corn: The Whole Plant. The Whole Season.' Bulletin.

Monsanto (1999). *Annual Report 1998*. Posted at http://www.monsanto.com. Accessed October 18.

Monsanto (2000). 'Scientists Achieve Major Breakthrough in Rice; Data to be Shared with Worldwide Research Community.' Posted at http://www.monsanto.com/monsanto/mediacenter/2000/00apr4_rice.html. Accessed July 11.

Mooney, H.A. (ed.) (1990). *Introduction of Genetically Modified Organisms into the Environment* (SCOPE 44). New York: Wiley.

Moore, G.C. *et al.* (1999) *Bt Cotton Technology in Texas: A Practical View*. Texas Agricultural Extension Service, L-5169.

Morris, M. and Hoisington, D. (1999). 'Bringing the Benefits of Biotechnology to the Poor: The Role of the CGIAR Centers.' Paper Presented at the Conference on Agricultural Biotechnology in Developing Countries: Toward Optimizing the Benefits for the Poor,' Bonn University, Bonn, Germany. 15–16 November.

Moschini, G. and Lapan, H. (1997). 'Intellectual Property Rights and the Welfare Effects on Agricultural R&D.' *American Journal of Agricultural Economics* 79(November): 1229–42.

Moschini, G., Lapan, H. and Sobolevsky, A. (2000). Roundup Ready ® soybeans and welfare effects in the soybean complex. *Agribusiness* 16: 33–55.

Mullins, J.W. and Mills, J.M. (1999). 'Economics of Bollgard Versus Non-Bollgard Cotton in 1998.' *1999 Proceedings Beltwide Cotton Conferences* (eds. P. Dugger and D. Richter), pp. 958–61. Memphis, TN: National Cotton Council of America.

Myers, N. (1998). 'Lifting the Veil on Perverse Subsidies.' *Nature* 392: March 26.

Myerson, A.R. (1997). 'Breeding Seeds of Discontent: Cotton Growers Say Strain Cuts Yields.' *New York Times* November 19.

National Academy of Sciences (1982). *Genetic Engineering of Plants*. Washington DC: National Academy of Sciences.

National Academy of Sciences (1987). *Introduction of Recombinant DNA-Engineered Organisms Into the Environment: Key Issues*. Washington DC: National Academy Press.

National Agricultural Statistics Service, United States Department of Agriculture (Various years). *Agricultural Statistics*. Washington DC: United States Government Printing Office.

National Center for Food and Agricultural Policy (1999). 'Agricultural Biotechnology: Insect Control Benefits,' July (http://www.bio.org/food&ag/slide31.htm).

National Center for Nutrition and Dietetics (1996). 'Food Biotechnology: Safe, Nutritious, Healthful, Abundant and Tasty Food.' Nutrition Fact Sheet.

National Public Radio (1999). Genetically-Engineered Plants and the Environment. Morning Edition. December 2. Available at http: //search.npr.org/cf/cmn/cmnps05fm.cfm?SegID=67342. Accessed June 24, 2000.

National Research Council, Committee on a National Strategy for Biotechnology in Agriculture, Board on Agriculture (1987). *Agriculture Biotechnology: Strategies for National Competitiveness*. Washington DC.

National Research Council (1995). *Nitrate and Nitrite in Drinking Water*. Washington DC: National Academy Press.

National Research Council (2000). *Genetically Modified Pest-Protected Plants: Science and Regulation*. Washington DC: National Academy Press.

Nelson, G.C., Josling, T., Bullock, D., Unnevehr L., Rosegrant, M. and Hill, L. (1999). *The Economics and Politics of Genetically Modified Organisms in Agriculture: Implications for WTO 2000*. University of Illinois at Urbana-Champaign. Champaign IL. November. http://web.aces.uiuc.edu/wf/GMO/GMO.pdf

Nestle, M. (1996). 'Allergies to Transgenic Foods – Questions of Policy.' *New England Journal of Medicine* 334(11): 726–7.

Newhouse, K., Singh, B., Shaner, D. and Stidham, M. (1991). 'Mutations in Corn (*Zea mays* L.) Conferring Resistance to Imidazolinone Herbicides.' *Theoretical and Applied Genetics* 83: 65–70.

New Zealand Ministry of Agriculture and Forestry (NZMAF) (2000). 'Labelling – The Issues.' GM Foods Information Kit, Part 5. Available at http://www.maf.govt.nz/MAFnet/index/GMF/sheet5.htm. Viewed July 20.

Nordlee, J.A., Taylor, S.L., Townsend, J.A., Thomas, L.A. and Bush, R.K. (1996). 'Identification of Brazil-Nut Allergen in Transgenic Soybeans.' *New England Journal of Medicine* 334(11): 688–92.

Nottingham, S. (1998). *Eat Your Genes: How Genetically Modified Food Is Entering Our Diet*. London: Zed Books.

Obrycki, J. and Hansen, L. (2000). 'Field Disposition of *Bt* Transgenic Corn Pollen: Lethal Effects on the Monarch Butterfly,' http://link.springer.de/link/service/journals/00442/instr.htm, *Oecologia*, August 19.

Oerke, E.-C., Dehne, H.-W., Schonbeck, F. and Weber, A. (1994). *Crop Production and Crop Protection: Estimated Losses in Major Food and Cash Crops*. Amsterdam: Elsevier.

Ohio State University. *Ohio Herbicide Selector Program*. Available at http://www2.ag.ohio-state.edu~ohioline/software/. Columbus, OH.

Onstad, D.W. and Guse, C.A. (Forthcoming). 'Economic Analysis of Transgenic Maize and Nontransgenic Refuges for Managing European Corn Borer (Lepidoptera: Pyralidae).' *Journal of Economic Entomology*.

Organisation for Economic Co-operation and Development (OECD) (1989). 'Biotechnology: Economic and Wider Impacts.' *Policy Brief*. Paris.

Organisation for Economic Co-operation and Development (OECD) (1992). 'Biotechnology, Agriculture, and Food.' *Policy Brief*. Paris.

Organisation for Economic Co-operation and Development (OECD) (1993). 'Safety Evaluation of Foods Derived by Modern Biotechnology: Concepts and Principles.' *Policy Brief*. Paris.

Organisation for Economic Co-operation and Development (OECD) (1994). 'Aquatic Biotechnology and Food Safety.' *Policy Brief*. Paris.

Organisation for Economic Co-operation and Development (OECD) (1996). 'Food Safety Evaluation.' *Policy Brief*. Paris.

Organisation for Economic Co-operation and Development (OECD) (1998). 'Biotechnology and the Changing Role of Government.' *Policy Brief*. Paris.

Organisation for Economic Co-operation and Development (OECD) (1999). 'Modern Technology and the OECD.' *Policy Brief*. Paris.

Organisation for Economic Co-operation and Development (OECD) (2000a). *GM Food Safety: Facts, Uncertainties, and Assessment*, Edinburgh, Scotland, February 28–March 1.

Organisation for Economic Co-operation and Development (OECD) (2000b). 'Biotechnology and Food Safety Web Site.' http://www.oecd.org/subject/biotech/.

Ostlie, K.R., Hutchison, W.D. and Hellmich, R.L. (eds.) (1997). 'Bt Corn and European Corn Borer: Long-Term Success Through Resistance Management.' NC-205, North Central Regional Publication. University of Minnesota Extension Service. Available at http: //www.extension.umn.edu/Documents/D/C/DC7055.html. (October 1998 supplement available at https//ent.agri.umn.edu/ecb/nc205doc.htm)

O'Sullivan, K. (1999). 'Claims on GM Crop Yields Rejected.' *The Irish Times* July 9.

Overseas Development Institute (ODI) (1999). 'The Debate on Genetically Modified Organisms: Relevance for the South.' *Briefing Paper*, January.

Panem, S. (ed.) (1985). *Biotechnology: Implications for Public Policy*. Washington DC: The Brookings Institution.

Pardey, P. and Alston, J.M. (1996). 'Revamping Agricultural R & D.' *2020 Brief* No. 24. Washington DC: IFPRI.

Parekh, R.B., Dwek, R.A., Thomas, J.R., Opdenakker, G., *et al.* (1989a). 'Cell-type-specific and Site-specific N-glycosylation of Type I and Type II Human Plasminogen Activator. *Biochemistry* 28(19): 7644–62.

Parekh, R.B., Dwek, R.A., Rudd, P.M., Thomas, J.R., *et al.* (1989b). 'N-glycosylation and in vitro Enzymatic Activity of Human Plasminogen Activator Expressed in Chinese Hamster Ovary Cells and a Murine Cell Line.' *Biochemistry*, 28(19): 7670–9.

Patterson, L.A. (2000). 'Biotechnology Policy: Regulating Risks and Risking Regulation. *Policy Making in the European Union*, 4th edn (eds. H. Wallace and W. Wallace). Oxford: Oxford University Press.

Pedigo, L. (1989). *Entomology and Pest Management*. New York: Macmillan.

Pelkmans, J. (1998). 'Atlantic Economic Cooperation: the Limits of Plurilateralism.' Paper presented at the HWWA Conference on Transatlantic Relations in a Global Economy, Hamburg, Germany, May.

Perdikis, N., Kerr, W.A. and Hobbs, J.E. (1999). 'Can the WTO/GATT Agreements on Sanitary and Phyto-Sanitary Measures and Technical Barriers to Trade Be Renegotiated to Accommodate Agricultural Biotechnology?' Paper presented at the NE-165 Conference, Transitions in Agbiotech: Economics of Strategy and Policy, Washington DC, June 24–25.

Perpich, J.G. (ed.). (1986). *Biotechnology in Society: Private Initiatives and Public Oversight*. New York: Pergamon Press.

Persley, G.J. (1990). *Beyond Mendel's Garden: Biotechnology in the Service of World Agriculture*. Wallingford, UK: CAB International.

Persley, G.J. and Doyle, J.J. (1999). *Biotechnology for Developing-Country Agriculture: Problems and Opportunities. Overview*. 2020 Vision. Washington DC: International Food Policy Research Institute (IFPRI). October.

Peterson, G.R. (ed.) (1993). *Understanding Biotechnology Law: Protection, Licensing, and Intellectual Property Policies*. New York: Marcel Dekker.

Phillips, R.L. and Vasil, I.K. (1994). *DNA-based Markers in Plants*. New York: Kluwer Academic Publishers.

Pike, D., Kirby, H.W. and Kamble, S.T. (1997). 'Distribution and Severity of Pests in the Midwest.' Dept. Crop Sci., University of Illinois at Urbana-Champaign.

Pilcher, C.D., Rice, M.E., Obrycki, J.J. and Lewis, L.C. (1997). 'Field and Laboratory Evaluations of Transgenic *Bacillus thuringiensis* Corn on Secondary Lepidopteran Pests (Lepidoptera: Noctuidae).' *Journal of Economic Entomology* 90(2): 669–78.

Pimentel, D., Lach, L., Zuniga, R. and Morrison, D. (1999a). 'Environmental and Economic Costs Associated with Non-Indigenous Species in the United States.' Cornell University, June 12. Available at http://www.news.cornell.edu/releases/Jan99/species_cost. Accessed July 2000.

Pimentel, D., *et al.* (1999b). *Will Limits of the Earth's Resources Control Human Numbers?* Cornell University College of Agriculture, February 25.

Pinstrup-Andersen, P. (1999). 'Modern Biotechnology and Tropical Agriculture: Summary.' Seminar on Agricultural Research in Africa, Washington DC; Center for International Development, Harvard University.

Pinstrup-Andersen, P. and Cohen, M.J. (2000a). 'Agricultural Biotechnology: Risks and Opportunities for Developing Country Food Security.' *International Journal of Biotechnology* 2 (1/2/3): 145–63.

Pinstrup-Andersen, P. and Cohen, M.J. (2000b). 'Biotechnology and the CGIAR.' *Proceedings of the International Conference on Sustainable Agriculture in the Next Millennium – The Impact of Modern Biotechnology on Developing Countries*. Brussels: Friends of the Earth Europe. Forthcoming. (Available at www.ifpri.org)

Pinstrup-Andersen, P., Ruiz de Londou, N. and Hoover, E. (1976). The impact of increasing food supply on human nutrition: implications for commodity priorities in agricultural research and policy. *American Journal of Agricultural Economics* **58**: 131–42.

Pinstrup-Andersen, P., Pandya-Lorch, R. and Rosegrant, M.W. (1999). *World Food Prospects: Critical Issues for the Early Twenty-First Century, 2020 Vision Food Policy Report*. Washington DC: IFPRI.

Pleasants, J.M., Hellmich, R.L. and Lewis, L.C. (1999). 'Deposition of Pollen on Milkweeds and Exposure Risk to Monarch Larvae in Iowa.' Monarch Butterfly Research Symposium, Rosemont, IL, November.

Plexus Marketing Group, Inc. and Timber Mill Research, Inc. (1998). *Enhanced Market Data Cotton 1997/1998 Report*. Indianapolis, IN: Timber Mill Research, Inc.

Plexus Marketing Group, Inc. and Timber Mill Research, Inc. (1998). *Enhanced Market Data Cotton 1996/1997 Report*. Indianapolis, IN: Timber Mill Research, Inc.

Plucknett, D.L. (1994). 'Sources of the Next Century's New Technology.' *Agricultural Technology* (ed. J.R. Anderson), pp. 147–77. Wallingford, UK: CAB International.

Pohl-Orf, M., Morak, C., Wehres, U., Saeglitz, C., Drießen, S., Lehnen, M., Hesse, P., Mücher, T., von Soosten, C., Schuphan, I. and Bortsch, D. (2000). 'The Environmental Impact of Gene Flow from Sugar Beet to Wild Beet: An Ecological Comparison of Transgenic and Natural Virus Tolerance Genes.' The Biosafety of Genetically Modified Organism (eds. C. Fairbairn, G. Scoles and A. McHughen), pp. 51–5. Proceedings of the 6th International Symposium, Saskatoon, Canada, July.

Pollan, M. (1998). 'Playing God in the Garden.' *New York Times Magazine* October 25.

Pollan, M. (1999). 'Feeding Frenzy.' *New York Times* December 12.

Poston, F.L., Pedigo, L.P. and Welch, S.M. (1983). 'Economic Injury Levels: Reality and Practicality.' *Bulletin of the Entomological Society of America* 29: 49–53.

Poulter, S. (1999). 'M&S Bows to Shoppers' Fears and Orders Ban on Frankenfoods.' *Daily Mail* March 16.

*Poultry Annual Report India* (1995). US Department of Agriculture/Foreign Agricultural Service, New Delhi.

Pratley, J., *et al.* (1996). 'Glyphosate Resistance in Annual Ryegrass.' 11th New South Wales Grasslands Conference. Wagga Wagga, Australia.

Pray, C.E. and Fuglie, K.O. (1997). 'The Private Sector and International Agricultural Technology Transfer.' Paper presented at ERS-Farm Foundation Workshop: Public-Private Collaboration in Agricultural Research, Rosslyn, VA.

Purrington, C.B. and Bergelson, J. (1995). 'Assessing Weediness of Transgenic Crops: Industry Plays Plant Ecologist.' *Trends in Evolution and Ecology* 10(8): August 8.

Pusztai, A., Ewen, S.W.B., Grant, G., *et al.* (1990). Relationship Between Survival and Binding of Plant Lectins During Small Intestinal Passage and Their Effectiveness as Growth Factors. *Digestion* 26 (suppl. 2): 306–16.

Qaim, M. (1998). 'Transgenic Virus-Resistant Potatoes in Mexico: Potential Socioeconomic Implications of North-South Biotechnology Transfer.' *ISAAA Briefs*. Ithaca, NY: International Service for the Acquisition of Agri-Biotech Applications.

Qaim, M. (1999a). 'Assessing the Impact of Banana Biotechnology in Kenya.' *ISAAA Briefs*. Ithaca, NY: International Service for the Acquisition of Agri-Biotech Applications.

Qaim, M. (1999b). 'Potential Benefits of Agricultural Biotechnology: An Example from the Mexican Potato Sector.' *Review of Agricultural Economics* 21:2(Fall/Winter): 390–408.

Raeburn, P. (1995). *The Last Harvest*. New York: Simon and Schuster.

Raybould, A.F. and Gray, A.J. (1994). 'Will Hybrids of Genetically Modified Crops Invade Natural Communities?' *Trends in Ecology and Evolution* 9: 85–89.

Raynor, G.S., Ogden, E.C. and Hayes, J.V. (1972). 'Dispersion and Deposition of Corn Pollen from Experimental Sources.' *Agronomy Journal* 64(4): 420–7.

Reddy, S.A. and Thomas, T.L. (1996). 'Expression of a Cyanobacterial Delta 6-Desaturase Gene Results in Gamma-linolenic Acid Production in Transgenic Plants.' *Nature Biotechnology* 14: 639–42.

Rejesus, R.M., Greene, J.K., Hamming, M.D. and Curtis, C.E. (1997). 'Economic Analysis of Insect Management Strategies for Transgenic *Bt* Cotton Production in South Carolina.' *1997 Proceedings Beltwide Cotton Conferences* (eds. P. Dugger and D. Richter), pp. 247–51. Memphis, TN: National Cotton Council of America.

Reuters (1999a). 'ADM to Offer Premium to Grow Non-Genetic Soy.' Available at http://biz.yahoo.com May 5.

Reuters (1999b). 'Europe Said Not Ready for Gene Crops.' *Science Headlines* [dailynews.yahoo.com/headlines/sc], July 8.

Reuters (1999c). 'US Official: EU Biotech Food Fear Threatens Trade.' *Science Headlines* [dailynews.yahoo.com/headlines/sc], June 29.

Rice, M. (1999). 'Corn Borers Should Increase in 1999.' *Integrated Crop Management*. Ames, IA: Department of Entomology, Iowa State University, April 12. (http://www.ipm.iastate.edu/ipm/icm/1999/4–12–1999/cborerinc.html).

Rice, M. and Pilcher, C.D. (1998). 'Potential Benefits and Limitations of Transgenic *Bt* Corn for Management of the European Corn Borer.' *American Entomologist*, Summer.

Rifkin, J. (1999). 'Unknown Risks of Genetically Engineered Crops.' *The Boston Globe* June 7 (A3).

Riley, P. (1999). 'Value-Enhanced Crops: Biotechnology's Next Stage'. *Agricultural Outlook*. Economic Research Service, US Department of Agriculture, March, pp. 23–30.

Riley, S.L. (1989). 'Pyrethroid Resistance in *Heliothis virescens*: Current U.S. Management Programs.' *Pesticide Science* 26: 411–21.

Rissler, J. and Mellon, M. (1995). 'Managing Resistance to *Bt*.' *The Gene Exchange* 6: 4–7.

Rissler, J. and Mellon, M. (1996). *The Ecological Risks of Engineered Crops*. Cambridge, MA: MIT Press.

Roberts, D. and DeRemer, K. (1997). 'An Overview of Technical Barriers to U.S.

Agricultural Exports.' Staff Paper, Economic Research Service Report, US Department of Agriculture. Washington DC: GPO.

Roberts, D., Josling, T. and Orden, D. (1997). 'Technical Barriers to Trade: An Analytic Framework.' Agricultural Economic Report, US Department of Agriculture. Washington DC: GPO, November.

Rockefeller Foundation, (1999a). 'Food Gains for the World's Poor Are Being Threatened by Furore Over Genetically Modified (GM) Foods.' June 24.

Rockefeller Foundation (1999b). *Annual Report 1998*. Posted at http://www.rockfound.org. Accessed October 18.

Rodriguez, S.N. (1997). 'Food Labeling Requirements'. *Fundamental of Law and Regulation: An In-depth Look at Foods, Veterinary Medicines, and Cosmetics* (eds. R.P. Brady, R.M. Cooper and R.S. Silverman), vol. 1, pp. 237–256. Washington DC: FDLI.

Roger, P.L., and Fleet, G.H. (eds.) (1989). *Biotechnology and the Food Industry*. New York: Gordon and Breach.

Roof, M.E. and Durant, J.A. (1998). 'Experiences With *Bt* Cotton Under Light to Medium Bollworm Infestations in South Carolina.' *1998 Proceedings Beltwide Cotton Conferences* (eds. P. Dugger and D. Richter), pp. 964–5. Memphis, TN: National Cotton Council of America.

Rosegrant, M.W., Agcaoili-Sombilla, M. and Perez, N.D. (1995). 'Global Food Projections to 2020: Implications for Investment.' *Food, Agriculture, and Environment Discussion Paper* No. 5. Washington DC: IFPRI.

Rosset, P. (1999). 'Why Genetically Altered Food Won't Conquer Hunger.' *New York Times*, September 1.

Rosset, P. and Altieri, M. (1999). 'The Productivity of Small Scale Agriculture.' IFA White Paper.

Roth, M.J. and Shear, R.H. (1999). 'Intellectual Property Law in the Protection of Plant Inventions.' Paper presented at the NE-165 Conference, Transitions in Agbiotech: Economics of Strategy and Policy. Washington DC, June 24–25.

Rotman, D. (1996). 'Monsanto Snaps Up Agracetus.' *Chemical Week* 158: 8.

Rural Advancement Foundation International (RAFI) (2000). Agbiotech's Five Jumbo Gene Giants: Just Five Survive Consolidation Squeeze. RAFI Geno-Types. January 7. Available at http://www.rafi.org. Accessed July 2000.

Saegusa, A. (1999). 'Japan Tightens Rules on GM Crops to Protect the Environment.' *Nature*, June.

Sainsbury. (1999). World Wide Web homepage. Available at http://www.sainsburys.co.uk.

Sampaio, M. (1999). 'Perspectives from National Agricultural Research Systems.' *Biotechnology and Biosafety* (eds. I. Serageldin and W. Collins). Washington DC: The World Bank.

Sampson, H.A., Mendelson, L. and Rosen, J.P. (1992). 'Fatal and Near-Fatal Anaphylactic Reactions to Food in Children and Adolescents.' *New England Journal of Medicine* 327: 380–4.

Sanford, J.C. (1988). 'The Biolistic Process.' *Trends in Biotechnology* 6: 229–302.

Sasson, A. and Costarini, V. (eds.) (1991). *Biotechnologies in Perspective: Socio-Economic Implications for Developing Countries*. Paris: UNESCO.

Saxena, D., Flores, S. and Stotzky, G. (1999). 'Insecticidal Toxin in Root Exudates from *Bt* Corn.' *Nature* 402: December 2.

Scheffler, J.A. and Dale, P.J. (1994). 'Opportunities for Gene Transfer From Transgenic Oilseed Rape (*Brassica napus*) to Related Species,' *Transgenic Research* 3(February): 263–78.

Scherpenbeg, J.V. (1998). 'Regulation and Market Access – Conflicts and Common Interests in Services Markets.' Paper presented at the HWWA Conference on Transatlantic Relations in a Global Economy, May. Hamburg, Germany.

Scherr, S.J. (1999). 'Soil Degradation: A Threat to Developing-Country Food Security by 2020?' *2020 Brief* No. 58. Washington DC: IFPRI.

Schioler, E. (1998). *Good News from Africa: Farmers, Agricultural Research, and Food in the Pantry*. Washington DC: IFPRI.

Schnieke A., Harbers K. and Jaenisch, R. (1983). 'Embryonic Lethal Mutation in Mice Induced by Retrovirus Insertion into the Alpha-1(I) Collagen Gene.' *Nature* 304: 315–20.

Schoelz, J. and Wintermantelk, W. (1993). 'Expansion of Viral Host Range Through Complementation and Recombination in Transgenic Plants.' *Plant Cell* 5: 1669–79.

Schomberg, R. Von, and Wheale, P. (ed.) (1996). *The Social Management of Biotechnology: Workshop Proceedings*. Tilburg University, The Netherlands, April.

Schubbert, R., Lettmann, C. and Doerfler, W. (1994). 'Ingested Foreign (Phage M13) DNA Survives Transiently in the Gastrointestinal Tract and Enters the Bloodstream of Mice.' *Molecules, Genes and Genetics* 242: 495–504.

Schulz, E. (2000). *The Opportunities and Hazards of Agricultural Biotechnology*. Economic Strategy Institute, May.

Scott, M.R., Will, R., Ironside, J., Nguyen, H.B., Tremblay, P., DeArmond, S.J. and Prusiner, S.B. (1999). 'Compelling Transgenetic Evidence for Transmission of Bovine Spongiform Encephalopathy Prions to Humans.' *Proceedings of the National Academy of Sciences*, 96(26): 15137–42.

Scott, S.E. and Wilkinson, M.J. (1999). 'Low Probability of Chloroplast Movement from Oilseed Rape (*Brassica napus*) into Wild *Brassica rapa*.' *Natural Biotechnology* 17: 390–2.

Sears, M. and Shelton, A. (2000). 'Questionable Conclusions from Latest Monarch Study,' http://www.agbioworld.org, August 29.

Sedjo, R. (1995). 'Global Forests Revisited.' *The State of Humanity* (ed. J. Simon), p. 336. Oxford: Blackwell.

Serageldin, I. (1999). 'Biotechnology and Food Security in the 21st Century.' *Science* 285 (16 July): 387–9.

Shaunak, R.K. (1995). 'Economic Implications of Adopting Genetically Engineered Heliothis Resistant *Bt* Cotton Cultivars in Mississippi.' PhD Dissertation, Mississippi State University.

Shemitt, L.W. (ed.) (1982). *Chemistry and World Food Supplies: The New Frontiers, CHEMRAWN II*. Canada: Pergamon Press.

Shiva, V. (1996). *Biodiversity-based Productivity*. New Delhi: RFSTE.

Silver, S. (ed.) (1986). *Biotechnology: Potentials and Limitations*. Berlin: Springer-Verlag.

Sindel, B. (1996). 'Glyphosate Resistance Discovered in Annual Ryegrass.' *Resistant Pest Management* 8: 5–6.

Skarzhinskaya, M., Fahleson, J., Glimelius, K. and Mouras, A. (1998). 'Genome Organization of *Brassica napus* and *Lesquerella fendleri* and Analysis of their Somatic Hybrids Using Genomic in situ Hybridization. *Genome* 41: 691–701.

Sloan, A.E. and Powers, M.E. (1986). 'A Perspective on Popular Perceptions of Adverse Reactions to Food.' *Journal of Allergy and Clinical Immunology* 78: 127–33.

Smith, R.H. (1997). 'An Extension Entomologist's 1996 Observations of Bollgard (*Bt*) Technology.' *1997 Proceedings Beltwide Cotton Conferences* (eds. P. Dugger and D. Richter), Memphis, TN: National Cotton Council of America.

Smith, R.H. (1998). 'Year Two of Bollgard Behind Boll Weevil Eradication – Alabama

Observations.' *1998 Proceedings Beltwide Cotton Conferences* (eds. P. Dugger and D. Richter), pp. 965–6. Memphis, TN: National Cotton Council of America.

Snow, A.A. and Jorgensen, R.B. (1998). 'Costs of Transgenic Glufosinate: Perspective on Popular Perceptions of Adverse Reactions to Foods.' *Journal of Allergy and Clinical Immunology* 78: 127–33.

Snow, A.A. and Jorgensen, R.B. (1999). 'Fitness Costs Associated with Transgenic Glufosinate Tolerance Introgressed From *Brassica napus* ssp *oleifera* (Oilseed Rape) into Weedy *Brassica rapa*.' BCPC Symposium Hearings No. 72: *Gene Flow and Agriculture: Relevance for Transgenic Crops*, pp. 137–42.

Snow, A.A. and Palma, P.M. (1997). 'Commercialization of Transgenic Plants: Potential Ecological Risks.' *Bioscience* 47: 86–96.

Snow, A.A., Andersen, B. and Jørgensen, R.B. (1999). 'Costs of Transgenic Herbicide Resistance Introgressed from *Brassica napus* into Weedy *B. Rapa*.' *Molecular Ecology* 8(4), April.

Song, H.-S., Brotherton, J.E., Gonzales, R.A. and Widholm, J.M. (1998). 'Tissue Culture Specific Expression of a Naturally Occurring Tobacco Feedback-Insensitive Anthranilate Synthase.' *Plant Physiology* 117: 533–43.

Sparks, A.N., Jr (1998). 'Texas Guide for Controlling Insects on Commercial Vegetable Crops.' Extension Bulletin B-1305, Texas A&M University.

Spillane, C. (1999). *Recent Developments in Biotechnology as They Relate to Plant Genetic Resources for Food and Agriculture*. Food and Agriculture Organization of the United Nations (FAO). Commission on Genetic Resources for Food and Agriculture. Background Study Paper No. 9. April.

Spring, S., *et al.* (1992). 'Phylogenetic Diversity and Identification of Nonculturable Mangeto-tactic Bacteria.' *Systematics and Applied Microbiology*, 15: 116–22.

Stark, C., Jr (1997). 'Economics of Transgenic Cotton: Some Indicators Based on Georgia Producers.' *1997 Proceedings Beltwide Cotton Conferences* (eds. P. Dugger and D. Richter), pp. 251–3. Memphis, TN: National Cotton Council of America.

Steffan, R.J., *et al.* (1988). 'Recovery of DNA from Soils and Sediments.' *Applied Environmental Microbiology*, 54: 2908–15.

Steffey, K. (1998). 'A Flurry of European Corn Borer Activity, and Bt-Corn for '99?' In: *Pest Management and Crop Bulletin* No. 22 (August 28). University of Illinois Cooperative Extension Service, pp. 200–2. (http://www.ag.uiuc.edu/cespubs/pest/articles/199822e.html).

Steffey, K. and Gray, M. (1997). 'Corn BorerDensities, 1987 to 1996.' In: *Pest Management and Crop Bulletin* No. 25 (December 5). University of Illinois Cooperative Extension Service. (http://www.ag.uiuc.edu/cespubs/pest/articles/v9725g.html).

Stephens, P.A., Nickell, C.D. and Widholm, J.M. (1991). 'Agronomic Evaluation of Tissue-culture-derived Soybean Plants.' *Theoretical and Applied Genetics* 82: 633–5.

Strong, D. (1997). 'Fear No Weevil.' *Science* 277(August 22): 1058–9.

Sugita, K., Matsunaga, E. and Ebinuma, H. (1999). 'Effective Selection System for Generating Marker-free Transgenic Plants Independent of Sexual Crossing.' *Plant Cell Reports* 18: 941–7.

Sullivan, J., Wainio, J. and Roningen, V. (1989). *A Database for Trade Liberalization Studies*. Washington DC: US Department of Agriculture, Agriculture and Trade Analysis Division.

Sumida, S. (1993/1994). 'Plant Biotechnology Comes of Age.' *OECD Observer* 1985: 9–11.

Sykes, A.O. (1995). *Product Standards for Internationally Integrated Goods Markets*. Washington DC: The Brookings Institution.

Tabashnik, B.E. (1994). 'Evolution of Resistance to *Bacillus thuringiensis.' Annual Review of Entomology* 39: 49–79.

Tabashnik, B.E., Cushing, N.L., Finson, N. and Johnson, M.W. (1990). 'Field Development of Resistance to *Bacillus thuringiensis* in Diamondback Moth.' *Journal of Economic Entomology* 83: 1671–6.

Tangermann, S. (1998). 'The Common and Uncommon Agricultural Policies: An Eternal Issue?' Paper presented at the HWWA Conference on Transatlantic Relations in a Global Economy, May. Hamburg, Germany.

Tangley, L. (2000). 'Of Genes, Grain, and Grocers: The Risks and Realities of Engineered Crops.' *US News and World Report.* April 10.

Tanksley, S. and McCouch, S. (1997). 'Seed Banks and Molecular Maps: Unlocking Genetic Potential from the Wild.' *Science* 277(August 22): 1063–6.

Taylor, C.R. (1993). 'AGSIM: Model Description and Documentation.' *Agricultural Sector Models for the United States* (eds. C.R. Taylor, K.H. Reichelderfer and S.R. Johnson). Ames, IA: Iowa State University Press.

Teich, A.H., Levin, M.A. and Pace, J.H. (1985). *Biotechnology and the Environment: Risk and Regulation.* Washington DC: American Association for the Advancement of Science.

TeKrony, D.M., Egli, D.B. and White, G.M. (1987). 'Seed Production and Technology.' *Soybeans: Improvement, Production, and Uses* (2nd edn) (ed. J.R. Wilcox), pp. 295–354. Madison, WI: American Society of Agronomy.

Tenenbaum, D. (1996). 'Weeds from Hell.' *Technology Review* 99: 32–40.

Thayer, A.M. (1999). 'Ag Biotech Food: Risky Or Risk Free?' *Chemical and Engineering News* November 1.

Thompson, P. (1997). *Food Biotechnology in Ethical Perspective.* London: Blackie Academic and Professional.

Timmons, A.M., Charters, Y.M., Crawford, J.W., *et al.* (1996). 'Risks From Transgenic Crops.' *Nature* 380: 487.

Tolbert, K. (2000). 'In Japan, it's Back to Nature; Consumers Add Non-modified Products to Shopping Cart.' *Washington Post* January 24, p. A08.

Tomlin, C. (ed.) (1994). *The Pesticide Manual: Incorporating the Agrochemicals Handbook.* (10th ed). Farnham, UK: British Crop Protection Council.

Traavik, T. (1998). 'Too Early May Be Too Late: Ecological Risks Associated with the Use of Naked DNA as a Tool for Research, Production and Therapy.' Directorate for Nature Research. Trondheim, Norway.

Trimble, S. (1999). 'Decreased Rates of Alluvial Sediment Storage in the Coon Creek Basin, Wisconsin, 1975–93.' *Science* 285(Aug. 20): 1244–6.

Tu, J., Ona, I., Zhang, Q., Mew, T.W., Khush, G.S. and Datta, S.K. (1998). 'Transgenic Rice Variety IR 72 with Xa21 Is Resistant to Bacterial Blight.' *Theoretical and Applied Genetics* 97: 31–6.

UN Development Program (1999). 'Demographic Trends, Total Fertility Rate, 1997, All Developing Countries (Table 16).' *Human Development Report 1999.* Oxford.

UN Population Division (1998). *World Population Prospects: The 1998 Revision.* New York: United Nations.

Union of Concerned Scientists (1999). 'New Worries About Moderate-Dose Bt Corn. The Gene Exchange.' May.

US Bureau of Labor Statistics (1999). Data posted at http://www.bls.gov. Accessed November 18.

US Census Bureau (1999). Data posted at http://www.census.gov. Accessed November 18.

US Congress, Office of Technology Assessment (1988). *US Investment in Biotechnology.* Washington DC.

US Congress, Office of Technology Assessment (1993). *Harmful Non-Indigenous Species in the United States.* Washington DC: GPO.

US Department of Agriculture. *Agricultural Chemical Usage: Field Crops Summary, 1995–1998.* National Agricultural Statistics Service (separate volumes).

US Department of Agriculture (1995). *APHIS Protecting US Agriculture from Noxious Weeds.* Washington DC: GPO.

US Department of Agriculture (1997a). *The Food Safety Enforcement Enhancement Act of 1997.* Washington DC: GPO.

US Department of Agriculture (1997b). *Biotechnology Permits Database. US Department of Agriculture, Animal and Plant Health Inspection Service.* Available at http: //www.aphis.usda.gov: 80/bbep/bp/.

US Department of Agriculture, Agricultural Marketing Service (1997). *Cotton Varieties Planted.* Washington DC: Government Printing Office.

US Department of Agriculture, Agricultural Marketing Service (1998). *Cotton Varieties Planted.* Washington DC: Government Printing Office.

US Department of Agriculture, Agricultural Marketing Service (1999). *Cotton Varieties Planted.* Washington DC: Government Printing Office.

US Department of Agriculture, Agricultural Marketing Service (2000). *Cotton Varieties Planted.* Washington DC: Government Printing Office.

US Department of Agriculture, Animal and Plant Health Inspection Service (1993) 'Genetically Engineered Organisms and Products: Notification Procedures for the Introduction of Certain Regulated Articles; and Petition for Nonregulated Status.' *Federal Register* 58 (60) 17044.

US Department of Agriculture, Animal and Plant Health Inspection Service (1997) 'Genetically Engineered Organisms and Products: Simplification of Requirements and Procedures for Genetically Engineered Organisms.' *Federal Register* 62 (79) 19903.

US Department of Agriculture, Economic Research Service (1996). *Cotton and Wool: Situation and Outlook.* Washington DC.

US Department of Agriculture (1998). *1998 Agriculture Fact Book.* http://www.usda.gov.

US Department of Agriculture Economic Research Service (1999). 'Impacts of Adopting Genetically Engineered Crops in the US – Preliminary Results. July 20, 1999.' Available at http: //www.econ.ag.gov/whatsnew/issues/gmo. Accessed July 2000.

US Department of Agriculture Economic Research Service (2000). 'Biotech Corn and Soybeans: Changing Markets and the Government's Role.' http://www.ERS.USDA.gov/whatsnew/issues/biotechmarkets/governmentrole.htm.

USDA/Grain Inspection Packers and Stockyards Administration (2000) USDA to Validate Tests for Biotech Grains, Accredit Labs. Press release viewed at http: //www.usda.gov/news/releases/2000/05/0147, July 20.

US Environmental Protection Agency (1993). 'R.E.D. FACTS: Glyphosate.' EPA 738-F-93-011. Available at http: /www.epa.gov: 80/oppsrrd1/REDs/factsheets/0178fact.pdf. September.

US Environmental Protection Agency (1995). 'EPA Gives Conditional Approval for Full Commercial Use of Cotton Plant Pesticide to Combat Cotton Bollworm, Tobacco Budworm, and Pink Bollworm.' EPA Press Release.

US Environmental Protection Agency (1996). 'EPA Issues Conditional Approval for Full Commercial Use of Field Corn Plant-Pesticide Targeting the European Corn Borer.' EPA Press Release.

US Environmental Protection Agency (1998a). 'White Paper on *Bt* Plant-Residue Resistance Management.' Available at http: //www.epa.gov/pesticides/ biopesticides/white_bt.pdf. May.

US Environmental Protection Agency (1998b). '*R.E.D. FACTS*: *Bacillus thuringiensis*.' EPA 738-F-98-001. Available at http: //www.epa.gov: 80/oppsrrd1/REDs/ factsheets/0247fact.pdf. March.

US Environmental Protection Agency (1998c). 'Transmittal of the Final Report of the FIFRA Scientific Advisory Panel on *Bacillus thuringiensis* (*Bt*) Plant-Pesticides and Resistance Management.' Available at http: //www.epa.gov/pesticides/SAP/ 1998/february/finalfeb.pdf. February 9–10.

US Environmental Protection Agency (1999). 'Pesticides Industry Sales and Usage: 1996 and 1997 Market Estimates.' November.

US Environmental Protection Agency and US Department of Agriculture (1999). 'EPA and USDA Position Paper on Insect Resistance Management in *Bt* Crops.' Available at http: //www.epa.gov/pesticides/biopesticides/otherdocs/ bt_position_paper_618.htm.

US Food and Drug Administration (1992). 'Statement of Policy: Foods Derived From New Plant Varieties.' *Federal Register* 57(104), May 29.

US Food and Drug Administration (1997). 'Food Safety from Farm to Table: A National Food Safety Initiative.' Report to the President. Washington DC, May.

US Food and Drug Administration (2000). 'FDA to Strengthen Pre-Market Review of Bioengineered Foods.' http://www.fda.gov/bbs/topics/NEWS/NEW00726.html.

US Food and Drug Administration Center for Food Safety and Applied Nutrition (1998). 'Guidance for Industry: Use of Antibiotic Resistance Marketer Genes in Transgenic Plants.' Available at http: //vm.cfsan.fda.gov/~dms/opa-armg.html. September 4.

US General Accounting Office (1997). *Agricultural Exports*: *US Needs a More Integrated Approach to Address Sanitary/Phytosanitary Issues*. Washington DC, December.

US National Agricultural Statistics Service (1999). Data posted at http://www.usda.gov/nass. Accessed November 18.

Van der Krol, Lenting, Veenstra, *et al.* (1988). 'An Anti-sense Chalcone Synthase Gene in Transgenic Plants Inhibits Flower Pigmentation.' *Nature* 333: 866–9.

Van Harten, A.M. (1998). *Mutation Breeding: Theory and Practical Applications*. New York: Cambridge University Press.

Vasil, I.K. (1990). *Biotechnology: Science, Education and Commercialization: An International Symposium*. New York: Elsevier.

Videla, G.W., Lorenz, E., Deaton, R., López-Mondo, E. and Torcasso, F. (1999). 'Efficacy of Biogodon (Bollgard) to Control Target Cotton Lepidopteran Pests in Argentina.' *1999 Proceedings Beltwide Cotton Conferences* (eds. P. Dugger and D. Richter), p. 1246. Memphis, TN: National Cotton Council of America.

Violand, B.N., Schlittler, M.R., Lawson, C.Q., Kane, J.F., Siegel, N.R., Smith, C.E., Kolodziej, E.W. and Duffin, K.L. (1994). 'Isolation of *Escherichia coli* Synthesized Recombinant Eukaryotic Proteins that Contain ε-N-acetyllysine.' *Protein Science* 3: 1089–97.

Virginia Cooperative Extension (1997). 'Outlook for Transgenic Soybeans.' Available at http: //fbox.vt.edu: 10021/cals/cses/chagedor/soy97.html.

Vogt, D. (1998). 'Food Safety Issues in the 105th Congress.' Washington DC: Congressional Research Service, January 30.

Vorman, J. (1999). 'European Biotech Fear Seen as Key US Trade Threat;' Reuters: Science Headlines, June 30 [dailynews.yahoo.com/headlines/sc].

Wade, N. (1999). 'Geneticists Tap Secrets of 'the Weed''. *New York Times*. Science Desk. December 21.

Wade, R. (1987). *Village Republics: Economic Conditions for Collective Action*. Cambridge: Cambridge University Press.

Walter, E. (2000). 'Botanist Develops Super Rice.' *Lewiston Morning Tribune* May 30.

Wan, Y. and Widholm, J.M. (1993). 'Anther Culture of Maize.' *Plant Breeding Reviews* 11: 199–224.

WARDA (West African Rice Development Association) (1999). *The Spark that Lit a Flame: Participatory Varietal Selection*. Bouaké, Côte D'Ivoire: WARDA.

Watal, J. (1999). 'Intellectual Property Rights and Agriculture: Interests of Developing Countries.' Paper presented at the World Bank and WTO Conference, Agriculture and the New Trade Agenda in the WTO 2000 Negotiations, October 1–2, Geneva.

Welsh, R. (1996). 'The Industrial Reorganization of Agriculture: An Overview and Background Report.' Policy Studies Report. Greenbelt MD; Washington DC: Henry A. Wallace Institute for Alternative Agriculture. April.

Wesseler, J. (1999). 'Temporal Uncertainty and Irreversibility – A Theoretical Framework for the Decision to Approve the Release of Transgenic Crops.' Paper presented at the NE-165 Conference, Transitions in Agbiotech: Economics of Strategy and Policy, Washington DC, June 24–25.

Wheale, P. and McNally, R. (1990). *The Bio-Revolution: Cornucopia or Pandora's Box?* London: Pluto Press.

White House Committee on Environment and Natural Resources (1999). *Report on Hypoxia in the Gulf of Mexico*, October 20. Washington DC: National Oceanic and Atmospheric Administration.

Wichmann, G., Purrington, C.B. and Bergelson, J. (1998) 'Male Promiscuity Is Increased in Transgenic Arabidopsis.' Available at http: //genome-www.stanford.edu/ Arabidopsis/madison98/abshtml/321.html. Accessed June 24, 2000.

Widholm, J.M. (1996). 'In Vitro Selection and Culture-induced Variation in Soybean.' *Soybean: Genetics, Molecular Biology and Biotechnology* (eds. D.P.S. Verma and R. Shoemaker), pp. 107–26. Wallingford, UK: CAB International.

Wiegele, T.C. (1991). *Biotechnology and International Relations: The Political Dimensions*. Gainesville, FL: University of Florida Press.

Wier, A.T., Mullins, J.W. and Mills, J.M. (1998). 'Bollgard Cotton Update and Economic Comparisons Including New Varieties.' *1998 Proceedings Beltwide Cotton Conferences* (eds. P. Dugger and D. Richter), pp. 1039–40. Memphis, TN: National Cotton Council of America.

Williams, M.W. (1996). 'Cotton Insect Losses 1995.' *1996 Proceedings Beltwide Cotton Conferences* (eds. P. Dugger and D. Richter), pp. 670–89. Memphis, TN: National Cotton Council of America.

Williams, M.W. (1997). 'Cotton Insect Losses 1996.' *1997 Proceedings Beltwide Cotton Conferences* (eds. P. Dugger and D. Richter), pp. 834–53. Memphis, TN: National Cotton Council of America.

Williams, M.W. (1998). 'Cotton Insect Losses 1997.' *1998 Proceedings Beltwide Cotton Conferences* (eds. P. Dugger and D. Richter), pp. 834–53. Memphis, TN: National Cotton Council of America.

Williams, M.W. (1999). 'Cotton Insect Losses 1998.' *1999 Proceedings Beltwide Cotton Conferences* (eds. P. Dugger and D. Richter), pp. 785–806. Memphis, TN: National Cotton Council of America.

Williams, M.W. (2000). 'Cotton Insect Loses 1999. ' *2000 Proceedings Beltwide Cotton Conferences* (eds. P. Dugger and D. Richter), pp. 887–911. Memphis, TN: National Cotton Council of America.

World Bank (1988). 'Demography and Fertility, 1965, Low-income Countries (Table 28).' *World Development Report*. Oxford.

World Bank (1999). Data posted at http://www.worldbank.org/poverty/data/trends/income.htm. Accessed October 15.

World Health Organization (1995). *Application of the Principles of Substantial Equivalence to the Safety Evaluation of Foods or Food Components From Plants Derived By Modern Biotechnology, Report of a WHO Workshop*. Geneva: WHO.

World Health Organization (WHO) (1999). 'Malnutrition Worldwide.' Posted at http://www.who.int/nut. Accessed September 29.

World Trade Organization (1995). *The Results of the Uruguay Round of Multilateral Trade Negotiations: The Legal Texts*. World Trade Organization.

Worthing, C.R. (1987). *The Pesticide Manual* (8th edn). UK: The British Crop Protection Council.

Wortman, S. and Cummings, R.W. Jr (1978). *To Feed this World: The Challenge and the Strategy*. Baltimore, MD: The Johns Hopkins University Press.

Wraight, C.L., Zangerl, A.R., Carroll, M.J. and Berenbaum, M.R. (2000). 'Absence of Toxicity of *Bacillus thuringiensis* Pollen to Black Swallowtails Under Field Conditions.' *Proceedings of the National Academy of Sciences of the USA* 97(14): 7700–3.

Wrubel, R.P. and Gressel, J. (1994). 'Are Herbicide Mixtures Useful for Delaying the Rapid Evolution of Resistance? A Case Study.' *Weed Technology* 8: 635–48.

Yoon, C.K. (1999a). 'No Consensus On the Effects Of Altered Corn On Butterflies.' *New York Times*. November 4.

Yoon, C.K. (1999b). 'Squash With Altered Genes Raises Fears of 'Superweeds."' *New York Times*. November 3.

Zelinski, L.J. and Kerby, T. (1997). *Comparison of NUCOTN33(B) to Best Adapted Varieties Throughout the US Cotton Belt*. Scott, MS: Delta and Pine Land Company.

Zhang, Q. (1999). 'Meeting the challenges of food production: the opportunities of agricultural biotechnology in China.' Paper presented at *Ensuring Food Security, Protecting the Environment, and Reducing Poverty in Developing Countries: Can Biotechnology Help? An International Conference on Biotechnology*. World Bank, Washington DC, October 21–22.

# Index

Page numbers in *italics* refer to figures and tables; those in **bold** to main discussions.